MW01493000

Fundamentals Of CNC

An Extended Introduction To
CNC Machining Center And Turning Center Usage

SECOND EDITION

by Mike Lynch

CONCEPTS, INC.

© 2013   CNC Concepts, Inc.
44 Little Cahill Road • Cary, IL 60013
Phone 847.639.8847 • Fax 847.639.8857
Internet: www.cncci.com
Email: lynch@cncci.com

Notice

The information in this book is meant to supplement, not replace, proper CNC machine usage training. In all cases, information contained in the machine tool builder's manual will supersede material presented in this book. Like any industrial equipment that requires skill to master, CNC machine tools pose some risk. The publisher advises readers to take full responsibility for their safety and know their limits, as well as the limits of the CNC machine tool being used. Before practicing the skills and techniques described in this book, read all safety advice presented in the machine tool builder's manuals, be sure that your equipment is well maintained, and do not take risks beyond your level of experience, aptitude, training, and comfort level.

Library of Congress Cataloging-in-Publication Data

Lynch, Mike

Fundamentals of CNC - 3rd edition   p.   cm.

Includes index.

ISBN 13: 978-1493509652

ISBN 10: 1493509659

Distributed to the book trade by CNC Concepts, Inc.

Table of contents

Turning Center Discussions Start Here

CNC Turning Centers 230

Know Your Machine From A Programmer's Viewpoint

It is from two distinctly different perspectives that you must come to know your CNC machine/s. Here in Key Concept Number One, we'll look at the machine from a programmer's viewpoint. Much later, during Key Concept Number Seven, we'll look at the machine from a setup person's or operator's viewpoint.

Key Concept Number One is the longest of the Key Concepts. It contains seven lessons:

    1: Machine configurations
    2: General flow of programming
    3: Visualizing program execution
    4: Understanding program zero
    5: Determining program zero assignment values
    6: How to assign program zero
    7: Introduction to programming words

A CNC programmer need not be nearly as intimate with a CNC machining center as a setup person or operator – but they must, of course, understand enough about the machine to create programs – to instruct setup people and operators – and to provide the related setup and production run documentation.

First and foremost, a CNC programmer must understand what the CNC machining center is designed to do. That is, they must understand the machining operations a machining center can perform. They must be able to develop a workable process (sequence of machining operations), select appropriate cutting tools for each machining operation, determine cutting conditions for each cutting tool, and design a workholding setup. All of these skills, of course, are related to basic machining practice – which as we state in the Preface – are beyond the scope of this text. For the most part, we'll be assuming you possess these important skills.

That said, we do include some important information about machining operations that can be performed on CNC machining centers throughout this text. For example, we provide a description of hole-machining operations in lessons eight and sixteen. We describe milling operations in lessons eight, nine, and twelve. And in general, and we provide suggestions about how machining operations can be programmed when it is appropriate. This information should be adequate to help you understand enough about machining operations to begin working with CNC machining centers.

If you've had experience with conventional (non-CNC) machine tools...
A CNC machining center can be compared to a conventional milling machine. Many of the same operations performed on a milling machine are performed on a CNC machining center. If you have experience with manually operated milling machines, you already have a good foundation on which to build your knowledge of CNC machining centers.

This is why machinists make the best CNC programmers. With a good understanding of basic machining practice, you can easily learn to program CNC equipment. You already know *what* you want the CNC machine to do. It is a relatively simple matter of learning *how to tell the CNC machine* to do it.

If you have experience with machining operations like face milling, end milling, drilling, tapping, boring, and reaming, and if you understand the processing of machined workpieces, believe it or not, you are well on your way understanding how to program a CNC machining center. Your previous experience has prepared you for learning to program a CNC machining center.

We can also compare the importance of knowing basic machining practice in order to write CNC programs to how important it is for a speaker to be well versed with the topic they will be presenting. If not well versed with their topic, the speaker will not make much sense during the presentation. In the same way, a CNC machining center programmer who is not well versed in basic machining practice will not be able to prepare a program that makes any sense to experienced machinists.

Lesson 1:
Machine Configurations

As a programmer, you must understand what makes up a CNC machining center. You must be able to identify its basic components – you must understand the moving components of the machine (called axes) – and you must know the various functions of your machine that are programmable.

Most beginners tend to be a little intimidated when they see a machining center in operation for the first time. Admittedly, there will be a number of new functions to learn. The first point to make is that you must not let the machine intimidate you. As you go along in this text, you will find that a machining center is very logical and is almost easy to understand with proper instruction.

You can think of any CNC machine as being nothing more than the standard type of equipment it is replacing with very sophisticated and automatic motion control added. Instead of activating things manually by hand-wheels and manual labor, you will be preparing a program that tells the machine what to do. Virtually anything that needs to be done on a true machining center can be activated through a program – meaning anything you need the machine to do can be commanded in a program.

There are two basic types of machining centers that we will be addressing in this text. They are vertical machining centers and horizontal machining centers. Let's start by describing the most common features of each.

Vertical machining centers

A vertical machining center has its spindle oriented in the vertical position. The cutting tool will be pointing downward toward the machine's table. This is a very popular type of CNC machining center because it closely resembles a popular type of conventional machine tool – the knee mill. For anyone having experience with a knee mill, a vertical machining center will appear quite familiar. Note that during machining, chips will tend to collect and build up on the workpiece, and may eventually interfere with machining operations.

C-frame style

Figure 1.1 shows a C-frame style vertical machining center.

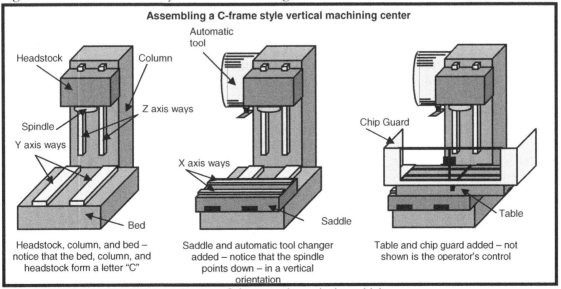

Assembling a C-frame style vertical machining center

Headstock, column, and bed – notice that the bed, column, and headstock form a letter "C"

Saddle and automatic tool changer added – notice that the spindle points down – in a vertical orientation

Table and chip guard added – not shown is the operator's control

Figure 1.1 – Components that make-up a C-frame style vertical machining center

This type of vertical machining center is called a C-frame style machining center because the bed, column, and headstock, when taken together, form the letter "C". An automatic tool changer is mounted to the machine (usually on the left side) to allow tools to be loaded into the spindle without operator intervention. The table has a series of tee-slots and/or location/clamping holes to allow workholding devices (like a table vise) to be mounted on the table.

Directions of motion (axes) for a C-frame style vertical machining center

Basic vertical machining centers allow three directions of motion, or *axes*. Each of these basic axes is linear – allowing motion along a straight line. With a C-frame style vertical machining center, the table can move left/right (the X axis) – the table can move fore/aft (the Y axis) – and the headstock or spindle can move up/down (the Z axis). Figure 1.2 shows the axes of a C-frame style vertical machining center.

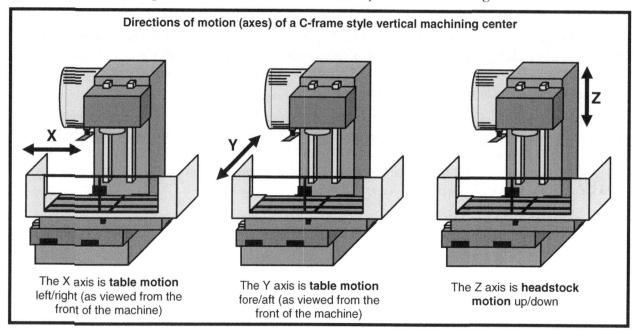

Directions of motion (axes) of a C-frame style vertical machining center

The X axis is **table motion** left/right (as viewed from the front of the machine)

The Y axis is **table motion** fore/aft (as viewed from the front of the machine)

The Z axis is **headstock motion** up/down

Figure 1.2 – Directions of motion (axes) of a C-frame style vertical machining center

With this kind of machine, notice that the cutting tool will not move along with the X and Y axis. Instead, the cutting tool will remain stationary in XY – it will only move along with the Z axis.

Axis polarity

Though not depicted in the previous illustrations, each axis has a polarity (plus and minus direction). As the table moves to the left, it is moving in the X plus direction. As it moves to the right, it is moving in the X minus direction. As the table moves toward you, it is moving in the Y plus direction. As it moves away from you, it is moving in the Y minus direction. As the headstock/cutting tool moves up, it is moving in the Z plus direction. As it moves down, it is moving in the Z minus direction.

For X and Y, since the cutting tool does not move along with every axis, it can be a little confusing (especially for programmers) to understand polarity by looking at table motion. From a programmer's viewpoint, we feel it is much easier to understand polarity if you imagine that the cutting tool *is* moving along with each axis. Figure 1.3 shows how to visualize polarity with this method.

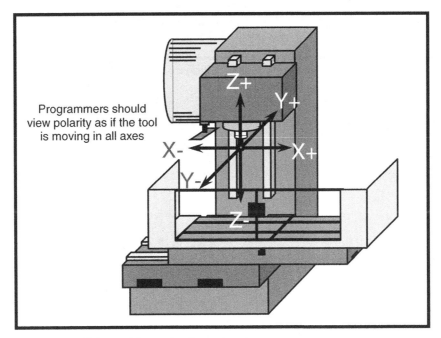

Figure 1.3 – Viewing polarity as if the tool is moving in each axis

If you imagine that the cutting tool can move in X and Y, determining polarity will be easier (especially when we introduce the coordinate system used with CNC programming in Lesson Four). As the cutting tool moves to the right, it is moving in the X plus direction. (But remember, the cutting tool cannot move to the right in X axis. Instead, the table moves to the left – which again, is the X plus direction.) As the cutting tool moves to the left, it is moving in the X minus direction. As the tool moves away from you, it is moving in the Y plus direction. As it moves toward you, it is moving in the Y minus direction. In Z, of course, the tool is moving with the axis, so polarity is much easier to understand. Up is Z plus, down is Z minus.

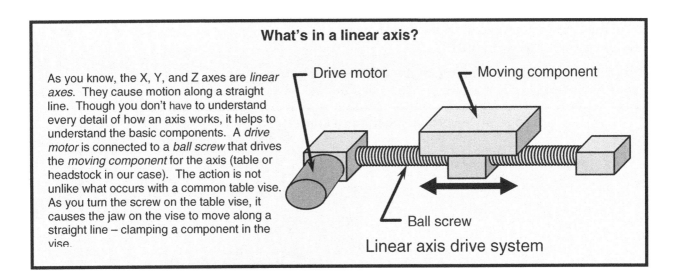

What's in a linear axis?

As you know, the X, Y, and Z axes are *linear axes*. They cause motion along a straight line. Though you don't have to understand every detail of how an axis works, it helps to understand the basic components. A *drive motor* is connected to a *ball screw* that drives the *moving component* for the axis (table or headstock in our case). The action is not unlike what occurs with a common table vise. As you turn the screw on the table vise, it causes the jaw on the vise to move along a straight line – clamping a component in the vise.

Drive motor

Moving component

Ball screw

Linear axis drive system

Knee style vertical CNC milling machines

While the C-frame style of vertical machining center is the most popular type of production vertical machining center, it is but one of several types. Another common type is called the *knee-style vertical CNC milling machine*. It is identical to a conventional knee-style milling machine with the addition of CNC motion control. Many knee-style vertical milling machines do not have a tool changer – which is why they're called CNC milling machines instead of CNC machining centers. If the machine does not have an automatic tool changer, cutting tools must be manually loaded during the CNC cycle. Figure 1.4 shows a typical knee-style CNC milling machine.

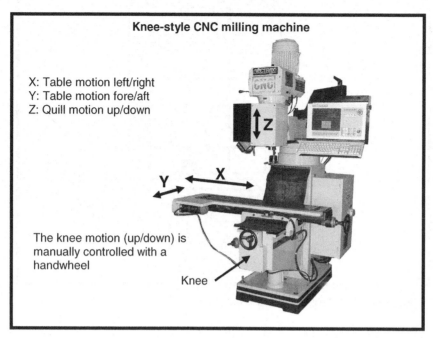

Figure 1.4 – Knee-style CNC milling machine with axes directions shown

Notice how similar the axes of this machine are to the C-frame style CNC machining center. The only difference is that a knee-style CNC milling machine has a quill that moves within the headstock to form the Z axis as opposed to the entire moving headstock of a C-frame style machining center.

The polarity for each axis is also the same. X plus is table motion to the left. X minus is table motion to the right. Y plus is table motion toward you. Y minus is table motion away from you. Z plus is quill motion up. Z minus is quill motion down. And since the tool does not move along with the X and Y axes, we recommend that you view polarity as if the tool is moving in X and Y – just as we suggested with a C-frame style machining center (see Figure 1.3).

Bridge-style vertical machining center (also called gantry-style)

Yet another type of vertical machining center is called the bridge-type machining center. Figure 1.5 shows one.

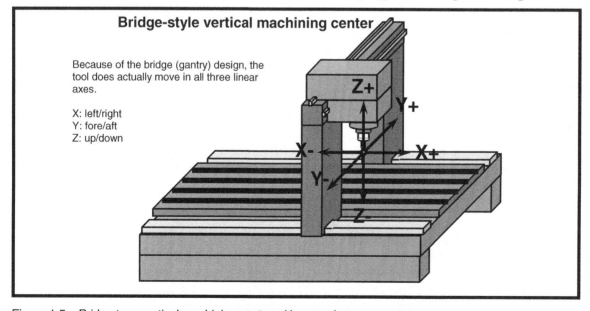

Figure 1.5 – Bridge type vertical machining center with axes shown

As you can see, the table remains stationary on this kind of machine, which is necessary when machining very large workpieces – like larger aircraft components. With this kind of machine, the tool *does* move along with each axis, and this makes understanding polarity much easier. X plus is tool movement to the right. X minus is tool movement to the left. Y plus is tool movement away from you. Y minus is tool movement toward you. Z plus is tool movement up. Z minus is tool movement down.

Horizontal machining centers

As the name implies, a horizontal machining center has its spindle oriented in a horizontal position. This provides one major advantage over a vertical machining center having to do with chip removal. With vertical machining centers, chips tend to gather on the workpiece right around the cutting tool during machining – and if they are not removed by some means (manually by the operator or flushing away by coolant or air blow), they will eventually interfere with the machining operation. By comparison, chips will naturally fall away from the machining operation with a horizontal machining center.

Another advantage of a horizontal machining center is that some kind of rotary device is usually incorporated in the table to allow the workpiece to be rotated during the machining cycle. This allows several surfaces of the workpiece to be exposed to the spindle for machining during the CNC cycle. The most common form of rotary device for current model machining centers is a rotary axis, which is called the B axis. Figure 1.6 shows the components making up a horizontal machining center.

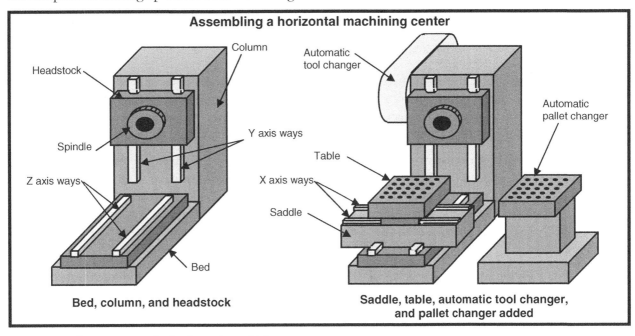

Figure 1.6 – Components of a horizontal machining center

Most horizontal machining centers sold today come with an automatic pallet changer (shown in the right-side illustration). This device (which can also be equipped with vertical machining centers) allows the operator to be loading a workpiece on one pallet while a workpiece on the other pallet is being machined. While it is not shown in the illustration above, most horizontal machining centers have guarding that completely surrounds the work area, which keeps the operator safe while loading workpieces.

Directions of motion (axes) for a horizontal machining center

Figure 1.7 shows the axes for a horizontal machining center.

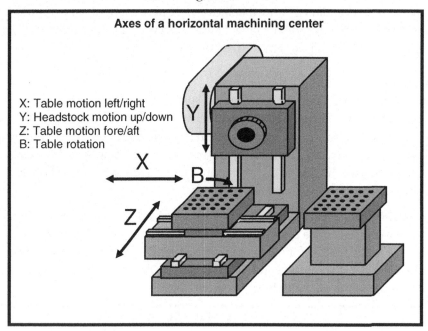

Axes of a horizontal machining center

X: Table motion left/right
Y: Headstock motion up/down
Z: Table motion fore/aft
B: Table rotation

Figure 1.7 – Axes for a horizontal machining center

At first glance, the axes of a horizontal machining center (X, Y, and Z) appear to be radically different from a vertical machining center. But if you look more closely, you'll see that a horizontal machining center is little more than a vertical machining center placed on its back. In fact, a program written for a vertical machining center will run properly in a horizontal machining center, at least with regard to axis movements.

Also notice the addition of the rotary axis (B) within the table. This allows the table to be rotated during the execution of a CNC program – and can expose different surfaces of the workpiece to the spindle for machining. Some horizontal machining centers do not have a full rotary axis within the table. Instead, they have a simple indexing device. A rotary axis will allow machining to occur *during* table rotation. With an indexer, machining can only occur *after* the table has rotated. We discuss both types of rotary devices during Lesson Nineteen.

There is a bit of a misconception that horizontal machining centers are harder to work with than vertical machining centers. Frankly speaking, the only major difference is that programs for horizontal machining centers tend to be longer since several surfaces of the workpiece are exposed to the spindle and more cutting tools will be required. While this will require a longer program, the program will not be more difficult to write. In fact, a vertical machining center will require several programs to machine the same number of workpiece surfaces.

Axis polarity

Notice in Figure 1.7 that the tool will not move along with the X and Z axes (it will move only with the Y axis). And as we said during the discussion of C-frame vertical machining centers, this tends to cause some confusion when it comes to determining the polarity of each axis. Again, we recommend that you view polarity as if the tool is actually moving along with each axis. Figure 1.8 shows this.

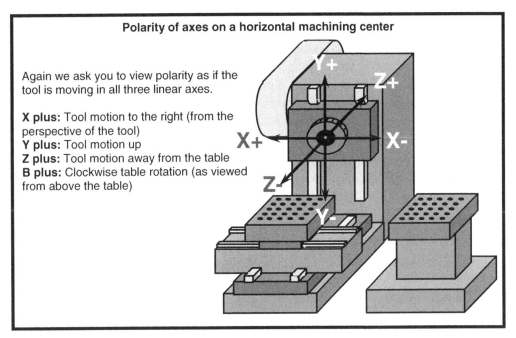

Polarity of axes on a horizontal machining center

Again we ask you to view polarity as if the tool is moving in all three linear axes.

X plus: Tool motion to the right (from the perspective of the tool)
Y plus: Tool motion up
Z plus: Tool motion away from the table
B plus: Clockwise table rotation (as viewed from above the table)

Figure 1.8 – Polarity of axes for a horizontal machining center

When it comes to polarity for the B axis, most machine tool builders will make clockwise table rotation (as viewed from above the table) the plus direction. But not all machine tool builders adhere to this standard.

Note: Since vertical machining centers are more popular than horizontal machining centers, most example programs in this text are shown in the format for a vertical machining center. During our discussion of rotary devices in Lesson Nineteen, we discuss special programming considerations for horizontal machining centers.

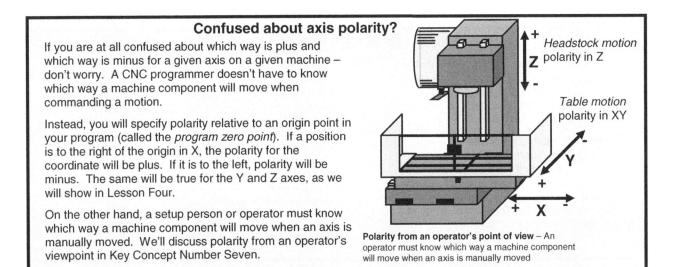

Confused about axis polarity?

If you are at all confused about which way is plus and which way is minus for a given axis on a given machine – don't worry. A CNC programmer doesn't have to know which way a machine component will move when commanding a motion.

Instead, you will specify polarity relative to an origin point in your program (called the *program zero point*). If a position is to the right of the origin in X, the polarity for the coordinate will be plus. If it is to the left, polarity will be minus. The same will be true for the Y and Z axes, as we will show in Lesson Four.

On the other hand, a setup person or operator must know which way a machine component will move when an axis is manually moved. We'll discuss polarity from an operator's viewpoint in Key Concept Number Seven.

Headstock motion polarity in Z

Table motion polarity in XY

Polarity from an operator's point of view – An operator must know which way a machine component will move when an axis is manually moved

Programmable functions of machining centers

A true CNC machining center will allow you to control just about any of its functions from within a program. There should be very little operator intervention during a CNC cycle. Here we list some common functions that can be programmed on all *true* machining centers. While we do show the related CNC words used to command these functions, our intention here is not to teach programming commands (yet). It is to simply make you aware of the kinds of things a programmer can control through a program.

Spindle

The spindle of all machining centers can be programmed in at least three ways, activation (start/stop), direction (forward/reverse), and speed (in rpm). Some machining centers additionally provide multiple power ranges (like the transmission of an automobile).

Spindle speed

You can precisely control how fast the spindle of a machining center rotates in one rpm increments. An *S word* is used for this purpose. If you want the spindle to rotate at 350 rpm, use the word **S350** to do so. Since spindle speed is specified in whole numbers, you must not include a decimal point with the S word. Also, the S word by itself does not actually activate the spindle.

Most cutting tool manufacturers recommend spindle speed in *surface feet per minute* (sfm), meaning you'll have to perform a calculation in order to determine the appropriate speed in rpm for every machining operation. Here is the rpm formula:

rpm = 3.82 times sfm divided by cutting tool diameter

Say, for example, you need to drill a 0.5 diameter hole in mild steel with a high speed steel twist drill. The cutting tool manufacturer recommends that you drill mild steel at 70 sfm. This requires a speed of 535 rpm (actually it comes out to 534.8, but round the result to the closest whole number). Again, 3.82 times 70 divided by 0.5 is 534.8. The word **S535** will be used to specify the spindle speed.

Spindle activation and direction

You can also control which direction the spindle rotates – forward or reverse. The forward direction is used for right hand tooling. It will appear as counter-clockwise when viewed from in front of (below) the spindle. The reverse direction is used for left hand tooling and will appear as clockwise when viewed from in front of the spindle.

Three *M codes* control spindle activation. **M03** turns the spindle on in the forward direction (used with right-hand tools). **M04** turns the spindle on in a reverse direction (for left-hand tools). **M05** turns the spindle off.

What is an M code?

M codes control many of the programmable functions on a machining center. In many cases, you can think of them as being like programmable on/off switches. "M" stands for "miscellaneous" or "machine" function.

M codes are created by the machine tool builder, and will often vary from one machining center to another. Here we show some common M codes, but you must look in your machine tool builder's programming manual to find the complete list of M codes for a given machining center.

These M codes don't vary:
M00: Program stop
M01: Optional stop
M03: Spindle on (forward)
M04: Spindle on (reverse)
M05: Spindle off
M06: Activate tool changer
M08: Flood coolant on
M09: Coolant off
M30: End of program
M60: Activate pallet changer

These M codes vary:
M___: Automatic door open
M___: Automatic door close
M___: Chip conveyor on
M___: Chip conveyor off
M___: _____
M___: _____
M___: _____
M___: _____
M___: _____
M___: _____

Spindle range

Some, especially larger machining centers, have two or more spindle ranges. Spindle ranges are like the gears in an automobile transmission. Lower ranges are used for power – higher ranges are used for speed. With most current model machining centers, spindle range selection is a bit *transparent*. Spindle range will be automatically selected when you specify a spindle speed (S word). For this reason, some programmers don't even know the machine that they are programming has two or more spindle ranges!

Figure 1.9 shows the power curve chart for a machining center that has two spindle ranges. If your machining center has more than one spindle range, you can find this kind of chart in your machine tool builder's programming manual.

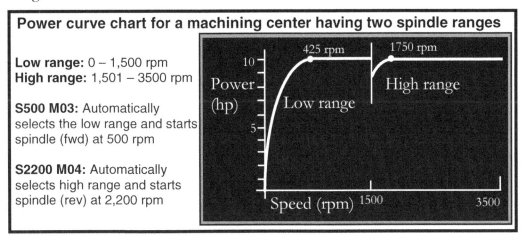

Power curve chart for a machining center having two spindle ranges

Low range: 0 – 1,500 rpm
High range: 1,501 – 3500 rpm

S500 M03: Automatically selects the low range and starts spindle (fwd) at 500 rpm

S2200 M04: Automatically selects high range and starts spindle (rev) at 2,200 rpm

Figure 1.9 – Typical power curve chart for a machining center having two spindle ranges.

For our example machining center, full power is achieved in the low spindle range at 425 rpm and in the high spindle range at 1,750 rpm. This is important information, especially when you are programming powerful machining operations. Consider for example, using a 2.0 inch face mill for a rough milling operation. Say the tooling manufacturer recommends using a speed of 800 surface feet per minute (sfm). This equates to 1,528 rpm (again, rpm equals 3.82 times sfm divided by tool diameter – 3.82 times 800 divided by 2.0 is 1,528). If S1528 is specified as the spindle speed, the machine will automatically select the *high* spindle range, and there may not be enough power to perform the powerful milling operation. The spindle may stall. In this case, it will be necessary to compromise the spindle speed down to S1499 to force the machine to select the low spindle range in order to achieve the needed power.

Feedrate

As you know, a machining center has three linear axes, X, Y, and Z. You must be able to control how quickly these axes move, especially when machining is being done. Feedrate is the motion rate at which the cutting tool will move during a machining operation. It is a programmable function of all machining centers. Feedrate is specified with an F word (obviously, F for feedrate). For most machining centers, feedrate can only be specified in per-minute fashion (inches per minute or millimeters per minute).

Note that most cutting tool manufacturers recommend feedrate in inches (or millimeters) per-revolution or inches (or millimeters) per-tooth/flute. To determine the per-minute feedrate, you must multiply the per-revolution value times the (previously calculated) rpm.

Again, say you need to drill a 0.5 diameter hole in mild steel. The previously calculated speed is 535 rpm (see the previous discussion of spindle speed). The cutting tool manufacturer recommends a feedrate of 0.007 inches per revolution. In this case, you need to program a feedrate of F3.74, since 535 times 0.007 is 3.74.

Since calculated feedrate often includes a decimal portion of a whole number, you *are* allowed to include a decimal point with the F word. Again, this will be F3.74 in the previous example.

Cutting tool manufacturers sometimes recommend feedrate with per-tooth values. This is commonly the case for milling cutters. To determine the per-revolution feedrate amount, multiply the per-tooth value times the number of teeth or flutes on the cutter.

There are some machine tool builders that do allow feedrate to be specified directly in per-revolution's fashion (which minimizes calculations). The vast majority do not. If both feedrate specifications are allowed (per-minute and per-revolution), two *G codes* are used to specify the feedrate type.

For machining centers that allow this, **G94** is commonly used to select feed-per-minute mode and **G95** is used to select feed-per-revolution mode. Any F word following a **G94** will be taken as a per-minute feedrate. Any F word following a **G95** will be taken as a per-revolution feedrate.

> Check with an experienced person in your company or school to find out whether or not your machining center allows feedrate specification in per-revolution fashion.

What is a G code?

G codes are called *preparatory functions.* They prepare the machine for what is coming up – in the current command and possibly in up-coming commands. In many cases, they set *modes,* meaning once a G code is *instated* it will remain in effect until the mode is changed or cancelled.

Here we list a few common G codes, but don't worry if they don't make much sense yet. Upcoming discussions will clarify.

Common G codes:

G00: Rapid motion
G01: Straight line motion
G02: Circular motion (CW)
G03: Circular motion (CCW)
G04: Dwell
G20: Inch mode selection
G21: Metric mode selection
G28: Zero return command
G40: Cancel cutter radius comp.
G41: Cutter comp. left

G42: Cutter comp. right
G43: Tool length comp.
G81: Drilling cycle
G83: Peck drilling cycle
G84: Tapping cycle
G90: Absolute mode
G91: Incremental mode
G98: Initial plane
G99: R plane

Look in your control manufacturer's manual for a full list of G codes.

Coolant

Coolant is the liquid used to cool and lubricate the machining operation. All current model machining centers provide flood coolant capability. Flood coolant is turned on with an **M08** and turned off with an **M09**.

Some machine tool builders provide special coolant capabilities, including high pressure coolant, through-the-tool coolant, and air-blast (to clear chips from the work area). These special coolant functions are almost always programmed with M codes, but the specific M code numbers will vary from one machine tool builder to another. You must look in your machine tool builder's programming manual to see if your machine has any special coolant capabilities, as well as to determine the relate M codes.

Automatic tool changer

All true machining centers have automatic tool changers that allow tools to be loaded into the spindle during the program's cycle. This, of course, allows several machining operations to be performed on a workpiece within one program cycle. Machine tool builders vary when it comes to how many cutting tools their machines can hold. They range from under ten to well over one hundred. A typical automatic tool changer magazine can hold about twenty tools.

The automatic tool changer design will vary from one machine tool builder to another, but programming methods remain amazingly similar, falling into one of two basic programming styles. We'll discuss the specific differences during Key Concept Number Five (program formatting).

For now, we'll just mention that a *T word* is used to rotate the automatic tool changer mechanism. With most machines, the T word does not actually cause a tool change to occur. Again, it just rotates the magazine to its ready position (also called the waiting position).

Magazine stations are numbered. A machine having a tool changer magazine that can hold twenty tools will commonly have T words ranging from **T01** to **T20**. And again, the T word is used to specify the desired tool station number that will be brought to the ready position.

An **M06** is used to actually make the tool change. For most current model machining centers, **M06** will cause the tool that is in the spindle to be placed back into the magazine – and the tool in the ready position will then be placed into the spindle. For example, the command

    T05 M06

---

will cause tool station number five to rotate to the ready station and then be placed into the spindle (whatever tool was currently in the spindle will be placed back into the magazine). Figure 1.10 illustrates the activation of a double-arm-style automatic tool changer.

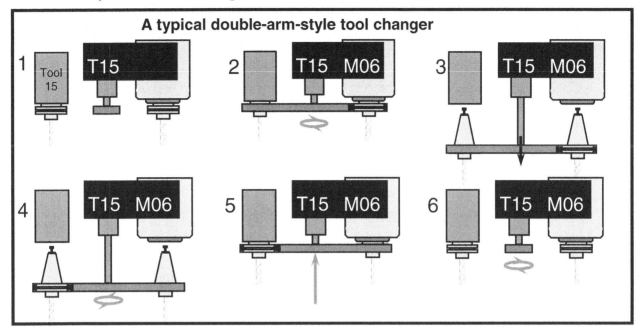

Figure 1.10 – Steps of a double arm automatic tool changer

In step 1 of Figure 1.10, the **T15** word will place tool station number fifteen in the ready position. Starting from step 2, the **M06** will cause the tool change. We're showing a double-arm automatic tool changer, which is the most efficient style of tool changer. Notice how the tool in the spindle is replaced with the tool in the ready station. We'll discuss automatic tool changers in more detail during Key Concept Number Five.

Measurement system mode (inch or metric)

All machining centers used in the United States allow you to work in the English (inch) system or the Metric system. Since this text has been prepared for use in the United States, and since most companies in the US (still) work in the English system, all examples in this text are provided in the English system. Note that for CNC use, we refer to the two measurement systems as inch mode or metric mode.

Two G codes are used to select you measurement system of choice. With most machining centers, **G20** is used to select the inch mode and **G21** is used to select the metric mode.

Many CNC functions are affected by your measurement system of choice, including two we've already discussed – axis coordinates and feedrate. In the inch mode, these values are specified in inches. In the metric mode, they are specified in millimeters.

For example, the command

G20

selects the inch mode. Any coordinate or feedrate following **G20** will be taken in inches. The command

X1.0 Y1.0

will specify a value of one inch in each axis (as opposed to one millimeter).

Feedrates specified after a **G20** will be taken in inches per minute. Feedrates after a **G21** will be taken in millimeters per minute. (Do remember that some machines, though not many, additionally allow feed per revolution. This will additionally allow feedrate to be specified in inches or millimeters per revolution.)

What else might be programmable?

While this text will acquaint you with the most common programmable functions of machining centers, you must be prepared for more. Other programmable devices that may be equipped on your machining center include pallet changer, programmable chip conveyer, spindle probe, tool length measuring probe, and a variety of other application-based features. If you have any of these features, you must reference your machine tool builder's programming manual to learn how these special features are programmed.

> Check with an experienced person in your company or school to find out what other programmable features are available on the machining center you will be working with.

Key points for Lesson One:

- There are two types of machining centers – vertical and horizontal.
- The C-frame style vertical machining center is the most popular type of production machining center.
- You should be able to recognize the most basic machine components.
- Every machining center has three linear axes, X, Y, and Z.
- Some machining centers have one or more rotary axes (B on a horizontal machining center).
- Each axis has polarity (plus and minus direction).
- For linear axes, it is best to view polarity as if the tool is moving along with the axis.
- You must understand the functions of your machining center that can be programmed.
- Spindle can be controlled in at least three ways (activation, direction, and speed).
- Feedrate specifies the motion rate for machining operations.
- Coolant can be activated to allow cooling and lubricating of the machining operation.
- All true machining centers have an automatic tool changer.
- Machines in the US allow the use of inch or metric mode.
- You must determine what else is programmable on your machining center/s.

Quiz

1) Which style of machining center is the most popular?

  A) C-frame style vertical   C) Bridge-style vertical
  B) Knee-style vertical   D) Horizontal

2) Specify the correct axis for each moving component below (C-frame style vertical machining center).

  a) _____Table motion in/out
  b) _____Headstock motion up/down
  c) _____Table motion left/right

3) Specify the correct axis and polarity for each tool motion below (C-frame style vertical machining center).

  a) _____Tool motion left
  b) _____Tool motion away from you
  c) _____Tool motion down

4) Provide the CNC word or command needed to activate the following:

  a) Start spindle (fwd) at 300 rpm: _____
  b) Start spindle (reverse) at 2,000 rpm: _____
  c) Stop spindle: _____
  d) Turn on the flood coolant: _____
  e) Turn off the coolant: _____
  f) Place tool station seven in spindle: _____
  g) Select inch mode: _____

Talk with experienced people in your company to learn more: Do your machining centers have more than one spindle range? If so, what are the cut-off points for each range? At what rpm does the spindle achieve maximum horsepower in each range? Do any of your machines have high pressure coolant systems? If so, what is the related M code? How many cutting tools can your machining centers hold?

Answers: 1: A, 2a: Y, 2b: Z, 2c: X, 3a: X-, 3b: Y+, 3c: Z-, 4a: S300 M03, 4b: S2000 M04, 4c: M05, 4d: M08, 4e: M09, 4f: T07 M06, 4g: G20

Lesson 2
General Flow Of The CNC Process

The tasks of programming, setup, and operation are but three of the things that must be done in order to get a CNC job up and running. It really helps to understand how these tasks fit into the bigger picture of a company's manufacturing environment.

CNC machine tools are being used by all sorts of companies. Indeed, if a company manufacturers anything, it's likely that they're using at least some CNC machine tools. With the diversity of applications, there comes diversity in what is expected of CNC people. Understanding where your company fits in to this diverse group should help you understand what will be expected of you.

Companies that use CNC machining centers

There are many factors that contribute to how a CNC-using company applies its CNC machining centers. These factors include (among others) lot sizes, lead times, percentage of new jobs, closeness of tolerances held, materials machined, and company type. The most important of these factors is company type.

When it comes right down to it, there are only four types of companies that use CNC machine tools:

- Product producing companies – get revenue from the sale of a product
- Workpiece producing companies – (also called job-shops or contract-shops) get revenue from the sale of production workpieces to product producing companies
- Tooling producing companies – get revenue from the sale of manufacturing tooling (fixtures, jigs, molds, dies, gauges, cutting tools, etc.) to product producing and workpiece producing companies
- Prototype producing companies – get revenue from the sale of prototypes to product producing companies

There are also overlaps in company type. For example, some product producing companies have a tool-room in which CNC machine tools are used – or they may have a research and development department that produces prototypes. Some workpiece producing or tooling producing companies have a product of their own.

While there will be exceptions to what we say here, some pretty good generalizations can be made based upon the company type alone, especially when it comes to what CNC people will be doing.

Product producing companies tend to have more resources than workpiece producing, tooling producing, and prototype producing companies. Since their profit is one step removed from manufacturing (a product won't come to market unless the company can make a profit), they tend to engineer all facets of the manufacturing environment. For this reason, they commonly break up the tasks related to CNC machine tool usage. People will specialize in the tasks they perform.

This will maximize machine tool utilization. It ensures that machines are running for as high a percentage of time as possible. Many CNC tasks will be performed while the machine is running production (like programming for upcoming jobs, gathering components needed for future setups, and assembling cutting tools, among others). So while a CNC operator is running a job on the CNC machine, other people are getting ready to run the next (and other upcoming) jobs. Again, this minimizes the amount of time that the CNC machining center is down between production runs.

On the other hand, workpiece producing, tooling producing, and prototype producing companies tend to charge an hourly rate for machine usage time. Their resources will be much smaller – and they will require much more of their CNC people. One person may be responsible for all of the tasks related to running a CNC job. They will sacrifice machine tool utilization to some extent in order to run jobs with a small number of CNC people. This means machines may sit idle while the person responsible for the machine performs all or some of the tasks related to CNC machine usage.

What will you be doing?

This text includes enough information to help you fully master CNC machining center usage, and you'll probably want to learn it all. But if you work for a product producing company, it's likely that you'll only have to master a portion of what we present in this text. If you work for a workpiece producing, tooling producing, or prototype producing company, you'll probably have to master just about everything shown.

Common job titles in the CNC environment include:

- Process engineer – This person (often the CNC programmer) determines which machine will be used to produce the workpiece and develops the sequence of machining operations, or *process*
- Tooling engineer – This person (often the CNC programmer) designs any special workholding devices (fixtures) needed for a job and determines which cutting tools will be used to machine the workpiece
- CNC programmer – This person creates the CNC program and develops the setup and production run documentation used by people that actually run the job on the CNC machine
- CNC setup person – This person gets the CNC machine ready to run a new job
- Inspector – Once a setup is made and the first workpiece is run, this person checks the workpiece to confirm that it is within specifications
- CNC operator – Once a setup is completed and the first workpiece passes inspection, this person runs out the job
- CNC helper – This person does as much as possible to help in the shop – gathering components (fixtures, cutting tools, inserts, etc.) needed by others in the CNC environment, cleaning machines, performing basic preventive maintenance, and in general, doing whatever it takes to minimize work for other CNC people
- Tool crib attendant – This person gathers and assembles tooling components for upcoming jobs and supplies perishable tooling (like inserts) to CNC operators currently running jobs

> Ask an experienced person in your company
> what your CNC-related responsibilities will be.

Flow of the CNC process

Here we give an example of how a typical shop will handle a CNC job. By no means will this flow be the same for all shops. As stated, workpiece producing companies tend to have one or two people handling all steps in this flow. Product producing companies will usually break these steps up into small tasks, each to be handled by a different person. We present the related tasks in the approximate order that jobs are run.

Study the workpiece drawing

All questions about how a workpiece must be machined are answered by the workpiece drawing. Everyone involved with a job must get acquainted with the workpiece to be produced.

Decision is made as to which CNC machine to use

If the company has more than one CNC machining center, there may be some question as to which one should be used to produce a given workpiece. Based on the required number of workpieces (lot size), the accuracy required of the workpiece, the material to be machined, the required surface finish, and the shop loading on any one CNC machining center (among other things), the decision is made as to which CNC machining center to use.

The machining process is developed

If more than one machining operation must be performed by the CNC machine, someone with machining practice background (commonly the programmer) must come up with a sequence of machining operations to be used to machine the part. The program will follow this machining process. It is during this step that cutting tools needed for the job will be determined. Cutting conditions (speeds, feeds, and depths-of-cut) must also be calculated.

Tooling is ordered and checked

Based on a previously developed process, the required tooling must be obtained. This includes workholding tooling, like fixtures, as well as cutting tools.

The program is developed

In this step, the programmer codes the program into a language that the CNC machine can understand. This step is, of course, one major focus of this text. We present six Key Concepts to help you understand this step.

Setup and production run documentation is made

Part of the programmer's responsibility is to make it clear as to how the setup is to be made at the machine – as well as how the production run is to be completed. Drawings are commonly made to describe the work holding setup. A cutting tool list is made, specifying the components needed for each tool, the offsets to be used for each tool, and the automatic tool changer magazine station number to be used for each tool. Production run documentation commonly includes workpiece loading instructions, and instructions related to holding size on the workpiece during the production run. Truly, anything that helps a setup person or operator should be documented.

Program is loaded into the CNC control's memory

Once the program is prepared, there are two common methods used to load it into the control's memory. One way is for someone to physically type it through the keyboard and display screen of the control panel. Frankly speaking, this is a rather cumbersome way to get the program into the control's memory. For one thing, the keyboard of the control panel is quite difficult with which to work. The keyboard is not oriented in a logical way (most are not like the keyboard of a computer) and usually the panel itself is mounted vertically, resulting in the person typing the program experiencing fatigue. In many cases, the machine will sit idle while the program is being typed. Even a relatively short program will take time to enter. A CNC machine makes a very expensive typewriter!

Another more popular way to enter programs into the control's memory is to use some outside device for typing programs. A personal computer could be used for this purpose. The software used for this computer application resembles a common word processor. In fact, most word processors can be used for the purpose of typing CNC programs. Once the program is entered through the computer's keyboard, it can be saved on the computer's hard drive. When the program is needed at the machine, it can be quickly transmitted to the machine from the computer. This transmission takes place almost instantaneously, even for lengthy programs. The software, computer, and cabling used to transfer CNC programs to and from the CNC machine is commonly called a distributive numerical control (DNC) system.

Again, the computer makes it much easier for the person typing the program. They can sit in a comfortable environment to type the program. Most importantly, the program can be typed *while* the machine continues to machine workpieces for the current production run, meaning almost no machine time is wasted while the program is loaded.

The setup is made

Before the program can be run, the setup must be made. Using the set-up instructions (commonly prepared by the programmer), the setup person makes the workholding setup. Cutting tools are assembled and loaded into the proper magazine tool stations. Measurements must also be made to determine the position of program zero and the length of each tool. Certain program zero assigning and cutting tool-related offsets must be entered.

The program is cautiously verified

It is very rare that a new program requires no modification at the machine. Even if the programmer does a good job programming the required cutting tool motions, there will almost always be some optimizing that is necessary to minimize program execution time (especially for larger lots).

Production is run

At this point the machine is turned over to the CNC operator to complete the production run. While the programmer's and setup person's job could be considered finished at this point, there may be some long term problems that do not present themselves until several workpieces are run. For example, cutting tool wear may be excessive. Tools may have to be replaced more often than the company would like. In this case, the speeds and feeds in the program may have to be adjusted.

Corrected version of the program is stored for future use

Most workpieces that run on CNC machines are run on a regular basis, especially in product producing companies. At some future date it will probably be necessary to run the workpiece again. If changes have been made during a new program's verification, it will be necessary to transmit the corrected CNC program back to the program storage device (DNC system). If this step is not done, the program will have to be verified again the next time the job is run.

Key points for Lesson Two:

- There are four types of companies that use CNC machine tools.
- The type of company dramatically affects what is expected of CNC people.
- You must know what you will be expected to do in the CNC environment.
- It helps to understand the bigger picture of how a job is processed through a CNC using company.

Talk to experienced people in your company...

... to learn more about how your company operates. If you work for a CNC-using company, here are questions you'll want to get answered.

1) What kind of company do you work for: product producing, workpiece producing, tooling producing, or prototype producing?

2) Who are the programmers? Who does setups? Is there a tool crib? If so, who's the attendant? Do setup people inspect their own workpieces or is there a special inspector that inspects the first workpiece? What will *you* be doing?

3) Who develops the machining process for jobs run on CNC machines? Is it the CNC programmer?

4) Do programmers write programs manually or do they use some kind of computer aided manufacturing (CAM) system?

5) Does the company have a distributive numerical control (DNC) system? If so, where is it and how is it used?

6) How often are new programs run? Does the setup person verify programs alone, or does the programmer come to the machine to help?

Lesson 3

Visualizing The Execution Of A CNC Program

A CNC programmer must possess the ability to visualize movements a CNC machine will make as it executes a program. The better a person can visualize what the machining center will be doing, the easier it will be to prepare a workable CNC program.

Once again, we stress the importance of basic machining practice as it applies to CNC machining center usage. A machinist that has experience running a conventional milling machine will have seen machining operations taking place many times. While this experience by itself does not guarantee the ability to *visualize* a machining operation (seeing it happen in your mind), it dramatically simplifies the task of learning how to visualize a CNC program's execution.

When a machinist prepares to machine a workpiece on a conventional milling machine, they will have all related components needed for the job right in front of them. The machine, cutting tools, workholding setup, and print are ready for immediate use. It is highly unlikely that the machinist will make a basic mistake like forgetting to start the spindle before trying to machine the workpiece.

On the other hand, a CNC machining center programmer will be writing the program with only the workpiece drawing to reference. No tooling – no machine – and no workholding setup will be available to them. For this reason, a programmer must be able to visualize just exactly what will happen during the execution of the program - and this can sometimes be difficult, since this visualization must take place in the programmer's mind. A beginning programmer will be prone to forget certain things - sometimes very basic things (like turning the spindle on prior to machining the workpiece).

In this lesson, we will acquaint you with those things a programmer must be able to visualize. We will also show the first (elementary) program example to stress the importance of visualization.

Visualization is necessary to develop a set of instructions

Consider the visualization it takes to write a set of travel instructions to get someone from the airport to your company. Before you can write the instructions, you must be able to visualize the path from the airport to your company. If you cannot visualize the path, you can't write the instructions. Worse, if you think you can visualize the path (but you're wrong), you'll write incorrect instructions and the person following your instructions will get lost!

In similar fashion, if you cannot visualize the path a cutter will take as it machines the workpiece, you cannot write the CNC commands that drive the cutter through this path.

1) Take airport exit to Highland Dr. Turn left.
2) Take Highland Dr. 4 mi to Elm St. Turn right.
3) Take Elm St. 1 mi to March Ave. Turn left.
4) Take March Ave 2 mi to Lance Dr. Turn right.
5) Take Lance Dr. 1/2 mi to company (on right).

Program make-up

Like the sentences making up a set of travel instructions, a CNC program is made up of *commands* (also called blocks). Within each command are *words*. Each word is made up of a *letter address* (N, X, Z, T, etc.) and a *numerical value*. Figure 1.11 shows the beginning of a CNC program that stresses program make-up.

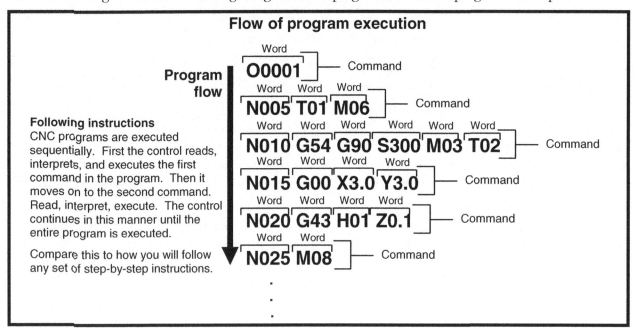

Figure 1.11   The flow of a CNC program

Method of program execution

You can also compare writing a CNC program to giving a set of step-by-step assembly instructions. For example, say you have just purchased a bookcase that requires assembly. The instructions you receive will be in sequential order. You will perform step number one before proceeding to step number two. Each step will include a sentence (or paragraph) explaining what it is you are supposed to do at the current time. As you follow each step, performing the given procedure, you are one step closer to finishing.

A CNC program will also be executed in sequential order. The CNC control will read, interpret, and execute the first command in the program. It will then go on to the next command. Read, interpret, execute. The control will continue this process for the balance of the program.

Keep in mind that the CNC control will execute each command *explicitly*. Compare this to the set of instructions for assembling the bookcase. In the set of assembly instructions, the manufacturer may be rather vague as to what it is you are supposed to do in a given step. They will assume certain things. They may assume, for example, that you have a screwdriver and that you know how to use it. Assembly instructions can be so vague as to be open to interpretation. A CNC control, by comparison, will make no assumptions. Each command will be very explicit - and any one CNC command will have only one resultant machine action or set of actions.

An example of program execution

To stress the sequential order of execution, and the visualization that is necessary to write programs, let's look at a very simple machining center example. We will first show the steps a machinist will perform to machine a very simple workpiece on a conventional milling machine. Then we will show the equivalent CNC program that will perform the same machining operation on a CNC machining center.

Figure 1.12 shows the print for this machining operation. In this case, we are simply drilling a 0.500 inch diameter hole to 0.75 deep.

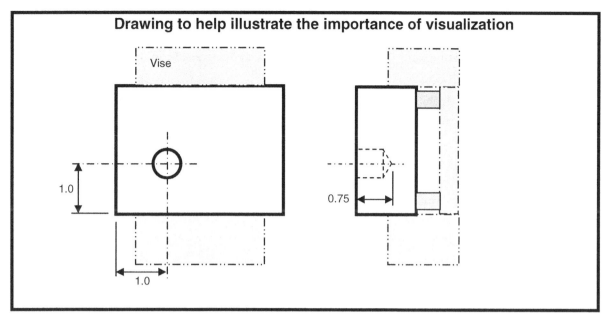

Figure 1.12   Drawing for example illustrating program execution flow

First let's look at the procedure a machinist will perform on a conventional milling machine or drill press. Note that we assume that the workpiece is already in a vise on the table and that the drill is already in the spindle.

Manual milling machine procedure:

1) Turn spindle on CW at 600 RPM

2) Move the workpiece under the drill

3) Move the drill up close to position

4) Turn on the coolant

5) At desired feedrate, drill the 0.500 hole

6) Retract the drill from the hole

7) Move tool away and turn off coolant and spindle

Again, this is a very simple example. But remember that we are stressing the sequential order by which a machinist will machine this workpiece and the visualization that is necessary to write a CNC program to perform this machining operation.

CNC program:

Now here is a CNC program to drill the 0.500 diameter hole in the workpiece on a CNC machining center. Again, we assume that the drill is in the spindle and that the workpiece is held in a vise on the table when this program begins.

```
O0001 (Program number)
N005 G54 G90 S600 M03 (Assign program zero, select absolute mode, turn spindle on CW at
   600 RPM)
N010 G00 X1.0 Y1.0 (Move the tool into position in X and Y)
N015 G43 H01 Z0.1 M08 (Rapid up to workpiece, instate tool length compensation, and turn on
   coolant)
N020 G01 Z-0.75 F3.5 (Drill hole at 3.5 IPM)
N025 G00 Z0.1 M08 (Rapid out of the hole, turn off coolant)
N030 G91 G28 X0 Y0 Z0 (Rapid to the machine's reference position)
N035 M30 (End of program, this command also turns off spindle)
```

Though the actual commands in this program probably don't make much sense, the messages within parentheses should nicely clarify what is happening in each command. (By the way, you can include messages within parentheses in your own programs - though in actual programs they must be in upper case letters).

Our intention is not to teach the programming words being used (yet). Instead, we want to stress two things. First is the sequential order by which the program will be executed. The control will execute line number N005 before moving on to line number N010. Then line number N015. And so on —until the entire program is activated.

Second is the visualization that is necessary in order to write programs. For this example, you must be able to visualize (see in your mind) the drill machining the workpiece before you can write this program. Indeed, if you cannot visualize how you will want a tool to move as it performs a machining operation, you will not be able to write the related CNC commands. For those machining operations with which you are not familiar, you will (at the very least) need the help of an experienced machinist.

Sequence numbers

Look at the example program again. Notice that each command of this program begins with an N *word*. N words are called *sequence numbers*. They are line identification numbers. For the most part, the CNC control will actually ignore them. They are just in the program to help you keep track of commands in the program. They really help us in this text, for example, to be able to point our specific things about a particular command – they confirm that you're looking at the command that's being discussed.

For longer programs, sequence numbers help you confirm that you are truly looking at the desired command (many commands look alike). This is extremely important if you are going to be making a modification to the program.

Again, the CNC control ignores sequence numbers. They can be left out of the program if you desire. Some programmers eliminate sequence numbers, for example, if the control's memory capacity is small. But we recommend that you include them in your programs – and place them in a logical order.

Throughout this text, every command of every program will have a sequence number. And you'll notice that we skip five numbers between sequence numbers. That is, we'll use N005, then N010, then N015, and so on. We recommend using this technique so you can add commands to your programs and still start them with sequence numbers.

A note about decimal point programming

As you have seen in the previous example program, certain CNC words allow a decimal point to be included within a numerical value. All current CNC controls allow a decimal point to be placed in CNC words which require *real numbers* (numbers that require a portion of a whole number). But most current controls do not allow a decimal point in words that always require integer values (whole numbers). All axis words (like X, Y, and Z) often require real numbers, so a decimal point is allowed. Spindle speed is specified with an integer value, so you are not allowed to include a decimal point in an S word.

Beginning programmers have the tendency to leave out decimal points. You must remember to include a decimal point within each word that allows a decimal point. If you do not, some very strange things can happen. Here's why.

Older controls (over thirty years old) do not allow a decimal point in any CNC word. These old controls require a *fixed format* for all real numbers needed in the program. Trailing zeros are required for these older controls to imply where the decimal point should be placed. For example, an X word of 5.0 in will be specified as "X50000" (note the four trailing zeros) if decimal point programming is not allowed.

Newer controls are *backward compatible*. This means programs written for older controls can still run in current controls. If a current CNC control sees a word that allows a decimal point, but the decimal point is not specified in the word, it will simply place the decimal point automatically – four places to the left of the right-most digit (in the inch mode).

Here's an example that illustrates the kind of mistake you can make. Say you intend to specify an X word of 3.0 in. The correct way to designate this movement is "**X3.0**" or "**X3.**". But say you make the mistake of omitting the decimal point. You specify "**X3**", leaving out the decimal point. In this case, the control will incorrectly interpret the X word. Instead of taking this word as 3.0 inches, the control will place the decimal point four places to the left of the right-most digit. In this case, "**X3**" will be taken as **X0.0003**, not 3.0 inches.

Concentrate on overcoming this tendency for making mistakes of omission. Get in the habit of including a decimal point within those words that allow it. These letter address words include: F for feedrate, I, J, & K, for circular motion commands, R for radius, and X, Y & Z for axis movements. Words that do not allow a decimal point and must be programmed as integer values include: N, G, H, D, L, M, S, T, O, and P.

A decimal point tip

When programming whole numbers in CNC words that allow a decimal point, be sure to carry the value out to the first zero after the decimal point. For example, write

> **X4.0**

Instead of

> **X4.**

This will force you to write/type the decimal point. In similar fashion, when specifying values under one, begin the value with the zero to the left of the decimal point. For example, write

> **X0.375**

instead of

> **X.375**

Again, this will force you to write/type the decimal point.

Other mistakes of omission

Knowing the mistakes beginners are prone to make may help you avoid them. Beginners tend to forget things in their programs. As stated, they forget to program a decimal point. They forget to turn on the spindle. They forget to turn on the coolant. They forget to drill a hole before tapping it. You will have to concentrate very hard on avoiding these mistakes. We're hoping that if you know about this tendency for making mistakes of omission, that it will help you avoid them.

Modal words

Many CNC words are *modal*. This means that the CNC word remains in effect until changed or canceled. In the previous example program, notice the **G00** command in line **N010**. This happens to be a rapid motion word. The positioning movement that results from line **N010** will be done at rapid. The very next (line **N015**) makes the movement in Z to approach the workpiece, but there is no motion type specification in this command. This movement will also occur at the machine's rapid rate because **G00** is modal. Modal words do not have to be repeated in every command.

Initialized words

Certain CNC words are initialized. This means the CNC control will automatically instate these words at power-up. For example, most machines used in the United States will be initialized in the inch mode (again, the instating word is **G20**). If a company will be using the inch mode exclusively, they can depend upon the CNC control to be in this state at all times (they'll never select the metric mode), meaning there is no need to actually include a **G20** in the program.

Letter O or number zero?

As you look at the example program in Figure 1.11, you see several zeros. But the *only* letter O in this program is the very first character (which happens to be the letter address that specifies a program number). All other characters that look like the letter O are actually *zeros*.

For example, notice the spindle-on-forward M code in line **N010** (**M03**). This is specified as M-zero-three, not M-oh-three. Again, the character in the middle is a number zero, not a letter O. This is very important.

The control will not behave properly if you use a letter O here. One very common beginner's mistake is to use the upper-case letter O instead of a number zero.

Word order in a command

With but one exception (that we'll show much later), the order in which CNC words appear in a command has no bearing upon how the command will be executed. For example, the command

N050 G00 X1.5 Y1.25 M08

will be executed in exactly the same way as

N050 M08 X1.5 G00 Y1.25

Key points for Lesson Three:

- You must be able to visualize the motions cutting tools will make as they machine workpieces.
- CNC programs are executed in sequential order, command by command.
- Commands are made up of words.
- Words are comprised of a letter address and a numerical value.
- You've seen your first complete CNC program. Though it was short, you've seen the sequential order by which CNC programs are executed – and you should understand the importance of being able to visualize a program's execution.
- We've introduced some program-structure-related points, including decimal point programming, modal words, and initialized words.
- You've seen some of the mistakes beginners are prone to making – like mistakes of omission and using an upper case letter O instead of a number zero.

Quiz

1) You must have basic machining practice experience in order to be able to visualize cutting tool motions.

        True   False

2) A CNC program is executed
   a) Randomly
   b) Sequentially
   c) Right-to-left
   d) None of the above

3) A decimal point must be included in all CNC words.

        True   False

4) A modal word
   a) is not allowed
   b) Is automatically instated at power up
   c) remains in effect until changed or cancelled
   d) only takes effect in the command in which it is included

4) An initialized word
   a) is not allowed
   b) Is automatically instated at power up
   c) remains in effect until changed or cancelled
   d) only takes effect in the command in which it is included

Talk with experienced people in your company to learn more: What kinds of machining operations does your company perform on its CNC machining centers? Ask to watch the execution of existing programs to see the motions for these machining operations.

Answers: 1: True, 2: b, 3: False, 4: c, 5: b

Fundamentals of CNC

Lesson 4

Program Zero And The Rectangular Coordinate System

A programmer must be able to specify positions through which cutting tools will move as they machine a workpiece. The easiest way to do this is to specify each position relative to a common origin point called program zero.

You know that machining centers have three linear axes – X, Y, and Z. You also know these axes move and that they have a polarity (plus versus minus). In order to machine a workpiece in the desired manner, each axis must, of course, be moved in a controlled manner. One of the ways you must be able to control each axis is with precise *positioning control*.

In the early days of NC (even before CNC, over forty years ago), a programmer was required to specify drive motor rotation in order to cause axis motion. This meant they had to know how many rotations of an axis drive motor equated to the desired amount of linear motion for the moving component (table or headstock, for example). As you can imagine, this was extremely difficult – it was not logical. There is no relationship between drive motor rotation and motion of the moving component. Today, thanks to program zero and the rectangular coordinate system, specifying positions through which cutting tools will move is much easier.

The rectangular coordinate system has an origin point that we'll be calling *program zero*. It allows you to specify all positions (we'll be calling *coordinates*) from this central location. As a programmer, *you* will be choosing the location for program zero – and if you choose it wisely, many of the coordinates you will use in the program will come directly from your workpiece drawing, meaning the number and difficulty of calculations required for your program can be reduced.

Graph analogy

We use a simple graph to help you understand the rectangular coordinate system as it applies to CNC. Since everyone has had to interpret a graph at one time or another, we should be able to easily relate what you already know to CNC coordinates. Figure 1.13 is a graph showing a company's productivity for last year.

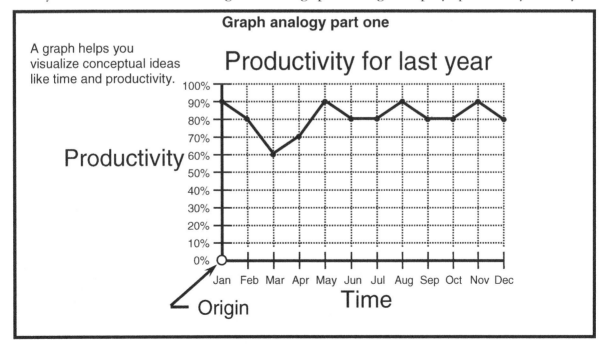

Figure 1.13 Graph example used to illustrate rectangular coordinate system

In figure 1.13, the *horizontal base line* represents time. The *increment* of the time baseline is specified in months. One whole year is the *range*, given from January through December. The *vertical base line* represents productivity. The *increment* for this base line is specified in ten percent increments and ranges from 0% to 100% productivity.

In order for a person to make this graph, they must have the productivity data for last year. They will plot a point along the vertical line corresponding to January (the vertical base line in this case) and along the horizontal line corresponding to the percentage of productivity (90% in our case). This plotting of points will be repeated for every month of the year. Once all of the points are plotted, a line or curve can be passed through each point to show anyone at a glance how the company did last year.

A graph is amazingly similar to the rectangular coordinate system used with CNC. Look at Figure 1.14.

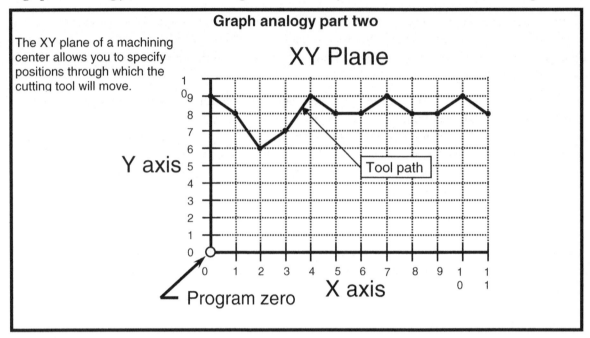

Figure 1.14 – The coordinate system of a machining center (XY plane)

For CNC machining centers, the *horizontal base line* represents the X axis. The *vertical base line* represents the Y axis. (The Z axis is at a right angle to this page, toward and away from you). For now, let's concentrate on the X and Y axis.) The *increment* of each base line is given in linear measurement. If you work in the inch mode (which we use throughout this text), each increment is given in inches. The smallest increment is 0.0001 inch, meaning each axis has a very fine grid. If you work in the metric mode, each increment will be in millimeters. In the metric mode, the smallest increment is 0.001 mm. The *range* for each axis is the amount of travel in the axis (from one over-travel limit to the other).

A metric advantage – Again, since most people in the US are accustomed to the inch mode, we use it for all examples in this text. However, you should know that there is an accuracy advantage with the metric mode. The advantage has to do with the least input increment – or resolution – for each axis. As stated, in the metric mode the least input increment is 0.001 mm, which is less than half of 0.0001 inch (0.001 mm is actually 0.000039 in). Think of it this way: A ten inch long linear axis has 100,000 programmable positions in the inch mode. The same ten inch long linear axis has 254,000 programmable positions in the metric mode!

What about the Z axis?
Figure 1.14 shows only two of the machining center's axes, X and Y. The Z axis behaves in exactly the same manner as X and Y. When taken all together, the X, Y, and Z provide you with a three dimensional grid. It is within this grid that you will be specifying positions (coordinates) that your tools will be passing through, as we show in Figure 1.15.

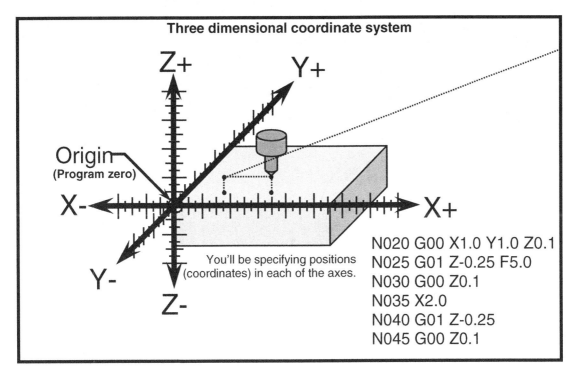

Three dimensional coordinate system

You'll be specifying positions (coordinates) in each of the axes.

N020 G00 X1.0 Y1.0 Z0.1
N025 G01 Z-0.25 F5.0
N030 G00 Z0.1
N035 X2.0
N040 G01 Z-0.25
N045 G00 Z0.1

Figure 1.15 – Three dimensional coordinate system of a machining center

Understanding polarity

In the graph example shown in Figure 1.13, notice that all points are plotted after January and above 0% productivity. The area up and to the right of the two base lines is called a *quadrant*. This particular quadrant is quadrant number one. The person creating the productivity graph intentionally planned for coordinates to fall in quadrant number one in order to make it easy to read the graph.

From Lesson Number One, you know that each axis has a polarity. You also know that since the cutting tool does not move along with every axis, it can be confusing to remember which way is plus and which way minus. (Consider the X and Y axes of a C-frame style vertical machining center, for example. As the table moves to the left, it is moving in the X plus direction.) To make things easier, we asked you to view polarity as if the tool is actually moving in each axis. Let's discuss the reason why a little more.

The rectangular coordinate system makes determining the polarity of coordinates used in a program *very* simple. Look at Figure 1.16, which shows the polarity for the X axis.

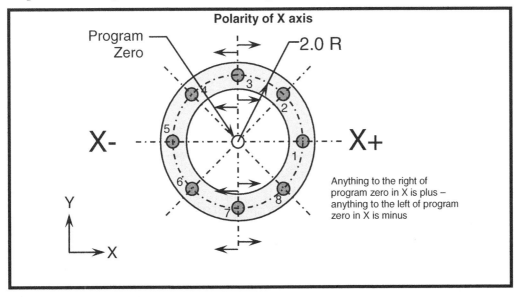

Polarity of X axis

Program Zero

2.0 R

Anything to the right of program zero in X is plus – anything to the left of program zero in X is minus

Figure 1.16 – X axis polarity

Notice that polarity is based upon the location of program zero – as it will be for all axes. For X, anything to the right of program zero is positive (plus). Anything to the left of program zero is negative (minus).

Now look at Figure 1.17, which shows the Y axis.

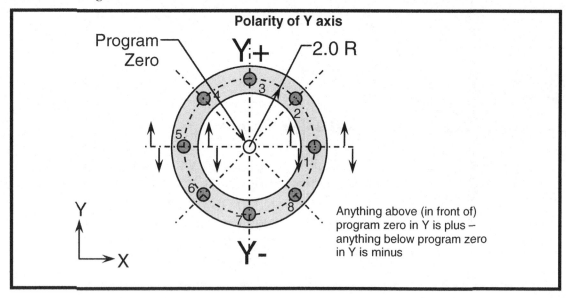

Figure 1.17 ~ Polarity for the Y axis

Anything above (in front of) program zero in Y is positive (plus). Anything below program zero in Y is negative (minus).

Here is an example showing coordinates in all four quadrants of the XY plane. Look at Figure 1.18.

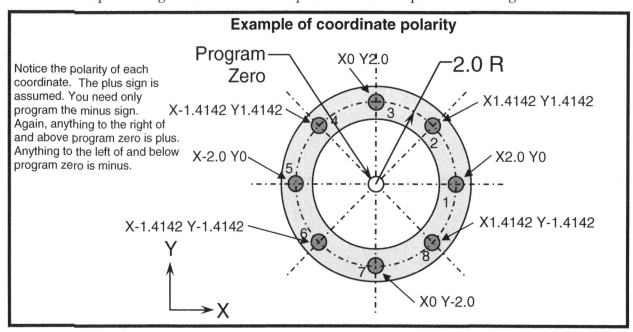

Figure 1.18 ~ Example of polarity for XY plane

Figure 1.18 shows a series of holes on a bolt pattern that must be specified relative to the center of the bolt pattern, which has been selected as program zero. As you can see, any hole to the left of program zero requires a negative X coordinate. Any hole below program zero requires a negative Y coordinate.

It just so happens that all CNC controls will *assume* that a coordinate is positive (plus) unless a minus sign is specified. The CNC word: X2.0, for example, specifies a position along the X axis of a *positive* two inches.

Some controls will actually generate an alarm if the plus sign is included within the word, meaning you must let the control assume positive values. Only include a polarity sign if it is negative (-).

Program zero must also be specified in the Z axis. Figure 1.19 shows the XZ plane (looking at a vertical machining center from the front).

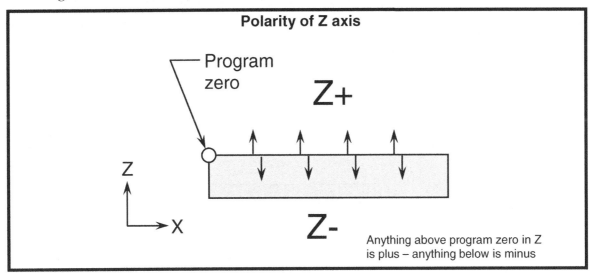

Figure 1.19 – Polarity for the Z axis

Anything above program zero in Z is positive (plus). Anything below program zero in Z is negative (minus). Figure 1.20 shows an example of polarity for the Z axis.

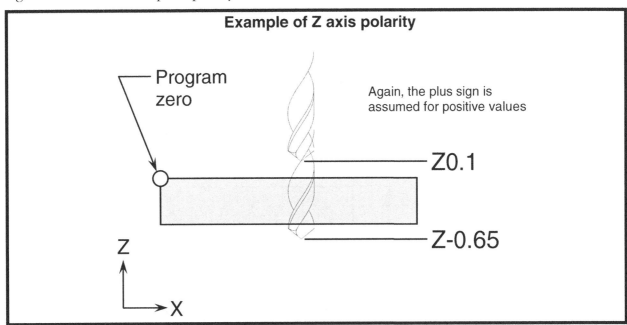

Figure 1.20 – Example of Z axis polarity

Notice in Figure 1.20, we've selected the top of the workpiece as program zero in Z (which is a common program zero point for vertical machining center applications). Any tool position above the top of the workpiece is positive (plus) in Z. Any position below the top of the workpiece is negative (minus) in Z.

Wisely choosing the program zero point location

As the programmer, *you* determine the program zero point location for every program you write. Frankly speaking, program zero could be placed in *any* location. As long as the coordinates used in your program are specified from the program zero point, the program will function properly. Though this is the case, the *wise* selection of the program zero point will make programming much easier. It may also make it easier for the setup person.

In X and Y

A good rule-of-thumb for selecting the program zero point location is to base your decision on how the workpiece drawing is dimensioned. Look for workpiece surfaces from which dimensions begin. Though design engineers vary with how obvious they make it, you should be able to find one surface in each axis from which dimensions along that axis start. Look at Figure 1.21, for example.

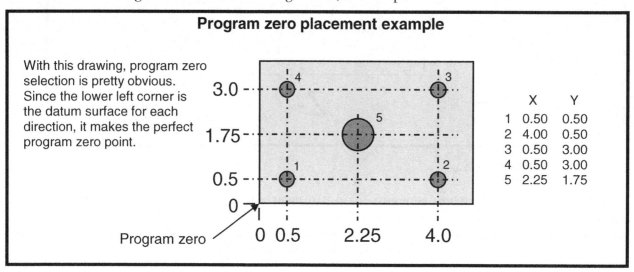

Figure 1.21 – Example of how to place the program zero point – Datum surface dimensioning

Design engineers don't always make it so obvious. Look at Figure 1.22.

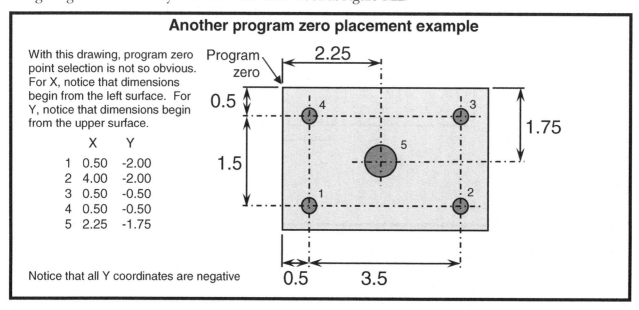

Figure 1.22 – The best location for program zero may be more difficult to determine

With the location of program zero determined, notice that all coordinates must be calculated from this point. With datum surface dimensioning (Figure 1.21), it is very easy. Coordinates used in your program will almost always match print dimensions. But if the design engineer doesn't use datum surface dimensioning techniques (Figure 1.22), you'll have some calculations to make.

Remember that you must determine how far it is from the program zero point to each coordinate you need in your program. As we show for the X and Y coordinates of point number two in Figure 1.22, you'll often have to do some arithmetic.

> For the X value of point two, you must add 0.5 plus 3.5, totaling X4.0

> For the Y value of point two, you must add 0.5 plus 1.5, totaling Y-2.0 (note that the result is negative, since it is below program zero)

And remember, anything to the right of program zero in X is plus. Anything to the left is minus. Anything above program zero in Y is plus. Anything below is minus.

Reminder about axis movement

Though we don't want to confuse you, we need to bring this up one more time. Remember, with a C-frame style vertical machining center, the tool does not move along with the X and Y axes. Instead, the table moves in these axes.

Consider once again the workpiece in Figure 1.21. Again, program zero is the lower left corner in X and Y. When you want the tool to move to point number five, what coordinates will you specify? You should easily agree that you'll send the tool to a position of

> X2.5 Y1.75

Notice that these coordinates are both specified as positive values (plus sign is assumed). At the completion of the motion, the tool will be resting over point number five. Here is the key question: Which way does the table have to move in each axis to bring the tool to this position? The answer is: *"Who cares?"*

This is the reason why we ask you to view polarity as if the tool is moving in all axes. You'll always be specifying polarity in your program relative to the program zero point (with one exception that we'll show shortly). You will never have to concern yourself about which way the table (or any axis) must move in order to get the tool to the desired position.

Say the tool is now resting at point number five (again, in Figure 1.21). You now want it to move to point number one (lower-left hole). What coordinates will you specify (be careful)? Hopefully, you agree that you will specify

> X0.5 Y0.5

Again, you specify *all* coordinates relative to program zero. Which way (plus or minus) will the machine move during this motion? Though again, you need not concern yourself with the direction of each motion, it just so happens that the tool will be moving in the negative direction in both axes during this motion – moving from a large coordinate in each axis (X2.5 Y1.75) to a smaller one (X0.5 Y0.5). And by the way, remember from Lesson Number One that the negative direction for X is table motion to the right. The negative Y direction is table motion away from you.

Hopefully, you are beginning to understand how much the rectangular coordinate system is doing for you. It makes it very simple to specify positions within your CNC program regardless of which way the tool, axis, or moving component of the machine must move to get there. You need only concern yourself with where you want the tool to move – and always specify this position relative to the program zero point.

In Z

As with XY, you must specify a program zero point in the Z axis. Though the location of Z axis program zero point can vary, most programmers will make the program zero surface in Z the top surface of the workpiece. See Figure 1.23. If this is done, any machining that occurs into the workpiece (below the top surface) will require a negative (-) Z axis coordinate. This is the Z axis program zero point location used for all examples shown in this text.

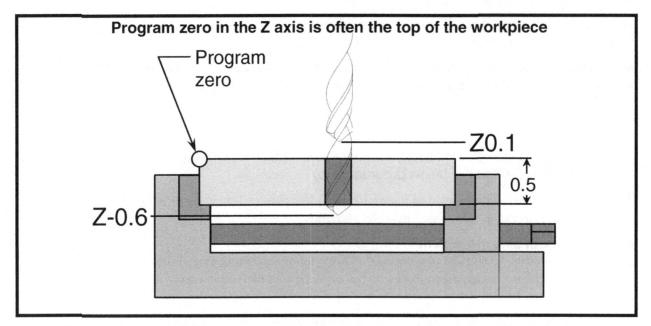

Figure 1.23 ~ Most programmers make program zero in Z the top of the workpiece for vertical machining center programs

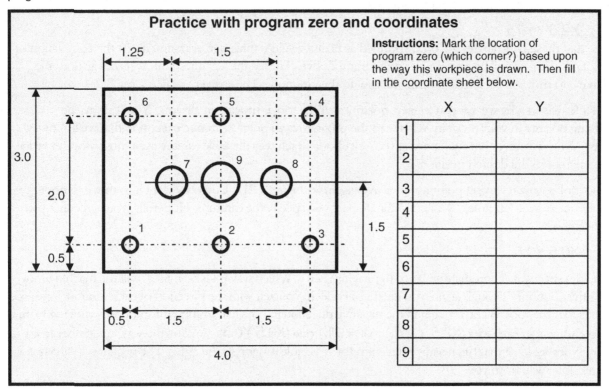

Answers:
Program zero is the lower left corner of the workpiece.
1: X0.5 Y0.5
2: X2.0 Y0.5
3: X3.5 Y0.5
4: X3.5 Y2.5
5: X2.0 Y2.5
6: X0.5 Y2.5
7: X1.25 Y1.5
8: X2.75 Y1.5
9: X2.0 Y1.5

What about tolerances?

Very few dimensions shown in this text specify a tolerance. But as you know – design engineers specify a tolerance for *every* dimension on a blueprint.

Which value do you program?
*Specify the **mean value** for every dimension (coordinate) you use in your program.* The mean value is, of course, right in the middle of the tolerance band. With some tolerances, it is more difficult to determine the mean value than with others.

Plus/minus tolerance: 4.0 +/- 0.002
The mean value is specified right in the dimension. 4.0 is the mean value of the dimension above

Uneven tolerance: 4.0 +0.003, -0.001
The mean value must be calculated. Divide the overall tolerance (0.004 above) by two (0.002) and subtract it from the high limit (4.003 is the high limit). The mean value of the dimension above is 4.001.

High and low limit specified: 4.004 / 4.001
The mean value must be calculated. Again, divide the overall tolerance (0.003 above) by two (0.0015) and subtract it from the high limit (4.004 is the high limit). The mean value of the dimension above is 4.0025.

Absolute versus incremental positioning modes

Though we have not actually said so yet, when you specify coordinates from the program zero point (as we have just introduced), it is called the *absolute positioning mode*. The absolute positioning mode is specified by a **G90** word. Once a **G90** is specified, all coordinates are taken to be from the program zero point.

Everything we've said to this point in Lesson Four has been related to the absolute positioning mode. And again, the point of reference for absolute positioning mode is the program zero point.

There is another method of positioning called the *incremental positioning mode*. **G91** is used to specify this mode. Unlike the absolute mode, once the control sees a **G91**, the point of reference for all specified positions will be the tool's *current position* – the location of the tool at the beginning of the motion.

In the incremental mode, each movement is specified as a distance and direction from the tool's current position. At first glance, it may seem easier to work in the incremental mode than in the absolute mode. But you will soon find that programming with incremental positioning is quite difficult and error-prone. And by the way, if you make a mistake in a series of incrementally specified motions, every movement from the mistake on will be incorrect.

While there are some excellent applications for incremental mode (we'll show them in Key Concept Number Six), beginning programmers should concentrate on working exclusively in the absolute mode. Note that all examples in this text (with the exception of some we show in Key Concept Number Six) will be shown using the absolute mode.

Any series of motions can be commanded in either the absolute or incremental mode. Look at Figure 1.24.

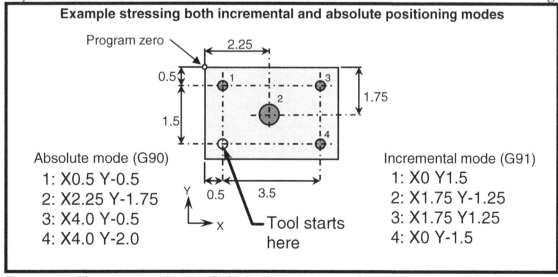

Figure 1.24 – Movements can be specified in both the absolute and incremental positioning mode.

As you can see, absolute positioning makes more sense. Coordinates often match print dimensions – but even when they don't – the point of reference for each position is the same – the program zero point. Incremental positioning doesn't make much sense. Positions are nothing more than a whole series of disjointed movements, each taken from the tool's previous position.

A decimal point reminder

As stated in Lesson Three, coordinates (X, Y, and Z) require real numbers. That is, values that often include a portion of a whole number. You must remember to include a decimal point with coordinates, even when you are specifying a whole number value. For example, if you want to specify an X coordinate of three inches, you must specify

> X3.0

not

> X3

If you specify "X3", the control will use the fixed format for the X word. It will automatically place the decimal point in a position four places to the left of the right-most digit (in inch mode). The value "X3" will be taken as X0.0003, and not X3.0.

We repeat our suggestion about how to specify any real number value:
When programming whole numbers in CNC words that allow a decimal point (real number values), be sure to carry the value out to the first zero after the decimal point. For example, use

> X4.0

Instead of

> X4.

This will force you to write/type the decimal point. In similar fashion, when specifying values under one, begin the value with the zero to the left of the decimal point. For example, use

> X0.375

instead of

> X.375

Again, this will force you to write/type the decimal point.

Most programmers prefer to use decimal point programming for obvious reasons. It doesn't make much sense to program using the fixed format. But be ready for some computer aided manufacturing (CAM) systems that automatically output G code level CNC programs in fixed format – without the decimal point. They can be a little tough to interpret.

Key points for Lesson Four:

- You must be able to specify positions (coordinates) within CNC programs.
- You know that the rectangular coordinate system of a CNC machine is very similar to that used for a graph.
- In CNC terms, the origin for a coordinate system is called the program zero point.
- From a programmer's viewpoint, polarity for each positioning movement is based upon the commanded position's relationship to program zero.
- The program zero point location is determined based upon how the workpiece drawing is dimensioned.
- When you specify coordinates relative to program zero, you're working in the absolute mode.

Lesson 5
Determining Program Zero Assignment Values

The programmer chooses the program zero point location. But the CNC machining center must also be told where program zero is located so it can move cutting tools accordingly.

You know from Lesson Four that the program zero point is the origin for your program. All coordinates specified in your program are taken from program zero (in the absolute positioning mode). You also know that the program zero point is determined based upon how the print is dimensioned. The program zero point is placed at the location where dimensions begin in each axis.

You must understand that just because you want the program zero point to be in a given location, doesn't mean the CNC machining center is automatically going to know where it is. A conscious effort must be made to *assign* program zero. Program zero assignment is the task of telling the CNC machining center where the program zero point is located. You can think of this task as marrying your program to the workholding setup that is made on the machine.

Much of what we present in lessons five and six is more related to setup than it is to programming. However, a CNC programmer must be able to *instruct and direct* setup people, providing instructions related to how a given setup must be made. This means they must understand almost as much about setups as setup people – and this includes an understanding of how program zero is assigned.

Program zero assignment values

Program zero assignment involves determining the distance between the program zero point and a special reference position in each axis. For the X and Y axes, the reference position is the spindle center. The program zero assignment values in X and Y are the distances between the program zero point and the centerline of the spindle. Figure 1.25 shows the program zero assignment values needed for the X and Y axes.

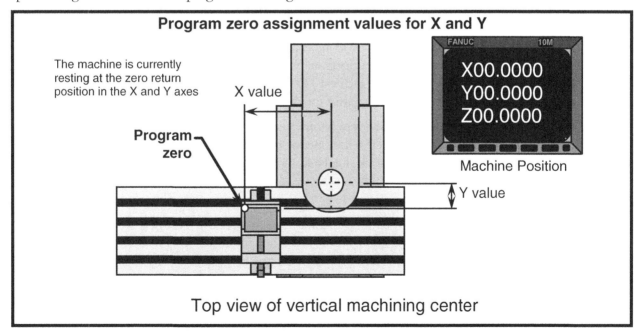

Figure 1.25 – You must determine the program zero assignment values for the X and Y axes

For the Z axis, the point of reference will be based upon how you choose to use a feature called *tool length compensation*. (We don't introduce tool length compensation until Lesson Eleven, so for now, we'll only show our recommended method.) Based upon our recommended method, the reference position for the Z axis will be the spindle face (the surface of the spindle itself – not the key that protrudes from it). The Z axis program zero assignment value is the distance between program zero in Z (commonly the top surface of the workpiece) and the spindle nose. Figure 1.26 shows the program zero assignment value needed for the Z axis.

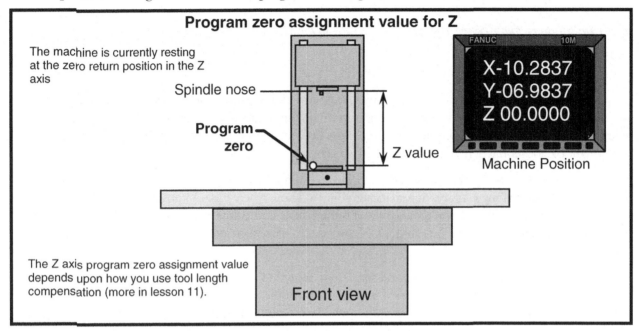

Figure 1.26 ~ You must determine the program zero assignment value for the Z axis

What is zero return position?

In Figures 1.25 and 1.26, the program zero assignment values are shown while the machine is resting at something called the *zero return position*. Some people call this position *home position* or *machine zero*. The zero return position is the point-of-reference for program zero assignment values. So there are actually two zero positions that are of extreme importance to you – the program zero position which is on the workpiece and the zero return position which is the point of reference for program zero assignment values.

The zero return position is a very accurate location along each axis. The machine start-up procedure actually requires that you send each axis to its zero return position. When the machine is sent to its zero return position, three *axis origin lights* come on (one for each axis) to indicate that the machine is truly resting at its zero return position. Additionally, the *machine position page* of the display screen will show zero in each axis.

The machine tool builder determines where the zero return position is placed. For vertical machining centers, the zero return position is almost always placed at the extreme plus limit of each axis. When a vertical machining center is resting at zero return position, the table is all the way to the left in X, all the way forward in Y (toward the operator), and the headstock of the machine is all the way up in Z. Figures 1.25 and 1.26 show this.

For horizontal machining centers, the zero return position for Y is usually when the headstock of the machine is all the way up. For Z it is when the table is as far away from the spindle nose as possible. Again, the Y and Z axes are at their extreme plus limits. But with the X axis, machine tool builders vary. Some place it at the center of the X axis travel (center of the table aligned with center of the spindle) while others place it at the extreme plus end of the X axis (table all the way to the left as viewed from the spindle).

Determining program zero assignment values

Again, the zero return position is the point-of-reference for program zero assignment values. By one means or another, you must be able to determine the distance in each axis between the program zero point and the zero return position (spindle center in X and Y – spindle nose in Z). Figures 1.25 and 1.26 show the needed values. There are two basic ways to determine program zero assignment values – physically *measuring* them during the setup and *calculating* them prior to making the setup. Which method you use is based upon whether or not your company makes *qualified workholding setups* and whether or not you use predictable workholding tools (fixtures).

How predictable are your workholding tools?
When an engineer designs a special fixture, they specify certain dimensions from workpiece location surfaces on the fixture to the components that locate the fixture to the table (commonly keys or pins). If the fixture is made accurately and these dimensions are correct, it will be possible to *calculate* the program zero assignment values before the setup is ever made. This also requires you to know some important distances on the machine itself – like the distances from the zero return position in each axis to key-slots or location holes on the machine table. Frankly speaking, not many programmers know the related dimensions or make the effort to find or measure them – so very few programmers calculate program zero assignment values.

What is a qualified workholding setup?
By qualified workholding setup, we mean one that can be replaced on the machine table in *exactly* the same location – over and over again. Product producing companies can often justify the higher fixture costs related to making qualified workholding setups. The fixture discussed above, for example, is qualified. It will be located in machine table tee-slots or pin holes. If a workholding setup is truly qualified, program zero assignment values will be *exactly the same* every time the setup is made. While it may not be possible to calculate program zero assignment values, it should not be necessary to measure them after the setup is made for the first time. Many companies run the same jobs over and over (lots of repeat business), and for them, qualifying workholding setups can save a lot of setup time.

Manually measuring the program zero point location on the machine

Again, workholding devices are seldom predictable enough to allow the calculating of program zero assignment values. This means another method of determining program zero assignment values must be used – *actually measuring them at the machine during setup*. Do keep in mind that if your company makes qualified workholding setups, this measurement should only be necessary the very first time each new job is run. If you document the program zero assignment values (and if the setup is truly qualified), you won't have to re-measure program zero assignment values every time the job is run.

Here we offer a general procedure that can be used to physically measure program zero assignment values at the machine. We'll be using the *position display screen* for the purpose of taking these measurements. With the *relative position display*, you are allowed to *reset* (set to zero) or *preset* (set to a specific value) any axis display at any time. This allows you to specify the starting point for a measurement. In essence, you'll be using the machine as a measuring tool to help you determine the program zero assignment values.

In X and Y for a square or rectangular workpiece

This procedure involves using an *edge finder* (also called a wiggler) and techniques similar to those used on manual milling machines when picking up an edge. As with many setup related tasks, this requires some basic machining practice experience. While you may be able to easily understand this presentation, if you do have questions, ask an experienced person in your company or school to demonstrate how an edge finder is used.

We'll be using an edge finder having a 0.200 inch diameter (0.100 inch radius). When the edge finder is flush with the edge of the workpiece, of course, the centerline of the spindle will be precisely 0.100 inch away from the surface being touched by the edge finder.

1) Load workpiece and edge finder
Begin by making the workholding setup and load a workpiece. Place the edge finder in the spindle and start the spindle at about 500 rpm. Push on the edge finder stylus to cause it to run out (it's now wiggling and ready to pick up a surface in X or Y).

2) Touch off the X program zero surface and reset the X axis display
Using the joystick and handwheel, move the table so that the edge finder comes flush with the X program zero surface – left side of our example workpiece – as shown in the drawing below. With the edge finder in this position, reset the X axis display (set it to zero).

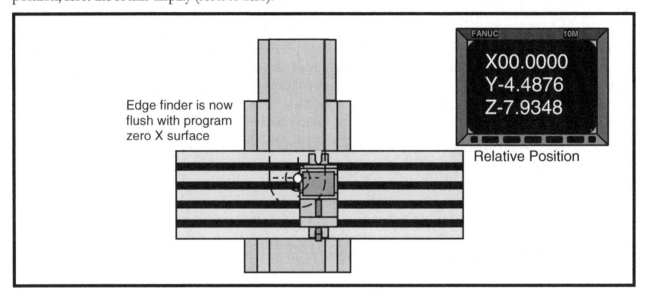

3) Back off in Z, move over 0.100 inches in X, and reset the X axis display again
Using the handwheel, move the edge finder up (plus) in the Z axis to clear the top of the workpiece. Then move the X axis precisely 0.100 in the plus direction. The centerline of the spindle will now be right over the top of the program zero surface in X as shown in the drawing below. Now reset the X axis display again.

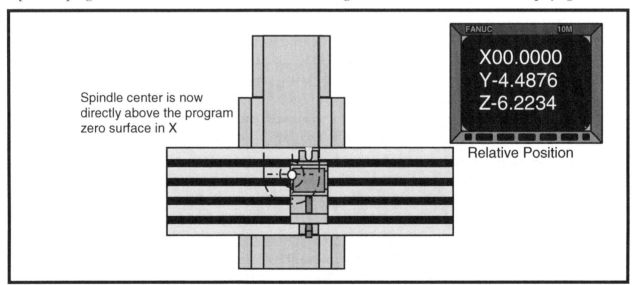

4) Touch off the Y program zero surface and reset the Y axis display
Now repeat for the Y axis. Using the joystick and handwheel, move the table so that the edge finder comes flush with the Y program zero surface – column side of our example workpiece – as shown in the drawing below. With the edge finder in this position, reset the Y axis display (set it to zero).

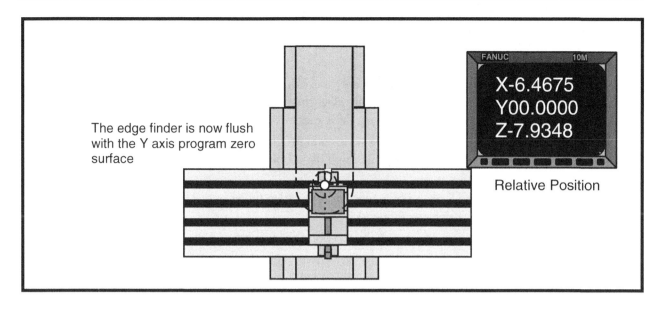

5) Back off in Z, move over 0.100 inch in Y, and reset the Y axis display again
Using the handwheel, move the edge finder up (plus) in the Z axis to clear the top of the workpiece. Then move the Y axis precisely 0.100 in the minus direction. The centerline of the spindle will now be right over the top of the program zero surface in Y as shown in the drawing below. Now reset the Y axis display again.

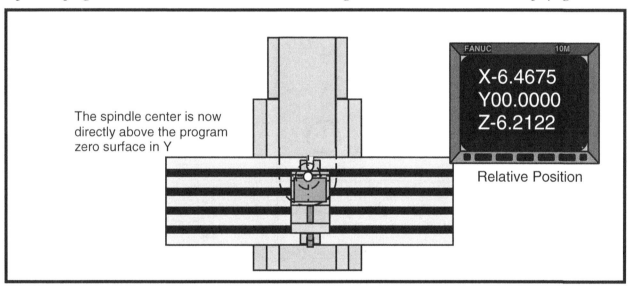

6) Send the X and Y axes to the zero return position
When you send the X and Y axes to the zero return position, the axis displays will follow along. When the X and Y axes are resting at the zero return position, the X and Y axis displays will be showing the program zero assignment values (distance from program zero to spindle center in X and Y while the machine is resting at the zero return position).

Key points for Lesson Five:
- Program zero assignment values must be determined before program zero can be assigned.
- Program zero assignment values are the distances between the program zero point and the machine's zero return position.
- There are only two ways to determine program zero assignment values – calculate them or measure them.

- Program zero assignment values can only be calculated if you make qualified workholding setups, if you use predictable workholding tools, and if you know some important dimensions on the machining center.

- If you make qualified workholding setups but your workholding tooling is not predictable, you will have to measure program zero assignment values the very first time a setup is made. If you document these values, you will not have to perform these measurements again.

- If you don't make qualified workholding setups and program zero assignment values must be measured for every setup, you should use a spindle probe to measure program zero assignment values and assign program zero.

- If you must manually measure program zero using an edge finder or dial indicator, you're using the machine as a very expensive measuring tool.

- You must consider the ease of setup as the top priority when it comes to deciding where the program zero point is placed.

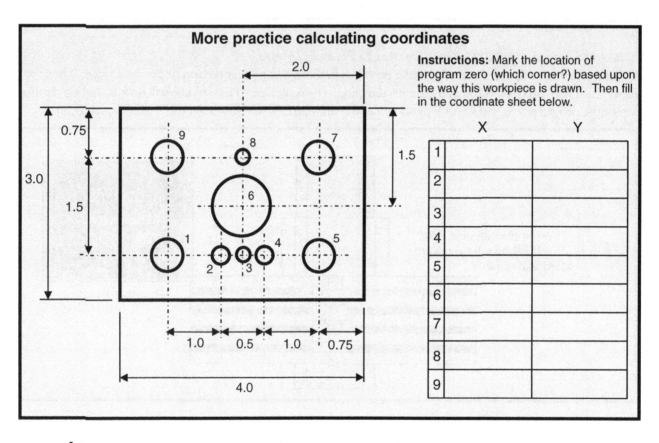

More practice calculating coordinates

Instructions: Mark the location of program zero (which corner?) based upon the way this workpiece is drawn. Then fill in the coordinate sheet below.

	X	Y
1		
2		
3		
4		
5		
6		
7		
8		
9		

Answers:
Program zero is the upper right corner of the workpiece.
1: X-3.25 Y-2.25
2: X-2.25 Y-2.25
3: X-2.0 Y-2.25
4: X-1.75 Y-2.25
5: X-0.75 Y-2.25
6: X-2.0 Y-1.5
7: X-0.75 Y-0.75
8: X-2.0 Y-0.75
9: X-3.25 Y-0.75

Lesson 6

Assigning Program Zero

Once program zero assignment values are determined as shown in Lesson Five, you must assign program zero. The most popular way to do so is with fixture offsets.

You now know how to determine program zero assignment values. As we discussed in Lesson Five, most setup people will measure program zero assignment values the very first time a new setup is made. If the workholding setup is qualified (it can be replaced on the table in exactly the same position), they will document these values so measurements need not be repeated every time the setup is made. If the workholding setup is not qualified, program zero assignment values must be re-measured every time the setup is made – which is an excellent application for a spindle probe.

In this lesson we're going to show how to *assign* program zero using the previously determined program zero assignment values. For the most part, we're going to assume that you *do not* have a spindle probe to measure and assign program zero. But even if you do, you must understand the points we make in this lesson.

As with Lesson Five, this lesson has more to do with setup than with operation. But again, programmers must know enough about making setups to instruct setup people. This includes knowing how program zero is assigned.

Understanding fixture offsets

Offsets are storage registers within the control. They are used to store numerical values – and will not be used until they are *invoked* by the program. In general, offsets are used to separate certain *tooling related values* from the CNC program – keeping you from having to know them when as program is written.

Fixture offsets are the registers used to assign the program zero point/s. In them, you'll be placing program zero assignment values. Most controls come with at least six sets of fixture offsets, meaning up to at least six different program zero points can be assigned and used by a program. (We haven't seen any applications for more than one program zero point in a program yet. We'll show some in Lesson Thirteen.)

Figure 1.29 shows the fixture offset display screen page of a popular CNC machining center control.

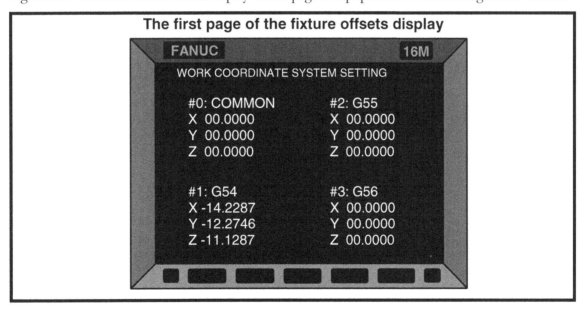

Figure 1.29 – First page of fixture offsets display screen

Notice that fixture offsets are organized by number. Figure 1.29 shows the *common offset* (#0 on the fixture offset display screen page) as well as fixture offsets one through three. The second page of fixture offsets (not shown) displays fixture offsets four through six.

For now, let's concentrate on assigning but one program zero point per program. We'll use fixture offset number one to do so. Again, notice that fixture offset number one is shown in the lower left corner of the fixture offset page (specified with #1). Also note the G code next to it: G54. G54 is the G code used to invoke fixture offset number one. When the control sees a G54 in the program, it uses the values stored in fixture offset number one's set of registers as the current set of program zero assignment values.

Fixture offset values have a *polarity*. These values are the distances *from* the zero return position *to* the program zero point in each axis. When the zero return position is very close to the plus limit for each axis (as it is for most machining centers), the polarity of fixture offset values will *always* be negative, as shown in Figure 1.29.

So to complete the task of program zero assignment, the setup person must enter the previously measured program zero assignment values into the appropriate fixture offset registers – with the required polarity (as negative values for most machines). As stated in Lesson Five, if a spindle probe is used to measure program zero assignment values, it will *automatically* enter the program zero assignment values into fixture offsets.

Key points for Lesson Six:

- If your machine has fixture offsets, use them to assign program zero.
- To assign program zero, the program zero assignment values shown in Lesson Five are placed into fixture offsets.
- You've seen a full example of how a program zero is assigned using fixture offsets.
- Again, fixture offsets offer numerous advantages over G92 – but if your machine does not have fixture offsets, you must use G92 to assign program zero. You must be very careful, since the G92 method of assigning program zero is not nearly as failsafe as when using fixture offsets.

Lesson 7
Introduction To Programming Words

As you know, all CNC words include a letter address and a numerical value. The letter address identifies the word type. You should be able to quickly recognize the most common ones.

You know that CNC programs are made up of commands – and that commands are made up of words. Words are categorized into types – and each word type has a special meaning to the control. Each word type is designated by a *letter address*. You already know a few the letter addresses, like N for sequence number, G for preparatory function, X, Y, and Z for axis designations, S for spindle speed, F for feedrate, and M for miscellaneous (or machine) functions. In this lesson, we're going to introduce the rest of the word types.

If you are a beginner looking at the word types for the first time, you may want to read this section a few times to get better acquainted with these word types. Note that we are *not* asking you to memorize the word types – just to get familiar with them. In Key Concept Number Five – program formatting – we will provide you with a way to remember each word's function.

Also, this lesson is only intended to *introduce* each word, not to give you an in-depth description. When appropriate, we'll point you to the lessons that discuss the word type in more detail.

You will find that certain words are seldom used, meaning you will have little or no need for them. Other words are constantly used, and you will soon have them memorized after writing a few programs.

Some CNC words have more than one function, depending on commanded format. We will be showing you the primary (most common) function of the word next to the "A" description and the secondary use for the word next to the "B" description.

Once you've seen a word type a few times, it should not be too difficult to remember its function. Again, most word types are aptly named with a logical letter address. Additionally, only about fifty words are used consistently when programming, so try to look at learning to program a CNC machining center as like learning a foreign language that contains only fifty words.

As you continue with this text, *use this lesson as a reference*. If you come across a word or word type you don't recognize, remember to come back to this lesson. You've probably already noticed that we provide a quick reference for CNC words on the inside front and back cover of this text, but information in this lesson is a little more detailed.

Words allowing a decimal point
Current CNC controls allow you to include a decimal point in those words that are used to specify *real numbers* (values that require a portion of a whole number). You must remember to include a decimal point with these words or the control will revert to the fixed format for the word (as discussed in lessons three and four). Word types that allow a decimal point include:

A, B, C, X, Y, Z, I, J, K, F, Q, and R

Certain CNC words are used to specify *integers* (whole numbers). These word types do *not* allow a decimal point:

O, N, G, P, L, S, T, M, D, and H

O

This is the word most controls use for a **program number**. All machines discussed in this text allow the user to store multiple programs in the memory of the control. The program number allows the specification of an individual program currently stored in the control. You will be assigning every program a program a number from 0001 through 9999 (**O0001** through **O9999**). The O word will be the *very first word* in the program. A decimal point is not allowed with the O word. Program numbers are discussed in more detail in Lesson Fifteen.

N

This word specifies a **sequence number**. Sequence numbers are used to identify commands in a CNC program. They allow you to organize each command in the program by number. This allows easier modification of the program at the machine. Sequence numbers are not required to be in any particular order and can even repeat in the program. Actually, they need not be in the program at all. But for the sake of organization purposes, we recommend that you include them in the program and place them in an understandable order. We use them for all examples in this text. We skip five numbers for each sequence number (going by fives). This allows extra commands to be added to the program while still using sequence numbers. A decimal point is not allowed with the N word. Sequence numbers are introduced in Lesson Three and discussed in more detail in Lesson Fifteen.

G

This word specifies a **preparatory function**. Preparatory functions prepare the control for what is coming – in the current command – and often in upcoming commands. They set modes (though some G codes are non-modal). There are many G words, but only a few that are used on a consistent basis. For a list of all G codes, see the list at the end of this lesson. A decimal point is not allowed with most G words. But do note that with Fanuc controls, some G codes do use a decimal point (like **G84.1** for rigid tapping). A decimal point used in this fashion is simply part of the G code's designation – and is not the true application for a decimal point. Preparatory functions are introduced in Lesson One and discussed often throughout this text.

X

A. The primary use for the X word is to designate a coordinate along the X axis. That is, it is the **X axis designator**. The X word allows a decimal point. An X position of 10 inches will be specified **X10.0**. Axis designators are introduced in Lesson One and discussed often during this text.

B. The secondary use for the X word is that it can be used to specify **dwell time** in seconds in a dwell command (**G04**). Dwell commands are used to make axis motion (for all axes) pause for a specified length of time. The dwell command is discussed in Lesson Eighteen.

Y

The Y word is the **Y axis designator**. The Y word allows a decimal point.

Z

A. The primary use for the Z word is as the **Z axis designator**. The Z word allows a decimal point.

B. The secondary use for the Z word is to specify the **hole-bottom position** in a canned cycle command. Canned cycles are discussed in Lesson Sixteen.

A

A. For machines that have a rotary axis mounted parallel to the X axis, letter address A is the **A axis designator**. By *parallel to the X axis*, we mean the *centerline* of the rotary axis is parallel to the X axis. Most vertical machining centers that are equipped with a rotary table have the rotary table mounted in this fashion. A decimal point is allowed with the A word when it is used as the A axis designator. The A word can be programmed to three places, meaning the rotary axis has 360,000 positions. A true rotary axis allows machining *during* rotary axis motion. Rotary axes are discussed in Lesson Nineteen.

B. For machines equipped with a one-degree indexer and when the indexer is mounted on the machine with its center parallel to the X axis, the A word is the **indexer activator**. You are not allowed to use a decimal point with the A word when it is used to activate an indexer. A50 specifies a fifty degree index. With an indexer, machining can only occur *after* the indexer rotates (not during rotation). Indexers are discussed in Lesson Nineteen.

B

A. For machines that have a rotary axis mounted parallel to the Y axis, B is the **B axis designator**. By *parallel to the Y axis*, we mean the *centerline* of the rotary axis is parallel to the Y axis. This is always the case when a rotary axis is mounted within the table of a horizontal machining center. A decimal point is allowed with the B word when it is used as the B axis designator. The B word can be programmed to three places, meaning the rotary axis has 360,000 positions. A true rotary axis allows machining *during* rotary axis motion. Rotary axes are discussed in Lesson Nineteen.

B. For machines equipped with a one-degree indexer and when the indexer is mounted on the machine with its center parallel to the Y axis, the B word is the **indexer activator**. You are not allowed to use a decimal point with the B word when it is used to activate an indexer. B50 specifies a fifty degree index. With an indexer, machining can only occur *after* the indexer rotates (not during rotation). Indexers are discussed in Lesson Nineteen.

C

A. For machines that have a rotary axis mounted parallel to the Z axis, C is the **C axis designator**. By *parallel to the Z axis*, we mean the *centerline* of the rotary axis is parallel to the Z axis. This may be the case when a machining center has *two* rotary axes (it is a five-axis machining center). A decimal point is allowed with the C word when it is used as the C axis designator. The C word can be programmed to three places, meaning the rotary axis has 360,000 positions. A true rotary axis allows machining *during* rotary axis motion. Rotary axes are discussed in Lesson Nineteen.

B. For machines equipped with a one-degree indexer and when the indexer is mounted on the machine with its center parallel to the Y axis, the C word is the **indexer activator**. You are not allowed to use a decimal point with the C word when it is used to activate an indexer. C50 specifies a fifty degree index. With an indexer, machining can only occur *after* the indexer rotates (not during rotation). Indexers are discussed in Lesson Nineteen.

Note about rotary axis designators and indexer activators
Not all machine tool builders adhere to the rotary axis and indexer activator naming conventions we've just introduced. What is more important than naming conventions, however, is that you know how the rotary axis or indexer is designated for the machine/s with which you must work. If your machining center is equipped with a rotary axis or one-degree indexer, consult your machine tool builder's programming manual to determine its designating letter address.

R
A. The primary use for the R word is to specify the **radius of a circular move**. The R word allows a decimal point. Circular motion is discussed in Lesson Nine.

B. The secondary use for the R word is to specify the **rapid plane** for a canned cycle command. Canned cycles are discussed in Lesson Sixteen.

I, J, K
A. I, J, and K words can be used to specify the **arc center point** of a circular motion. While they are still effective, we strongly recommend that beginners concentrate on using the R *word* to specify the arc size in a

circular move because it is much easier. I, J and K allow a decimal point. Circular motion (including the use of directional vectors) is discussed in Lesson Nine.

B. The secondary function for I and J is with canned cycles to specify the **move over amount** at the bottom of a fine boring cycle (**G76**). I and J used in this fashion allow a decimal point. Canned cycles are discussed in Lesson Sixteen.

Q

The Q word is used with the two peck drilling canned cycles (**G73** and **G83**) to specify **peck depth**. The Q word allows a decimal point. Canned cycles are discussed in Lesson Sixteen.

P

A. The P word can be used to specify **dwell time** in seconds for a dwell command (**G04**). Dwell commands are used to make axis motion (for all axes) pause for a specified length of time. A time of three seconds is specified as **P3000** (a decimal point is *not* allowed). Note the fixed format of the P word. There are three places to the right of the automatically placed decimal point position. Other examples: **P2500** is 2.5 seconds, **P500** is 0.5 second, and **P10000** is 10 seconds. Note that the X word can also be used to specify the time for a dwell command – and since it allows a decimal point – most programmers prefer using it instead of the P word. Again, a decimal point is not allowed with the P word. The dwell command is discussed in Lesson Eighteen.

B. The secondary use for the P word is with sub-programming to specify the **subprogram program** number of the program to be called. A decimal point is not allowed with the P word. Sub-programming is discussed in Lesson Seventeen.

L

A. The L word is used with sub-programming to specify the **number of executions** for the subprogram. A decimal point is not allowed with the L word. Subprograms are discussed in Lesson Seventeen.

B. The L word can be used with canned to specify the **number of holes** to machine. A decimal point is not allowed with the L word. Canned cycles are discussed in Lesson Sixteen.

F

The F word specifies **feedrate** – which is the motion rate for machining operations. It is used with straight line and circular motion commands (see **G01**, **G02**, and **G03**), along with any other interpolation types equipped with your machining center/s. Feedrate is affected by the currently instated measurement system mode (inch or metric). Inch mode is specified with **G20** and metric mode is specified with **G21**. With most machining centers, feedrate can only be specified in *per-minute* fashion – either inches- or millimeters-per-minute. Some, though not many machining centers, additionally allow feedrate to be specified in per-revolution fashion (inches- or millimeters-per-revolution). If per-revolution is additionally available, two G codes, usually **G94** and **G95**, allow the selection of per-minute and per-revolution feedrate modes respectively. The F word allows a decimal point. A feedrate of 3-1/2 inches per minute (assuming inch mode is instated) is specified as **F3.5**. Feedrate is introduced in Lesson One and discussed in more detail in Lessons Nine and Nineteen.

S

The S word specifies **spindle speed**. Most current machines allow you to specify spindle speed in one rpm increments. A spindle speed of 350 rpm is specified with **S350**. A decimal point is *not* allowed with the S word. The spindle is activated with M codes. **M03** turns the spindle on in a forward direction. **M04** turns the spindle on in a reverse direction. **M05** turns the spindle off. Spindle control is introduced in Lesson One and discussed in more detail during lessons fourteen and fifteen.

T

With most machining centers, the T word specifies the **ready position tool station**. When a T word is commanded, the automatic tool changer magazine will rotate, bringing the specified tool station to the *ready* or *waiting* position. This allows one tool to be cutting while the tool changer magazine is getting ready with the next. With these machines, an M06 word is used to command the tool change. The command T05 M06, for example, will first cause the machine to bring tool station number five to the ready position – then make the tool change – placing tool number five in the spindle. There are machining centers, however, with which the T word by itself commands the entire tool change. A decimal point is not allowed with the T word. Automatic tool changers are introduced in Lesson One and discussed in more detail in lessons fourteen and fifteen.

M

An M word specifies a **miscellaneous function** (also called a **machine function**). You can think of M words as programmable on/off switches that control functions like coolant and spindle activation. For a list of all M words, see the list at the end of this lesson. Note that machine tool manufacturers will select their own set of M words. While there are many standard M word numbers, you must consult your own machine tool builder's programming manual to find the exact list for your particular machine/s. A decimal point is not allowed with the M word. Miscellaneous functions are introduced in Lesson One and discussed numerous times throughout this text.

D

The D word specifies the **cutter radius compensation offset number**. In this offset, the setup person store's the milling cutter's radius. A decimal point is not allowed with the D word. Cutter radius compensation is discussed in Lesson Twelve.

H

The H word specifies the **tool length compensation offset number**. In this offset, the setup person will store the cutting tool's tool length compensation value. A decimal point is not allowed with the H word. Tool length compensation is discussed in Lesson Eleven.

EOB (end of block character)

EOB stands for **end-of-block**. Though it is not actually a CNC word, it is an important part of a CNC program. It is called a *command terminator*. On a machining center's display screen, it usually appears as a semicolon (;). When you type a program using a computer, it is automatically inserted at the end of each command when you press the *Enter key* (though it will not appear on the computer monitor). Though it is automatically inserted into your program when typing programs on a computer, it is not automatically entered if you type (or modify) programs through the keyboard and display of the CNC machining center. A special key on the control's keyboard labeled EOB must be pressed to insert the end-of-block character into the program. This must be done at the end of every command. Program structure is discussed in Lessons Fourteen and Fifteen.

/ (slash code)

This is called the **block delete** word (also called **optional block skip**). It is the slash character on your keyboard (under the question mark on most keyboards). It works in conjunction with an on/off switch on the control panel (labeled block delete or optional block skip). If the switch is on when the control reads the slash code, the control will ignore the words to the right of the slash code. If the switch is off, the control will execute the command in the normal manner. Block delete, including several applications for this feature, is discussed in Lesson Eighteen.

G and M codes

Here we list most of the G and M codes that can be used in programming, providing little more than the name for each word. Rest assured that the most often used G and M codes are discussed in detail in this text. You can find documentation for lesser used G and M codes in your control manufacturer's programming manual.

G codes

As you know, G codes specify preparatory functions. They prepare the machine for what is to come – in the current command – and possibly in up-coming commands. They set modes.

G code limitation:

With most controls, *only three compatible G codes are allowed per command.* If you exceed this limitation, most controls will *not* generate an alarm. They will simply execute the last three G codes in the command – ignoring those prior to the last three. For example, in the command

> G90 G80 G40 G20 (Select absolute mode, cancel canned cycles, cancel cutter radius compensation, select metric mode)

Only the G80, G40, and G20 codes will be executed. The G90 code will be ignored. If needed in a program, these four G codes must be broken into two commands.

By *compatible* G codes, we mean G codes that work together. For example, in the command

> G20 G90 G00 (select the inch mode, absolute positioning mode, and rapid mode)

all G codes are compatible. They work together. But you cannot, of course, specify G20 and G21 (selecting inch and metric mode) in the same command.

Option G codes

Some of the G codes in the up-coming list are *option G codes.* It is impossible to tell whether a given option G code is included in your control or not by just looking at our list since most machine tool builders include a standard package of options when they purchase controls from the control manufacturer. Our list shows what one popular control manufacturer specifies as options (Fanuc). Probably when your company purchased the machining center from your machine tool builder, other (option) G codes came with the machine. If there is any question as to whether your machining center has any a particular option G code, you can perform a simple test at the machine to find out if the G code is available to you (or you can call your builder to find out if the G code was included).

To make the test for an option G code, simply command the G code in the MDI mode (techniques given in Lesson Twenty-Three). You need not even specify the correct format for the G word. If you receive the alarm Unusable G code or G code not available, your machine does not have the G code. If you receive no alarm or if the alarm is related to the format of the G code, the option G code should be available for you to use.

What does initialized mean?

In the up-coming list, we also state whether the G code is *initialized*, meaning whether it is automatically instated the machine's power is turned on.

What does modal mean?

Most G codes are *modal*, meaning once they are invoked, they remain in effect until they are changed or cancelled. This means you do not have to keep repeating modal G codes in every command. There are some G codes that are non-modal (also referred to as *one-shot G codes*). Non-modal G codes only have an effect on the command in which they are included.

The most popular G codes

As you look at this list, it may seem a little intimidating. There are a lot of G codes. Here are the most popular G codes (about thirty), along with where they are discussed in this text.

> G00, G01, G02, G03: motion types – discussed in Lesson Nine
>
> G04: dwell command – discussed in Lesson Eighteen

G20, G21: inch and metric mode – discussed in lessons one, four, and fourteen

G28: zero return command – introduced in Lesson Five and discussed in more detail in Lessons Fourteen and Fifteen

G40, G41, G42 – cutter radius compensation – discussed in Lesson Twelve

G43: tool length compensation – discussed in Lesson Eleven

G54: invoke fixture offset number one – introduced in Lesson Six, discussed in more detail in Lesson Thirteen

G55 though G59: invoke other fixture offsets – discussed in Lesson Thirteen

G73 through G89: hole-machining canned cycles – discussed in Lesson Sixteen

G90, G91: absolute and incremental positioning modes – discussed in Lesson Four

G98, G99: initial and rapid plane used with hole-machining canned cycles – discussed in Lesson Sixteen

Common M codes used on a CNC machining center

Again, machine tool builders vary with regard to the M codes they provide – so this is just a partial list. You must reference your machine tool builder's programming manual for a complete list of the M codes available for a particular CNC machining center.

Note that some CNC control models (like many supplied by Fanuc) allow but one M code per command. If you include more than one M code, it's hard to predict what will happen for all machines. Some machines will stop executing the program while others will generate an alarm. Yet others will continue executing the program, ignoring all but the last M code in the command.

M CODE	DESCRIPTION
M00	Program stop
M01	Optional stop
M02	End of program (does not rewind memory for most machines – use M30)
M03	Spindle on in a clockwise direction
M04	Spindle on in a counter clockwise direction
M05	Spindle stop
M06	Tool change command
M07	Mist coolant on (option)
M08	Flood coolant on
M09	Coolant off
M30	End of program (rewinds memory)
M98	Sub program call
M99	End of sub program

Other M Codes for your machine (found in your machine tool builder's manuals)

_____ _____

_____ _____

_____ _____

_____ _____

_____ _____

_____ _____

_____ _____

Key points for Lesson Seven:

- While there are many CNC words available to programmers, there are only about fifty words that are used on a regular basis. Look at learning CNC programming as like learning a foreign language that has only fifty words.

- The letter addresses for many CNC word types are easy to associate with their usage (F for feedrate, T for tool, S for spindle, etc.). But other word types are more difficult to remember.

- You've been exposed to all of the word types used with CNC machining center programming in this lesson – as well as where (in this text) you can find more information about the most often-used words.

- There are only about thirty G words used on a regular basis.

- Only three compatible G codes are allowed per command with most CNC controls.

- Only one M code is allowed per command with some CNC controls (including Fanuc).

- You must reference your machine tool builders programming manual to find the complete list of M codes for your machining center/s.

You Must Prepare To Write Programs

While this Key Concept does not involve any programming words or commands, it is among the most important of the Key Concepts. The better prepared you are to write a CNC program, the easier it will be to develop a workable program.

Key Concept Number Two is made up of but one lesson:

8: Preparation for programming

Frankly speaking, this Key Concept can be applied to any CNC related task, including setup and operation. Truly, the better prepared you are to perform any task, the easier it will be to correctly complete the task. When it comes to making setups, for example, if you have all of the needed components to make the setup at hand (fixtures, cutting tools, cutting tool components, program, documentation, etc.), making the setup will be much easier and go faster. So *gathering needed components* is always a great preparation step to perform.

The same goes for completing a production run. If the CNC operator has all needed components (raw material, inserts for dull tool replacement, a place to store completed workpieces, gauging tools, etc.), they will be able to smoothly complete the production run.

In this Key Concept, however, we're going to limit our discussion to preparation steps that you can perform to get ready to write a CNC program. Remember, the actual task of *writing* a CNC program is but part of what you must do. Certain things must be done *before* you're adequately prepared to write the program.

Preparation for programming is especially important for entry-level programmers. For the first few programs you write, you will have trouble enough remembering the various CNC words – remembering how to structure the program correctly – and in general – you'll have trouble getting familiar with the entire programming process. The task of programming is infinitely more complicated if you are not truly ready to write the program in the first place.

Preparation and time

Without adequate preparation, writing a CNC program can be compared to working on a jigsaw puzzle. A person doing the puzzle has no idea where each individual piece will eventually fit. The worker makes a guess and attempts to fit the pieces together. Since the worker has no idea as to whether pieces will fit together, it is next to impossible to predict how long it will take to finish the puzzle.

In similar fashion, if you attempt to write a CNC program without adequate preparation, you will have a tendency to piece-meal the program together in much the same way as a person doing a jig saw puzzle. You will not be sure that anything will work until it is tried. The program may be half finished before it becomes obvious that something is seriously wrong with the process. Worse, the program may be completed and being verified on the CNC machining center before some critical error is found.

CNC machine time is much more expensive than your own time. There is no excuse for wasting precious machine time for something as avoidable as a lack of preparation.

You can also liken the preparation that is required to write a CNC program to the preparation needed for giving a speech. The better prepared the speaker, the easier it will be to make the presentation, and the more effective the speech. Truly, the speaker must think through the entire presentation (probably several times) before the speech can be presented. Similarly, the CNC programmer must think through the entire CNC process, setup, and program before the program can be written.

With adequate preparation, writing the program will be *much* easier. Most experienced programmers will agree that the actual task of writing a CNC program is the *easy part* of the programming process. The real work is done in the preparation stages. If preparation is done properly, writing the program will be a simple matter of translating what you want the machine to do (from English) into the language a CNC machine can understand and execute.

Though preparation is so very important, it is amazing to see how many so-called *expert* programmers muddle through the writing of a program without preparing at all. While an experienced person may be able to write workable programs for simple applications with minimal preparation (and even then they only gain this ability through trial-and-error practice), even the so-called expert programmer have problems with more complex programs.

Preparation and safety

Wasted time is but one of the symptoms of poor preparation. Indeed, it may be the least severe one. Poor preparation will often result in all kinds of mistakes.

A CNC machining center will follow a CNC program's instructions *to the letter*. While the control may go into an alarm state if it cannot recognize a given command, it will give absolutely no special consideration to motion mistakes. Indeed, a machining center cannot detect motion mistakes. The level of problem encountered because of motion mistakes ranges from minor to catastrophic.

Minor motion mistakes usually do not result in any damage to the machine or tooling, and the operator is not exposed to a dangerous situation. However, the workpiece will not be correctly machined. For example, say you intend to drill a hole at five inches per minute with a feedrate word of **F5.0** in the drilling command. But, you place the decimal point in the wrong location. You specify a feedrate of **F0.5** instead of **F5.0**. In this case, the control is being told to run the drill at a much slower feedrate than you intend. No damage to the tool or machine will result – but machining time will be much longer than it should be.

This mistake will be more serious, of course, if the programmer incorrectly specifies a feedrate of **F50.0** instead of **F5.0**. This time the tool will probably break – and the workpiece may be damaged. If the machine does not have a fully enclosed work area, debris flying out of the work area could injure the operator.

Catastrophic mistakes can result in damage to the machine and injury to the operator. For example, say you intend to position a cutting tool at the machine's *rapid rate* (the machine's fastest motion rate – which for some machines is well over 1,000 inches per minute). The correct position for your command is **Z0.1**, which is 0.100 inch above the work surface. But you make a mistake and include a minus sign for this motion word (**Z-0.1** instead of **Z0.1**). The tool is being told to crash into the part (at rapid). Depending on what kind of tooling is being used, this will, at the very least, cause the tool to break. Worse, the workpiece could be pushed out of the setup. Possibly, if the setup is very sturdy and the tool is very rigid, damage to the machine's way system and/or axis drive system will result. If the tool breaks and parts fly out of the work area, the operator could be injured.

We cannot overstress the importance of preparation. Just as the well-prepared speaker is less apt to make mistakes during his or her presentation, so will the well-prepared CNC programmer be less apt to make mistakes while writing a CNC program.

Lesson 8
Preparation Steps For Programming

Any complex project can be simplified by breaking it down into small pieces. This can make seemingly insurmountable tasks much easier to handle. CNC machining center programming is no exception. Learning how to break up this complex task will be the primary focus of Lesson Eight.

As you now know, preparation will make programming easier, safer, and less error-prone. Now let's look at some specific steps you can perform to prepare to write CNC programs.

Prepare the machining process

Process sheets, also called routing sheets, are used by most manufacturing companies to specify the sequence of machining operations that must be performed on a workpiece during the manufacturing process. The person who actually prepares the process sheet must, of course, have a good understanding of machining practice, and must be well acquainted with the various machine tools the company owns. This person determines the best way to produce the workpiece in the most efficient and inexpensive possible way, given the company's available resources.

In most manufacturing companies, this involves *routing* the workpiece through a series of different machine tools and processes. Each machine tool along the way will perform only those operations the process planner intends, as specified on the routing sheet. This commonly means that non-CNC machine tools are needed to complete a given workpiece. For example, the square workpiece shown in practice exercise at the end of Lesson Six might have the following routing sheet.

Op. #:	Operation:	Machine:
10	Procure material	Vender ID #12322
20	Cut bar stock to 3.2 long	Cut off saw
30	Clean and de-burr	Cleaning tanks
40	Mill contour, drill (9) holes	CNC machining center
50	Clean and de-burr	Cleaning tanks
60	Plate with nickel	Finishing tanks

Note that only operation 40 of this process requires a CNC machine tool. When a CNC machine is involved in the process plan, often the CNC machine will be required to perform several machining operations on the workpiece (as is the case in our example). As you know, all true CNC machining centers are multi-tool machines, meaning several tools can be used during one program. In some companies, the process sheet will clearly specify the order of machining operations that must be performed by the CNC machine. However, the vast majority of companies do not get so specific with their routing sheets. Instead, the sequence of machining operations to be performed on the CNC machine is left completely to the CNC programmer. If the programmer must develop the machining order, they must possess a good knowledge of basic machining practice.

In any event, the step-by-step machining order required to machine a workpiece on the CNC machine must be developed before the CNC program can be written. With a simple process, an experienced programmer may elect to develop the process as the program is being written. While some experienced programmers have the ability to do this, beginning programmers will find it necessary to plan the machining process first.

The process used to machine the workpiece will have a dramatic impact on the success of the program. If the process is correct, the workpiece will be machined efficiently and pass inspection. If the process is poor, the workpiece will not be machined correctly no matter how well the program is written. If you are new to the shop environment, you should seek help whenever there is a question as to whether your intended machining process will work.

Developing your machining process before the program is written will serve several purposes. First (as just stated), it will allow you to check the process for errors in basic machining practice before writing the program. You will be forced to think through the entire process before the first CNC command is written. You will have the opportunity to spot a problem with the process that will be difficult to spot and/or repair if the process is developed while the program is being written.

Second, developing the machining process prior to writing the program allows you *concentrate on your machining practice skills separate from your CNC programming skills*. While developing the machining process, you concentrate on the machining practice in order to develop a workable process. Your mind is occupied only with the task at hand. While programming, you concentrate on programming skills, translating your process into a language that the CNC machine can understand.

Third, if you spot an error in your thinking, you will be able correct the mistake before sending the program out to the machine. Indeed, you should spot the mistake before you write the program. Remember, there is never an excuse for wasting machine time for something as basic as inadequate preparation.

Fourth, developing the machining process prior to programming will provide you with documentation. You'll be able to use your process as a checklist while programming to ensure that you don't forget a machining operation. And anyone that must work on your program in the future will be able to quickly familiarize themselves with your machining process if they have this documentation. Figure 2.1 shows an example planning form that you can use to develop and document your machining process.

Part no.:	Date:	**Machining Process Planning Form**				
Part name:	Programmer:					
Machine:	Material:					
Seq.	Operation description	Tool	Station	Speed	Feed	Note

Figure 2.1   Example sequence of operations form

Notice how well this form allows you to document the process your program will use. Months or years after a CNC program is developed, there may be a need to revise it. If the person doing the revision can view the completed process planning form shown in Figure 2.1, it will be much easier to make the necessary changes.

The last reason we will give to plan the process first is to simply help you remember the operations to perform during programming. Remember, beginners tend to make mistakes of omission. You will have enough to think about when it comes to remembering the various commands needed in the program. The process planning form can be used as a kind of step-by-step set of instructions by which you machine the workpiece. It can be used as a check-list. Without this form, you will be prone to omitting important machining operations from the CNC program.

Develop the needed cutting conditions

Before a program can be completed, cutting conditions must be determined for the various cutting tools used in the program. All cutting tools will need a *spindle speed* in revolutions per minute (rpm) and a *feedrate* in inches per minute (ipm). For roughing tools, like rough milling cutters and rough boring bars, you must also determine a *depth-of-cut* for the tool – as well as how much *finishing stock* you will leave for the finishing tool. You must also determine whether or not to use coolant (and, if so, what kind – flood, mist, through-the-tool, or high pressure) based upon the workpiece and cutting tool material.

It is helpful to come up with the various cutting conditions needed in the program while developing the machining process – and again – *before* you write the program. This will keep you from having to break out of your train of thought while programming. If you use a planning form like the one in Figure 2.1, you will even be able to document the speeds and feeds needed for programming right on the form.

Here are a few terms and formulae used when calculating cutting conditions:

sfm (surface feet per minute) – this is the linear amount of material that will pass by the cutting tool's cutting edges during one minute. You find this recommended speed in reference books related to cutting conditions like tooling manufacturers' technical books (possibly the cutting tool manufacturer's catalog).

rpm (revolutions per minute) – this is the number of spindle revolutions that occur in one minute.

ipt (inches per tooth) – for milling cutters, this is the distance a cutting tool will move for each tooth/insert/flute of the cutting tool. Like sfm, this value is found in reference handbooks related to cutting conditions.

ipr (inches per revolution) – for any cutting tool, this is the distance a cutting tool will move during one revolution of the tool. For milling cutters, this value is determined by multiplying the number of flutes/teeth/inserts times the ipt value. For most other cutting tools, this value can be found in reference handbooks related to cutting conditions.

ipm (inches per minute feedrate) – this is the desired amount of cutting motion during one minute.

Some formulae:

ipr = ipt * number of teeth

rpm = 3.82 * sfm / cutting tool diameter

ipm = ipr * rpm

pitch (for tapping) = 1 / number of threads per inch

The data provided by cutting tool manufacturers includes the speed (in sfm) and feedrate (in either ipr or ipt). This information is based upon the cutting tool material (high speed steel, carbide, ceramic, etc.) and the workpiece material (mild steel, medium carbon steel, high carbon steel, stainless steel, aluminum, etc.). When appropriate, cutting tool manufacturers will also specify whether or not you should use coolant – as well as the recommended depth of cut for a roughing tool. In some cases, they will even provide recommendations about how the cutting tool should move as it machines the workpiece.

For machining center programs, you must of course, calculate the speed in rpm and feedrate in ipm. For rpm, you must know the recommended speed in sfm and the cutting tool diameter. For ipm, you must know the rpm and ipr feedrate.

An example

Say you must machine a mild steel workpiece with this machining process:

Operation:	Tool:
1) Rough face mill the top surface	4.0" face mill
2) Rough mill the two ends	1.0" end mill
3) Finish face mill the top surface	4.0" face mill
4) Finish mill the two ends	1.0" end mill
5) Center drill five holes	#3 center drill
6) Drill two 1/4 inch diameter holes	1/4 drill
7) Drill three 1/2 inch diameter holes	1/2 drill

You must first determine some important information about your cutting tools, including what they are made of (cutting edge material) and how many flutes, inserts, or teeth they have. We'll say the two 4.0 inch face mills have 8 carbide inserts each. The two 1.0 inch end mills have four flutes and are made of high speed steel (hss). The center-drill and the two drills are all made of high speed steel and have two flutes each (most drills have two flutes).

Say the face mills' manufacturer recommends a roughing speed of 400 sfm and a finishing speed of 500 sfm. They also specify a roughing feedrate of 0.004 ipt and a finishing feedrate of 0.0025 ipt when machining mild steel. The roughing speed in rpm will be 3.82 times 400 divided by 4.0, or 382 rpm. The roughing feedrate will be 382 times 8 (inserts) times 0.004 ipt, or 12.224 ipm. The finishing speed will be 3.82 times 500 divided by 4.0, or 477 rpm. The finishing feedrate will be 477 times 8 (inserts) times 0.0025 ipt, or 9.54 ipm.

The manufacturer of the two end mills recommends a roughing speed of 70 sfm and a finishing speed of 85 sfm. They specify a roughing feedrate of 0.005 ipt and a finishing feedrate of 0.004 ipt when machining mild steel. The roughing speed in rpm will be 3.82 times 70 divided by 1.0, or 267 rpm. The roughing feedrate will be 267 times 4 (flutes) times 0.005 ipt, or 5.34 ipm. The finishing speed will be 3.82 times 85 divided by 1.0, or 324 rpm. The finishing feedrate will be 324 times 4 (flutes) times 0.004 ipt, or 5.184 ipm.

The center-drill and drill manufacturer recommends a speed of 70 sfm and a feedrate of 0.004 for drills under 0.437 in diameter and 0.007 for drills larger than 0.437 in diameter.

For the center-drill, the speed in rpm will be 3.82 times 70 divided by 0.187 (the diameter of the center-drill's largest diameter), or 1,430 rpm. The feedrate in ipm for the center drill will be 1,430 times 0.004, or 5.72 ipm.

The speed in rpm for the 1/4 inch drill will be 3.82 times 70 divided by 0.25, or 1,069 rpm. The feedrate for the 1/4 inch drill will be 1,069 times 0.004, or 4.276 ipm.

The speed for the 1/2 inch drill will be 3.82 times 70 divided by 0.5, or 535 rpm. The feedrate in ipm for the 1/2 inch drill will be 535 times 0.007, or 3.74 ipm.

Cutting conditions can be subjective

Many programmers, indeed many machinists, choose feeds and speeds using a *seat-of-their-pants* approach. They do so by watching and listening to the cutting operation – and manually overriding the programmed spindle speed and feedrate until the machining operation *looks right*. Admittedly, expert machinists can do this pretty well, and it's hard to argue with success, but the majority of people we see using these techniques don't even come close to efficient cutting conditions. Instead, they tend to be well under the cutting tool manufacturer's recommendations. While many factors determine how quickly a cutting tool can machine (like rigidity of the setup, length of the cutting tool, and sharpness of the cutting tool), we urge beginners to at least *start with* the recommendations made by the manufacturer of the cutting tools they use.

Practice calculating speeds and feeds

Using the formulae shown previously, calculate the speed (in rpm) and feedrate (in ipm) for these tools based upon the recommendations in sfm and ipt/ipr.

No.	Tool description	Speed	Feedrate	Speed in rpm	Feedrate in ipm
1)	3.0 rough face mill having 6 carbide inserts	450 sfm	0.005 ipt	_____	_____
2)	3.0 finish face mill having 6 carbide inserts	525 sfm	0.004 ipt	_____	_____
3)	0.75 hss rough end mill having 4 flutes	70 sfm	0.006 ipt	_____	_____
4)	0.75 hss finish end mill having 4 flutes	90 sfm	0.0045 ipt	_____	_____
5)	0.375 hss twist drill	75 sfm	0.005 ipr	_____	_____
6)	0.625 hss twist drill	75 sfm	0.007 ipr	_____	_____
7)	0.875 hss twist drill	75 sfm	0.010 ipr	_____	_____

Answers:
1: 516 rpm, 15.48 ipm
2: 668 rpm, 16.032 ipm
3: 356 rpm, 8.544 ipm
4: 458 rpm, 8.244 ipm
5: 764 rpm, 3.82 ipm
6: 458 rpm, 3.208 ipm
7: 327 rpm, 3.274 ipm

Do the required math and mark-up the print

As stated in the Preface of this text, the word *numerical* in computer numerical control implies a strong emphasis on numbers and math. Most college curriculums related to CNC do require a strong math background. However, most forms of CNC equipment require less math than you might think. Believe it or not, many CNC machining center programs can be completely prepared solely with simple addition and subtraction. A basic knowledge of right angle trigonometry is also helpful, but not always mandatory.

While there are times when a manual programmer must apply trigonometry, there are many reference books that give all the formulae related to right-angle trigonometry in a very simple format. This makes it relatively easy to solve trig problems, even for a person who knows little about trigonometry

The reason why you should do the math needed to calculate coordinates before attempting to write the program is the same as the reason why you should first come up with a machining process. With the coordinates needed in your CNC program calculated, you won't have to break out of your train of thought while programming. Again, you'll be able to concentrate on programming *separately* from calculating coordinates.

Documenting your math also helps when mistakes are made. When you go back to check your mistake, if you have your math documented, you may be able to easily determine how the mistake was made to keep you from repeating the mistake in the future.

Marking up the print

You must have a good copy of the print. You should have your own working copy, and be allowed to do whatever you need to with the print to help you with the programming task. Your copy of the print should be kept with the program as part of the documentation for the program.

Depending on the complexity of the workpiece to be produced, interpreting a workpiece drawing can range from quite simple to very difficult. Once you study the print and understand the machining operations that must be performed, you should mark-up the print in any way that makes programming easier. The first thing we recommend is to take a high-lighting pen of a bright color and mark those surfaces on the print that require machining operations by your CNC program. Especially helpful for complicated workpieces, this helps you narrow down just what your program must do.

Additionally, you should indicate any information required for programming on the print. The location of program zero, the placement of workholding devices (fixtures, vises, etc.), and clamps should be included on the marked-up print. If there is room on the print, you can also include coordinates needed for programming.

Figure 2.4 shows an example of a marked-up print

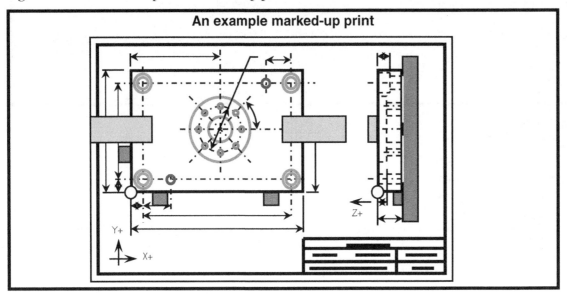

An example marked-up print

Figure 2.4 – A marked-up print showing the location of program zero, the position of clamps, the machining operations to be performed, and how the workpiece is oriented in the fixture

Doing the math

How dimensions are described on the part print will have a great deal to do with how much math is required for your program. In progressive companies, design engineers use *datum surface dimensioning* techniques. When datum surface dimensioning is used, each dimension on the print will be specified from one surface in each axis (the datum surface). This dramatically reduces the amount of math a programmer must do.

Unfortunately, not all design engineers use datum surface dimensioning techniques. You may be expected to do a great deal of math in order to calculate the coordinates required in your CNC programs.

When doing the required math for a program, of course, you will be calculating the coordinates to be used in the program. As discussed in Key Concept Number One, each coordinate is the distance from the program zero point to a position through which a cutting tool must move. Any single cutting tool position will have *three* values that comprise the programmed coordinates for the position (X, Y, and Z).

How the programmer documents the coordinates needed in a program depends on the print and how many coordinates must be calculated. If the print is large and roomy, and if there are only a few coordinates to be calculated, often you can easily write all of the coordinates right on the print close to the cutting tool position required in the program.

On the other hand, if the print is small and crowded, and/or if many coordinates are required, you should use a *coordinate sheet* with which to document the needed coordinates. On the print itself, draw a dot and place a number near each position to be included in the program. Each number represents a *point number* to be filled in on the coordinate sheet. The columns in the coordinate sheet will include point number and each axis letter address (X, Y, & Z). We have been using this technique throughout this text so it should be familiar to you.

For the Z axis, some points will require multiple positions. Consider, for example, any drilling operation. The drill must first move to the hole position in X and Y. Then it will move to an approach position above the hole in Z. It must then machine the hole, feeding to another Z position. It will then retract from the hole back to the approach position in Z. So, two Z positions are needed for a drill (the approach position and the hole-bottom position).

Also, certain points will be used by several cutting tools. Again consider hole-machining operations. A single hole may be center-drilled, drilled, and counter-bored. The X and Y coordinates will be the same for each tool, but each tool will require its own Z positions. Though the approach position in Z will usually be the same for each hole-machining tool (usually 0.1 inch above the work surface), each tool will require its own hole-bottom position.

While it may be enough for experienced programmers to calculate just the X and Y positions prior to programming (they'll determine Z positions as they write the program), we recommend that you calculate *all* coordinates needed in the program before you start writing the program. This includes all of the Z positions needed by hole-machining tools.

To document these Z values in your coordinate sheet, we recommend writing them in the same order you will need them in a program. Separate them with a comma. If the approach position is the same for all tools, just write it once (first). For example, here are the Z coordinates needed for a hole that must be center-drilled, drilled, and counter-bored.

0.1, -0.14, -1.23, -0.25

Again, these values are written in the Z column for any point (hole) requiring these three machining operations. 0.1 is the approach position for all three tools. -0.14 is the hole-bottom position for the center-drill. -1.23 is the hole-bottom position for the drill. -0.25 is the hole-bottom position for the counter-bore (end mill). Figure 2.5 shows a more elaborate example.

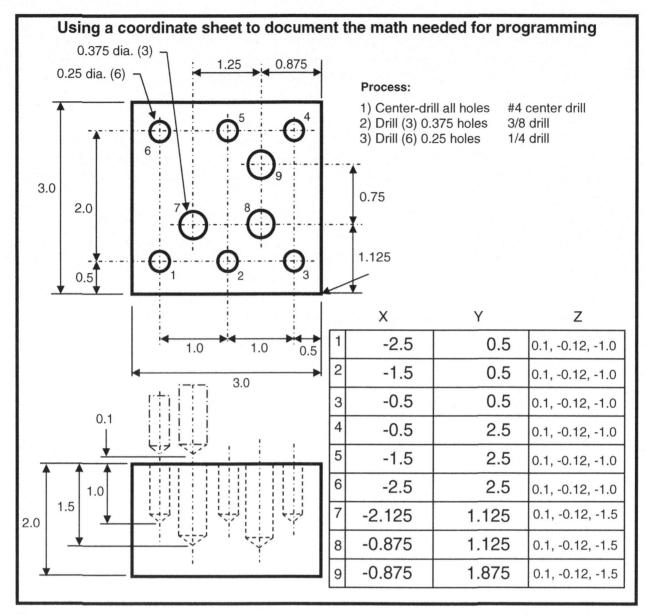

Using a coordinate sheet to document the math needed for programming

Process:

1) Center-drill all holes   #4 center drill
2) Drill (3) 0.375 holes   3/8 drill
3) Drill (6) 0.25 holes   1/4 drill

	X	Y	Z
1	-2.5	0.5	0.1, -0.12, -1.0
2	-1.5	0.5	0.1, -0.12, -1.0
3	-0.5	0.5	0.1, -0.12, -1.0
4	-0.5	2.5	0.1, -0.12, -1.0
5	-1.5	2.5	0.1, -0.12, -1.0
6	-2.5	2.5	0.1, -0.12, -1.0
7	-2.125	1.125	0.1, -0.12, -1.5
8	-0.875	1.125	0.1, -0.12, -1.5
9	-0.875	1.875	0.1, -0.12, -1.5

Figure 2.5 – Example showing how to document the math needed in a program

In the Z column of the coordinate sheet in Figure 2.5, notice the three values. The first (0.1) is the approach position of all tools. The second (-0.12) is the hole bottom position for the center drill. The third (-1.0 or -1.5) is the hole bottom position for the drill).

Our suggestions for documenting the math you do prior to programming are only *recommendations*. Frankly speaking, you can use any documentation methods that make sense to you as long as you get the desired result – calculating *all* coordinates needed for you program. You may be able to improve upon what we've shown. You'll quickly know how well you've done as you write your program. If you are constantly calculating (more) coordinates while you write the program, you haven't done very well.

What about milling operations?

To this point, we have only discussed how to calculate coordinates for hole-machining operations. But you must also have the ability to do so for milling operations. This means, of course, that you must be able to plan the tool path for the milling cutter. It also means you must know *which* path the tool will use, the centerline path or the work-surface path. At this early point in the text, we can only show how to plan the milling cutter's centerline tool path. Later, in Lesson Twelve, we'll discuss a feature called cutter radius compensation, and we'll show how to plan the (simpler) work-surface tool path.

Planning the milling cutter's centerline tool path

This involves determining every position the cutter must move through as it mills the workpiece. And the contour being milled, of course, determines where these positions will be. Again, this requires basic machining practice experience. Consider the drawing shown in Figure 2.6.

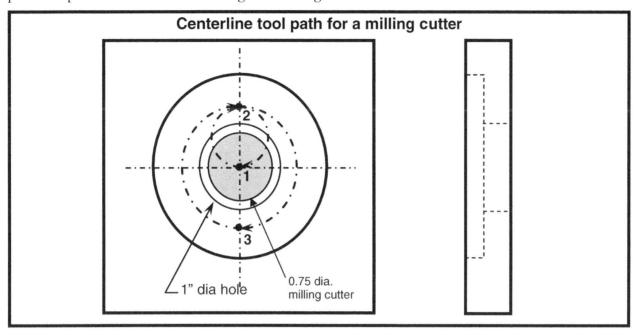

Figure 2.6 – Calculating coordinates for a milling cutter's centerline tool path

The 0.75 diameter milling cutter is milling a round pocket. The tool path being used first sends the tool to point number one. The tool will then approach in Z – into the previously drilled (1.0 diameter) hole and to the work surface. It will then form a half circle to point two – arcing into the pocket edge. Next it will form another half circle to point three milling the right half of the pocket. Next, the cutter will make a half circle back to point two, milling the left half of the pocket. Finally the cutter will make a half circle motion back to point one, arcing off the surface being milled and getting back to a clearance position in XY.

Admittedly, planning a milling cutter's tool path requires an understanding of the kinds of motions that the machining center can make – and we haven't talked about motion types yet (motion types are discussed in Lesson Nine). For now, just keep in mind that you must be able to determine the tool path and come up with coordinates for *all* cutting tools you use in your program.

Check the required tooling

The next preparation step is related to the cutting- and workholding- tools used by your program. Tooling problems can cause even a perfectly written CNC program to fail. You can avoid production delays by considering potential tooling problems while you prepare for programming. Again, there is no excuse for machine down-time for as avoidable a reason as poor preparation.

First, you will need to confirm that the various cutting tools to be used by your program are *available*. You may incorrectly assume, for instance, that common tools are always in your company's inventory. It is always wise to double-check, even with relatively common tools. And always be sure your company has lesser-used tools in stock.

If a given tool is not in stock, of course, it must be ordered. Or you may have to make do with what your company does have available. If this is the case, your entire machining process may have to be changed (another reason to check *before* you begin programming).

Second, you must confirm that each cutting tool is capable of *reaching* the surfaces to be machined. Castings can present special problems in this regard. Say for example, a hole must be machined very close to a tall wall of a casting, as Figure 2.7 shows.

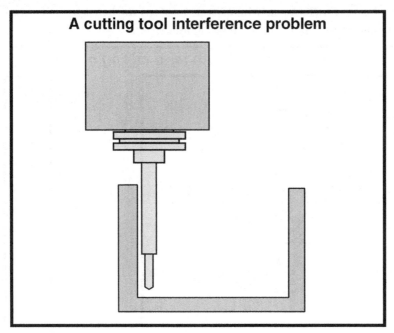

Figure 2.7 – Tooling related problem requiring a special extension to be used.

In this case, the tool must be long enough to reach the surface being machined. Additionally, the tool extension must be of a diameter small enough to clear the wall of the workpiece. Without being told how to assemble this cutting tool, most setup people will keep it as short and rigid as possible. For this reason, you must ensure that each cutting tool is made correctly by documenting special cutting tool considerations in the setup documentation. There is a column in the machining process planning form (Figure 2.1) named *Note* in which you can document special tooling considerations.

Another example of a tooling problem that can cause hold-ups during program verification has to do with the tool's *minimum length*. All machining centers have a Z axis travel limitation. On a large vertical machining center, for example, the spindle nose of the machine may only come down in Z to within about nine inches of the table top. With this machine, consider a very short setup – possibly a thin plate resting on short parallels. The top of the workpiece may only be about two inches above the table top.

In this situation, a cutting tool must be at least seven inches long just to reach the *top* of the workpiece. If the cutting tool is not long enough, the Z axis will over-travel when the tool attempts to approach the surface. If this problem isn't found until the program is being run on the machine for the first time, machine time will be wasted while the setup person remakes all of the cutting tools. This is yet another example of a down-time-causing problem that can be avoided with adequate preparation.

Plan the work holding set-up

The programmer is usually responsible for developing the workholding setup required to hold the workpiece during machining. Even for simple work holding setups, you should make a drawing or sketch indicating how the setup is to be made. For example, a sketch showing where a vise is placed on the machine's table may adequately instruct the setup person.

For more complicated setups, you may not be intimately involved in special fixture design. In most cases, a *tool designer* will actually design the workholding fixture and supervise its construction. Once the fixture drawing is made, you will need it to determine how the workpiece will be held, and will write the program accordingly. Though you may not actually design the fixture, you *will be responsible* for instructing the setup person with regard to how the fixture will be mounted on the machine's table.

Most companies use a *setup sheet* to help the setup person understand everything they need to know about how a given setup must be made. Most setup sheets will include a sketch of the setup (possibly even a photograph of the setup once it has been made), the location of program zero, a list of cutting tools (including a list of components needed for each tool), and in general, any other instructions necessary for getting the job up and running. Figure 2.8 shows an example of a *universal setup sheet*. We call it a *universal* setup sheet because this form is used for all setups made on a given CNC machine tool.

Part no.:	Date:	**Setup Sheet**		
Part name:	Programmer:			
Machine:	Program no:			
Stat.	Tool description	Offset	Insert	Instructions:
				Sketch of setup:
Fixture/vise:				
Clamps:				
Other notes:				

Figure 2.8 – An example universal setup sheet

This universal setup sheet does not include a detailed list of components for the cutting tools used in the program. Many companies do include a more complete tool list, possibly on a separate page. In similar fashion, this setup sheet does not include a complete list of the workholding tools. Again, many companies do include this kind of information so someone (probably *other* than the setup person) can be gathering all needed components even before the setup is made.

You should plan the setup and create the setup sheet *before* you write the CNC program. How the setup is made, of course, has a dramatic impact on how the program must be written. While you write the program, you'll need to know the position of workholding clamps and other obstructions in the setup so you can make sure that cutting tools don't hit them during their motions. You'll also need to know the orientation of the workpiece in the workholding devices. A workpiece is not always oriented in the same position as you see it in the top view of a workpiece drawing. When it comes to cutting tools, you'll need to know where they are placed in the machine's automatic tool changer magazine (station numbers). *Without having planed the setup, you really can't write the CNC program.*

Other documentation needed for the job

The programmer is responsible for providing all documentation needed for the jobs they program. And documentation should be aimed at the lowest skill level of people performing the related tasks. As with setup documentation, we recommend that *all* documentation related to a job be done *prior to* writing the program. You never know when you'll come across something that affects the way the program must be written.

Production run documentation

With longer production runs, many companies turn the machine over to another person, the CNC operator, once the setup is made and the first workpiece passes inspection. In some cases, more than one operator will complete the job (first and second shift operators, for example). Production run documentation should be aimed at the CNC operator, providing instructions for how to load and unload workpieces, how often to take sampling measurements, approximately how long cutting tools will last before they must be replaced, and a list of all perishable tools (inserts, twist drills, taps, etc.) used in the job so they can be quickly found and replaced when they get dull.

Program listing

Most companies supply a printed copy of the program with the documentation that goes out to the setup person and operator who will be running the job. This will help them make modifications to the program if they are required.

Is it all worth it?

You have seen that there is a great amount of work involved with *preparing* to write a CNC program. Again, experienced programmers will agree that this is where the *real work* lies. Once preparation is done, you will have a clear understanding of what the program must do. There will be no questions left to ponder while you write the program – and writing the program will be relatively easy – especially after you have written a few programs. Also, the work you do in preparation to write programs will pay special dividends when it comes time to run your program on the machine – there will be fewer mistakes to find and correct.

Key points for Lesson Eight:

- The better prepared you are, the easier it will be to write the program – and the fewer mistakes you will make.
- There is no excuse for wasting machine time for something avoidable as lack of preparation.
- Beginners are prone to making mistakes in four categories: syntax mistakes, motion mistakes, mistakes of omission, and process mistakes.
- First, study and mark up the print to become familiar with what must be done.
- Second develop a machining process, including cutting tools and cutting conditions to be used.
- Third, do the math calculations, coming up with all coordinates needed in the program.
- Fourth, check that cutting tools are available and capable of performing their machining operations.
- Fifth, plan the workholding setup.
- Sixth, create any other documentation that is required for the job (like production run documentation.

Understand The Motion Types

Motion control is at the heart of any CNC machine tool. CNC machining centers have at least three ways that motion can be commanded. Understanding the motion types you can use in a program will be the focus of Key Concept Number Three.

Key Concept Number Three is another one-lesson Key Concept:

9: Motion types

You know that all CNC machining centers have at least three axes (X, Y, and Z). You also know that you can specify positions (coordinates) along each axis relative to the program zero point. These coordinates will be positions through which your cutting tools will move. In Key Concept Number Three, we're going to discuss the ways to specify *how* cutting tools move from one position to another.

What is interpolation?

If a single linear axis is moving (X, Y, or Z), the motion will, of course, be along a perfectly straight line. For example, look at Figure 3.1.

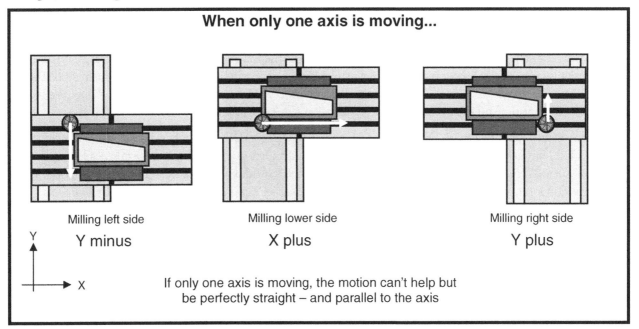

Figure 3.1 – A perfectly straight motion will occur if only one axis is moving

When milling the left side of the workpiece (left view in Figure 3.1), only the Y axis is moving. And since the Y axis is a *linear* axis, this will force the motion to be perfectly straight – and parallel to the Y axis. In like fashion, when milling the lower surface (middle view), only the X axis is moving – and again – this motion will be perfectly straight – parallel to the X axis. The same goes for milling the right side (right view) – only the Y axis is moving and the motion cannot help but be perfectly straight.

But notice that the upper side of this workpiece is *tapered*. It will require that both the X and Y axes move in a controlled manner, as shown in Figure 3.2.

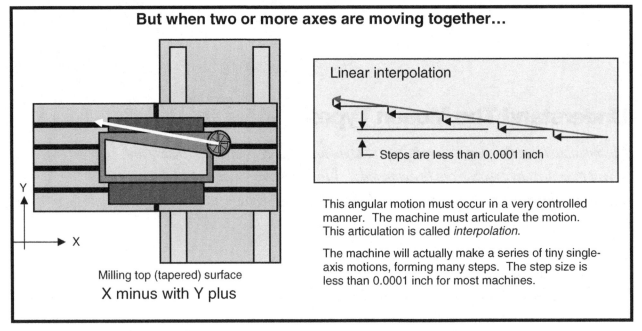

But when two or more axes are moving together...

Linear interpolation

Steps are less than 0.0001 inch

This angular motion must occur in a very controlled manner. The machine must articulate the motion. This articulation is called *interpolation*.

The machine will actually make a series of tiny single-axis motions, forming many steps. The step size is less than 0.0001 inch for most machines.

Milling top (tapered) surface
X minus with Y plus

Figure 3.2 – Milling the upper side of this workpiece requires that both the X and Y axes move in a controlled manner

The machine must articulate the motion shown in Figure 3.2. In CNC terms, this kind of articulation is called *interpolation*. Look at the drawing in the right side of Figure 3.2. As you can see, the machine will actually break the two-axis motion up in to a series of very tiny single-axis steps. The step size will be under 0.0001 inch for most machines. These steps are so small that you will not be able to see them – or measure them – with most measuring devices. For all intents-and-purposes, all machined surfaces will appear to be perfectly straight and without steps.

(By the way, the step size during interpolation is referred to as the machine's *resolution*. With many CNC machines, it is equal to the least-input-increment available in the measurement system you are using – 0.0001 inches in the inch mode or 0.001 millimeters in the metric mode. The smaller the step size, the finer the machine's resolution. And the more precisely it will follow your commanded motions.)

Machining center manufacturers offer a variety of *interpolation types*, based upon what their customers will be doing with their machines. *All* offer at least two types of interpolation – *linear interpolation* (also called straight-line motion) and *circular interpolation* (also called circular motion). Additionally, all machining centers come with a third motion type, called *rapid motion* – though we don't consider rapid motion to be a true form of interpolation since most machines do not articulate tool path during rapid motion.

Depending upon your company's specific needs, there may be other interpolation types available with your machining centers. However, these additional interpolation types will only be used for special applications. If, for example, you will be performing thread milling operations (not simple tapping), you'll need an interpolation type called *helical interpolation* (we discuss thread milling, including helical interpolation, in Lesson Eighteen). If your machining center has a rotary axis, and if you will be machining contours around the outside of a round cylinder or tube, you will probably need *cylindrical interpolation*. But again, these special interpolation types are only used in special applications. The bulk of your programming will not require them.

Actually, it should be refreshing to know that there are *only* three common ways to cause axis motion – rapid, straight-line, and circular. Just about every motion a CNC machining center makes can be divided into one of these categories. Once you master these three motion commands, you will be able to generate the motions required to machine a workpiece. Figure 3.3 illustrates them.

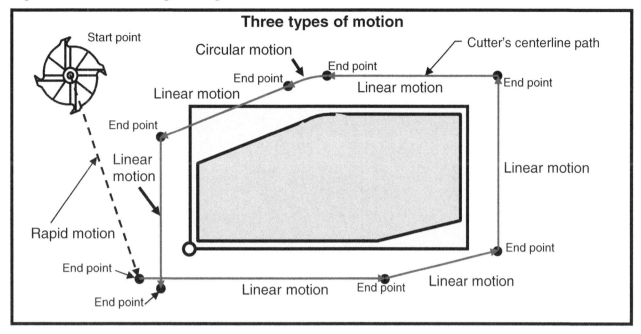

Figure 3.3 – The three kinds of motion available with all CNC machining centers

As you will see in Lesson Nine, it is actually quite easy to specify motion commands within a CNC program. In general, each motion will require you to specify the kind of motion (rapid, straight line, or circular) along with the motion's end point (coordinates at the end of the motion). Linear motion and circular motion additionally require that you specify the motion rate (feedrate) for the motion. Circular motion additionally requires that you specify the size of the arc being generated in the motion.

Answer program for the exercise at the end of lesson nine

O0001
N005 T01 M06 (Center drill)
N010 G54 G90 **S2000** M03 T02
N015 G00 **X0.75 Y0.75** (pt 11)
N020 G43 H01 **Z0.1** M08
N023 **G01 Z-0.12 F5.0**
N025 **G00 Z0.1**
N030 **X3.25** (pt 12)
N035 **G01 Z-0.12**
N040 **G00 Z0.1**
N045 **Y3.25** (pt 13)
N050 **G01 Z-0.12**
N055 **G00 Z0.1**
N060 **X0.75** (pt 14)
N065 **G01 Z-0.12**
N070 **G00 Z0.1** M09
N075 G91 G28 Z0 M19
N080 M01
N085 T02 M06 (1/2 drill)
N090 G54 G90 **S400** M03 T03

N095 G00 **X0.75 Y0.75** (pt 11)
N100 G43 H02 **Z0.1** M08
N105 **G01** Z-1.05 **F7.5**
N110 **G00 Z0.1**
N115 **X3.25** (pt 12)
N120 **G01** Z-1.05
N125 **G00 Z0.1**
N130 **Y3.25** (pt 13)
N135 **G01** Z-1.05
N140 **G00 Z0.1**
N145 **X0.75** (pt 14)
N150 **G01 Z-1.05**
N155 **G00 Z0.1** M09
N160 G91 G28 Z0 M19
N165 M01
N170 T03 M06 (3/4 end mill)
N175 G54 G90 **S450** M03 T01
N180 **G00 X-0.475 Y-0.125** (pt 1)
N185 G43 H03 Z0.1 M08
N190 G01 Z-0.25 F50.0

N195 **X3.25 F6.0** (pt 2)
N200 **G03 X4.125 Y0.75 R0.875** (pt 3)
N205 **G01 Y3.25** (pt 4)
N210 **G03 X3.25 Y4.125 R0.875** (pt 5)
N215 **G01 X0.75** (pt 6)
N220 **G03 X-0.125 Y3.25 R0.875** (pt 7)
N225 **G01 Y0.75** (pt 8)
N230 **G03 X0.75 Y-0.125 R0.875** (pt 9)
N235 **G02 X1.25 Y-0.625 R0.5** (pt 10)
N240 G00 Z0.1 M09
N245 G91 G28 Z0 M19
N250 M30

Lesson 9

Programming The Three Most Basic Motion Types

There are only three motion types used in CNC machining center programs on a regular basis – rapid, straight-line, and circular motion. You must understand how they are commanded.

While it helps to understand how the machining center will interpolate motion, it is not as important as *knowing how to specify motion commands in a program.* This is the focus of Lesson Nine. Let's begin our discussion by showing those things that all motion types share in common.

Motion commonalties

All motion types share five things in common:

First, they are all modal, meaning a motion type will remain in effect until it is changed. If more than one consecutive movement of the same type must be made, you need only include the motion type G code in the *first* command of the series of movements.

Second, each motion type command requires the *end point* of the motion. The control will assume the tool is positioned at the starting point of the motion prior to the motion command. Think of motion commands that form a tool path as being like a series of connect-the-dots.

Third, all motion commands are affected by whether or not you specify coordinates in the absolute or incremental positioning mode. In the absolute positioning mode (specified by G90), specified end points will be relative to the program zero point. In the incremental positioning mode (specified by G91), specified end points will be relative to the tool's current position. As stated in Lesson Four, you should concentrate on specifying coordinates in the absolute positioning mode.

Fourth, each motion command requires *only* the moving axes. If specifying a motion in only one axis, only one axis specification (X, Y, or Z) need be included in the motion command. *Axes that are not moving can be (and should be) left out of the command.*

Fifth, leading zeros can be left out of the G codes related to motion types. This means the actual G codes used to instate the motion types can be programmed in one of two ways. G00 and G0 (stated G zero-zero and G zero) mean exactly the same thing to the control, as do G01 and G1, G02 and G2, G03 and G3. All examples in this text do include the leading zero.

Understanding the programmed point of each cutting tool

In order to generate correct motions with your cutting tools, you must understand the position on the tool that you are programming. In many cases, you will need to calculate programmed coordinates – not just from the blueprint – but based on some cutting tool criteria as well. Figure 3.4 shows a series of center-cutting tools used to perform hole-machining operations.

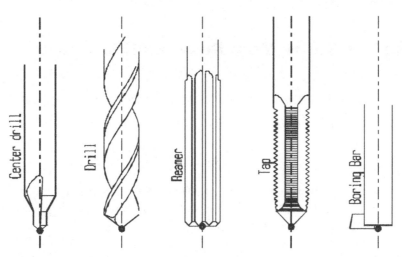

Figure 3.4 – The point you must program on each hole-machining tool

For all hole-machining tools shown in Figure 3.4, you program the *center of the tool* in X and Y. This makes programming hole-locations quite easy, since all holes are dimensioned on the workpiece drawing to their center points. For the Z axis, you'll always be programming the extreme end (tip) of the tool. However, as shown for the tools in Figure 3.4, you must often consider the *lead* of the tool when calculating hole-depth positions. This means your Z axis programmed coordinates may not match print dimensions. Here are some suggestions for determining the amount of lead for the most common hole-machining tools.

Center drill

Most programmers only center drill deep enough to make clearance for the web of the up-coming drill. That is, they do not try to center drill deep enough to form a chamfer on the drilled hole. If a chamfer is required on the up-coming hole, most programmers will use a *spot drill* (not shown above) instead of a center drill to start the hole.

Spot drill (not shown above)

Spot drills are used to form a chamfer on a hole that will eventually be drilled. Since spot drills have a ninety degree point angle, the depth of the spot drill will be half the size of the chamfer to be machined. If machining a 0.562 diameter chamfer for a 0.500 diameter hole (actually a 1/32" chamfer), the depth for the spot drill must be 0.281 (half of 0.562).

Drill

Most high speed steel drills (including twist drills) have a 118 degree point angle. The lead for these drills is calculated by multiplying 0.3 times the drill diameter. A 0.5 inch diameter twist drill, for example, has a 0.15 inch lead.

Reamer

The lead for a reamer is a small forty-five degree chamfer on the reamer's end. For reamers under 0.5 in diameter, the lead is usually about 0.03 inches. For reamers over 0.5 in diameter, the lead is larger – usually about 0.06 inches.

Tap

The lead for a tap is usually specified with the number of *imperfect threads* on the end of the tap. The plug tap shown above, for example, may have as many as six imperfect threads. To calculate a tap's lead, you must multiply the number of imperfect threads time the pitch of the tap (pitch is equal to one divided by the number of threads per inch). For a 1/2-13 tap having a lead of four imperfect threads, for instance, the lead will be 0.3079 inch (four times 0.0769).

Boring bar

Notice in Figure 3.4 that most boring bars do not have a lead, meaning you will be specifying the hole-depth directly from the print dimension.

What about milling cutters?

When performing milling operations, there are actually two possible sets of positions from which you may work. If performing hole-machining operations with an end mill (using an end mill for counter-boring, for example), you will still work from the tool's center line position. If you are performing face milling operations, you will also work from the cutter's centerline.

If you are performing *contour milling operations* (milling on the periphery of the milling cutter), you can still specify coordinates based upon the cutter's centerline (called the *centerline tool path*). But it is not very convenient to work from the cutter's centerline, especially with complicated contours. As you will see in Lesson Twelve, there is a feature called *cutter radius compensation*. This feature makes it possible to use coordinates that are on the *work surface* (right on the workpiece). This is called the *work surface tool path*. Since we have not yet introduced cutter radius compensation, contour milling examples shown in this lesson will be based upon the milling cutter's centerline tool path.

G00 – Rapid motion (also called positioning)

Rapid motion is used to position a cutting tool to and from machining positions. Under normal conditions, G00 (stated G-*zero-zero*) will cause the machine to move at its fastest possible rate – which is called the machine's *rapid rate*. The rapid rate will vary from one machine to another – but it is always very fast. Several current CNC machining centers boast rapid rates well over 1,500 inches-per-minute.

Due to this very fast – and somewhat scary – motion rate, most machine tool builders will allow you to override the machine's rapid rate during a program's verification using a multi-position switch called *rapid traverse override*. Though this feature varies from machine to machine, most machining centers allow you to slow the rapid rate to a crawl. This ability will relieve some of the stress of running a program for the first time, and minimize the potential for problems if a mistake has been made in a rapid motion command.

Figure 3.5 shows an example of rapid motion.

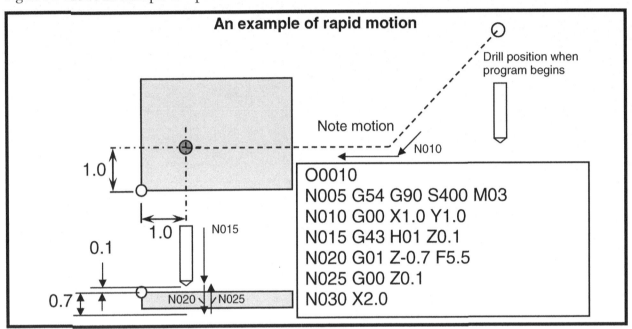

Figure 3.5 – Rapid motion example

Program with comments:

O00010 (Program number)

N005 G54 G90 S400 M03 (Select fixture offset #1, absolute mode, start spindle fwd at 400 rpm)

N010 G00 X1.0 Y1.0 (Rapid to hole-location)

N015 G43 H01 Z0.1 (Instate tool length compensation, rapid to just above work surface)

N020 G01 Z-0.7 F5.5 (Straight motion to hole-bottom at 5.5 ipm)

N025 G00 Z0.1 (Rapid out of hole)

N030 X2.0 (Rapid to next hole)

.

.

The tool is well away from the workpiece when the program in Figure 3.5 runs (possibly at the zero return position). In line N005, the G54 word tells the control to look in fixture offset number one to find the program zero assignment values. G90 tells the control that all up-coming coordinates will be specified from the program zero point. S400 and M03 start the spindle at 400 rpm in the forward direction.

Notice the G00 word in line N010. It specifies the rapid mode, so all motions from this point will be at rapid until the motion type is changed. In this command, the drill will rapid to the hole-position in XY (X1.0 Y1.0).

With most machining centers, this motion will *not* be along a straight line. Instead, most machines will allow all axes (X and Y in our case) to move as fast as they can. The rapid rate is the same for all axes (with most machining centers). One of the axes will arrive at its destination point first (the Y axis in our example). The other axes (just X in our example) will continue to move at rapid until they arrive at their destination points. Some programmers refer to this motion as a *dog-leg motion* (taken from the game of golf).

Now look at line N015. Though we haven't addressed it yet, G43 and H01 are invoking something called *tool length compensation*. (This feature is discussed in Lesson Eleven.) Notice the Z position of Z0.1. Also notice that this command does not include a motion type designation. Since G00 is *modal*, the motion to Z0.1 will be done at the machine's rapid rate. Again, G00 is modal. While it wouldn't hurt to include a G00 in line N015, it won't help either. We recommend leaving out redundant words to make your programs shorter and minimize the potential for writing/typing mistakes.

While we haven't discussed it yet, the G01 in line N020 specifies a straight-line motion. It changes the motion type from rapid to straight-line motion. The tool will now move at a feedrate of 5.5 inches-per-minute in during this motion, drilling the hole to a Z position of Z-0.7.

In line N025, the tool will once again rapid (note the G00 in this command) out of the hole and back to a position of Z0.1.

Again notice that in line N030, there is no motion type designation. So the machine will retain the rapid mode. The tool will rapid to the next hole-location (at X2.0).

How many axes can be included in a rapid motion command?

One, two, or three (even four if the machine has a rotary axis) axes can be included in the G00 command. Line N010 in Figure 3.5, for example, causes a two-axis rapid motion (in X and Y). The Z axis will remain stationary during this command. (Remember, you need only include the moving axes in each motion command – X and Y in our example).

While we recommend approaching in XY first, *then* Z, you can approach in all three axes *simultaneously*. To do so in our example, replace lines N010 and N015 with:

    N010 G00 X1.0 Y1.0 G43 H01 Z0.1

While this will minimize cycle time, it will be much scarier for the operator. If you do elect to use this method for approaching, be sure that the Z position you include in this command is above *all* obstructions in the setup. Most programmers that use this technique will keep Z higher than just 0.100 above obstructions. They'll use a distance of at least two inches above all obstructions (Z2.0 in our example) to keep from frightening the operator during the approach move. Once the cutting tool is in position, they'll rapid the it to its final approach position in Z, as shown in this new version of the program:

    O0010 (Program number)
    N005 G54 G90 S400 M03 (Select fixture offset #1, absolute mode, start spindle fwd at 400 rpm)
    N010 G00 X1.0 Y1.0 G43 H01 Z2.0 (Rapid to hole location, instate tool length compensation,
        rapid to two inches above work surface)
    N015 Z0.1 (Rapid to just above work surface)
    N020 G01 Z-0.7 F5.5 (Straight motion to hole-bottom at 5.5 ipm)
    N025 G00 Z0.1 (Rapid out of hole)
    N030 X2.0 (Rapid to next hole)
    ..

About the dog-leg motion…

It is very important that you understand the dog-leg style of motion most machines will use when two or more axes are included in a rapid motion command (as shown in line N010 in the example). You must always be concerned with obstructions between the starting and ending point for the rapid motion. The dog-leg motion will sometimes cause a tool to contact an obstruction that it would not if the motion occurs along a straight line. In these cases, you must break the rapid motion into two or more commands to go *around* the obstruction.

When do you use rapid motion?

Though it may seem obvious, you should use the rapid motion command whenever the cutting tool is not machining the workpiece during the motion. In this way, you can minimize a program's *air-cutting time* (reducing cycle time). This includes approaching surfaces to be machined, retracting tools to the machine's tool changing position, and any non-cutting operation that occurs during the motions of a cutting tool (getting from one cutting position to another). A good rule-of-thumb is *"If the tool is not cutting, it should be moving at rapid"*.

Certain commands *automatically* cause the machine to move at its rapid rate. During a G28 command, which sends the axes to the zero return position, the machine will automatically invoke the rapid mode. The command,

N075 G91 G28 Z0 M19

for example, will send the machine to its Z axis zero return position. This is the tool change position for most vertical machining centers. Though a G00 is not included in this command, the machine will move at its rapid rate during this motion.

What is a safe approach distance?

As stated, the rapid motion command minimizes motion time during your program. However, as the programmer, *you* must determine how closely you position each tool relative to a surface to be machined. Note that some machining operations occur at a very slow feedrate (at least when compared to the rapid rate), meaning a great deal of cycle time will be taken if you rapid a tool too far away from the work surface before the tool starts its cutting motion.

Though beginning programmers will need to be extra careful with rapid approach positioning movements, you must also be concerned with creating efficient programs. In the beginning of your programming career, the priority must be creating safe and easy-to-use programs. But as you gain experience, you'll need to create more efficient programs (safety and efficiency seldom go together when it comes to CNC machine tool usage).

Consider the effect that rapid approach distance has on cycle time. Most programmers use a 0.100 in (about 3.0 mm) approach distance when tools approach a *qualified surface*. By qualified surface, we mean a surface that will not vary by more than about 0.010 in from one workpiece to the next. Examples of qualified surfaces include surfaces machined during or prior to the CNC machining center operation, cold finished surfaces, precision die cast surfaces, and other accurate surfaces formed prior to the CNC operation.

On the other hand, if the surface is not qualified (it varies widely from one workpiece to another), the rapid approach distance must be increased. For sand castings, forgings, and other raw materials that are prone to variance, an approach distance of at least 0.25 in is recommended.

A good rule-of-thumb is to make the approach distance about ten times the amount that a surface is varying from one workpiece to the next (if the surface varies about 0.01 inch, use an approach distance of 0.100 inch). Even with this rule-of-thumb, remember that as you continue your career in programming, you may need to consider reducing rapid approach distance (at least for qualified surfaces) as production quantities – and your skills – grow.

Consider this example: You have fifty holes to drill into a qualified surface. You are using a high speed steel drill, requiring a feedrate of just 5.0 ipm. If using a rapid approach distance of 0.100 inch, the cycle will experience five inches of air cutting motion (50 holes times 0.1 in approach). At 5 ipm, that equates to *one minute* of cycle time. By reducing the rapid approach distance to 0.050 inch, thirty seconds of cycle time can be saved. Note that we do not encourage beginners to reduce rapid approach distance. Only consider doing so after you have written and verified several programs.

G01 – Linear interpolation (straight-line motion)

This motion type causes the machine to move along a perfectly straight path in one, two, or three axes. The control will calculate the path between the start point and the end point of the motion automatically, no matter what angle of motion is required. So you simply specify the *end point* for the motion. The G code used to command this motion type is **G01** (stated G-*zero*-one).

The motion rate for a straight-line motion is programmable. It is specified with an F word (specifying a feedrate for the motion). With almost all machining centers, the feedrate is specified only in per-minute fashion (inches-per-minute in the inch mode or millimeters-per-minute in the metric mode). Note that there are some machining centers (though not many) that additionally allow feedrate to be specified in per-revolution fashion (inches-per-revolution or millimeters-per-revolution). If your machining centers allow both feedrate types, two G codes will be used to specify which feedrate mode you desire (**G94** for per-minute, **G95** for per-revolution).

Like motion types, the feedrate word is *modal*. If a series of cutting motions will be done at the same feedrate, the F word need only be included in the *first* cutting motion command.

The straight-line motion command is used primarily to machine straight surfaces. Examples of when a **G01** command can be used include drilling a hole and milling a straight or angular surface.

As with rapid motion, you will be able to override the motion rate for all cutting motions, including straight-line motions. A special multi-position switch, called *the feedrate override switch*, allows you to do so. With most machines, you'll be able to stop the motion rate (0%) and increase it to double the programmed feedrate (200%) in ten percent increments on the switch. This switch will help you verify the cutting feedrate used for each cutting tool.

Figure 3.6 shows an example of straight-line cutting commands.

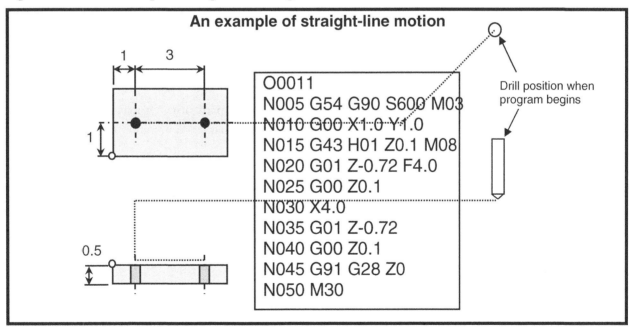

Figure 3.6 – Straight-line cutting command example

Program with comments:

O0011 (Program number)

N005 G54 G90 S600 M03 (Select fixture offset #1, absolute mode, start spindle fwd at 600 rpm)

N010 G00 X1.0 Y1.0 (Rapid to first hole-location)

N015 G43 H01 Z0.1 M08 (Instate tool length compensation, rapid to just above work surface, start coolant)

```
N020 G01 Z-0.72 F4.0 (Drill hole to bottom at 4.0 ipm)
N025 G00 Z0.1 (Retract from hole)
N030 X4.0 (Rapid to second hole)
N035 G01 Z-0.72 (Drill second hole at 4.0 ipm)
N040 G00 Z0.1 (Retract from hole)
N045 G91 G28 Z0 (Rapid to Z axis zero return position)
N050 M30 (End of program)
```

Though this example is similar to the one shown for rapid motion, notice that it is a complete program.

In line N005, the G54 word tells the control to look in fixture offset number one to find the program zero assignment values. G90 tells the control that all up-coming coordinates will be specified from the program zero point. S600 and M03 start the spindle at 600 rpm in the forward direction.

The G00 in line N010 specifies the rapid mode, so motions will be at rapid until the motion mode is changed. In this command, the drill will rapid to the hole-position in XY (X1.0 Y1.0). Again, notice the dog-leg style of motion caused by this G00 command.

In line N015, the drill will approach in Z (still at rapid), instating tool length compensation and turning on the coolant (M08) along the way.

Line N020 is the first straight-line cutting command, causing the drill to machine the hole at a feedrate of 4.0 ipm.

Line N025 retracts the drill from the hole at rapid back to a Z position of Z0.1.

Line N030 retains the rapid mode, moving the drill to its new hole position in X and Y.

Line N035 machines the second hole. The G01 in this command is very important. Don't you agree that if it is left out, the drill will rapid into the workpiece? This of course, would case the drill to break.

Notice that there is no feedrate word in line N035. Again, feedrate is modal. It will even be retained (at 4.0 ipm in our case) even after motion mode changes have been made. So the second hole will be machined at the same feedrate as the first, 4.0 ipm.

Line N040 retracts the drill from the second hole (at rapid) back to a position of Z0.1.

Line N045 sends the machine to its Z axis zero return position. In this particular example, the machine happens to be in the rapid mode at this point from line N040. But even if it is not, remember that G28 will automatically invoke the rapid mode.

The M30 in line N050 is an *end-of-program* word. It will cause the machine to turn off anything that is still running (spindle and coolant in our example) – then rewind to the beginning of the program and stop the executing the program.

Using G01 for a fast-feed approach

For the most part, you will use the straight line motion command whenever you are machining a straight surface. However, there may be times when you may elect to use G01 even when the tool is not actually machining. For example, many programmers are reluctant to rapid a tool past the top surface of the workpiece. They may be afraid that doing so will scare the person running the machine. Instead, they program approach movements below the work surface in Z with a *fast feedrate*. This relieves fear and still allows relatively efficient movements. Here is the beginning a program that uses the fast feedrate technique. In this program, the fast feedrate is 50 ipm (line N025).

```
O0004 (Program number)
N005 G54 G90 S350 M03 (Select fixture offset #1, absolute mode, start spindle fwd at 350 RPM)
N010 G00 X-0.25 Y-0.6 (Rapid to XY position)
N015 G43 H01 Z0.1 (Instate tool length compensation, rapid to just above workpiece)
N025 G01 Z-0.25 F50.0 (Fast feed to work surface – tool is clear of workpiece at this position)
```

N030 G01 Y3.25 F3.5 (Mill to point 2)

N035 X4.5 (Mill to point 3)

.

Just as you when you transition from rapid motion to straight-line cutting motion (and you must specify the G01 for the cutting motion), so must you remember to specify the cutting feedrate when transitioning from the fast feedrate to the cutting feedrate– as this program shows in line N030 with F3.5.

A milling example

Figure 3.7 shows another example using G01 – this time for a milling operation. It uses a fast-feed approach.

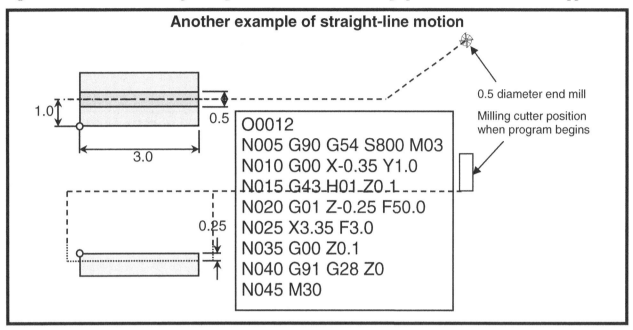

Figure 3.7 – Another G01 example

Program with comments:

O0012 (Program number)

N005 G54 G90 S800 M03 (Select fixture offset #1, absolute mode, start spindle fwd at 800 rpm)

N010 G00 X-0.35 Y1.0 (Rapid to XY approach position)

N015 G43 H01 Z0.1 (Instate tool length compensation, rapid to just above workpiece)

N020 G01 Z-0.25 F50.0 (Fast feed to work surface at 50 ipm)

N025 X3.35 F3.0 (Mill slot at 3.0 ipm)

N035 G00 Z0.1 (Rapid to just above workpiece)

N040 G91 G28 Z0 (Rapid to the Z axis zero return position)

N045 M30 (End of program)

Again, notice the fast feed approach in line N020. Also notice that in line N025, the milling cutter is milling the slot in the center of the workpiece. The motion type for line N025 will still be straight-line motion, carrying over from line N020. Notice the new feedrate in line N025 (F3.0), which is *very important*. Without this feedrate, the motion in line N025 will occur at 50.0 ipm, which will probably break the cutter.

Drill holes with G01?

We have already mentioned that the straight line motion command can be used for drilling holes. While this will be demonstrated several times during the course (including in your exercises and programming activities), there are special programming features, called *canned cycles*, that dramatically simplify the programming of hole-machining operations, including drilling, tapping, reaming and boring. Canned cycles are discussed in Key Concept Number Six.

G02 and G03 – Circular interpolation (circular motion)

Milling operations commonly require the machining of circular workpiece attributes. Consider, for example, the milling of a circular pocket. Frankly speaking, just about the *only* time you'll need to command a circular motion is when *contour milling*. When circular motion is commanded, two axes (usually X and Y) will be moving together to form the motion.

Circular motion can be either *clockwise* or *counter-clockwise*. Two G codes are involved – **G02** specifies clockwise motion while **G03** specifies counter-clockwise motion.

To determine whether a given XY circular motion is clockwise or counter-clockwise (**G02** or **G03**), view the motion from the perspective of the cutter. In most cases, this means viewing the motion from *above the print*. See Figure 3.8 for an example of clockwise and counter-clockwise motion. Notice that we're simply viewing the motion from above the print.

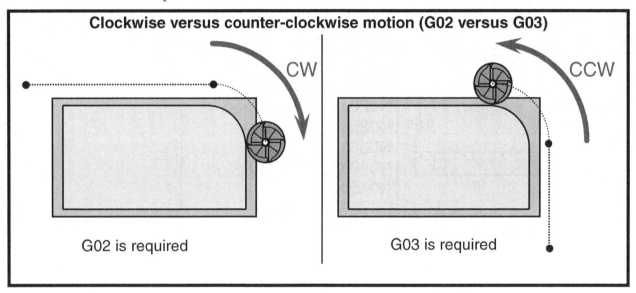

Figure 3.8 – Clockwise versus counter-clockwise circular motion

Like straight-line motion, circular motion requires that a feedrate be specified (with an F word). Again, feedrate is specified in per-minute fashion (inches- or millimeters-per-minute). And, feedrate is modal. Even if a feedrate is originally specified in a straight-line motion command, it will remain effective during subsequent circular motion commands.

Also as with straight line motion, circular motion commands require that the *end point* of the circular motion be specified. The tool position prior to the circular motion will be assumed as the starting point for the circular motion.

Which positions to program

As stated at the beginning of this lesson, you can either program the shape of the work surface being machined (work surface coordinates) or the centerline path of the milling cutter (centerline coordinates). And also as stated, work surface coordinates are easier for a manual programmer to calculate. The feature *cutter radius compensation* – not discussed until Key Concept Number Four – makes it possible for manual programmers to specify work surface coordinates. But until Key Concept Number Four, we must show all examples using the cutter's centerline tool path.

Specifying arc size with the R word

Circular motion commands also require that you specify the arc size of the circular path you are commanding. Almost all current controls allow you to do so with a simple *R word*. With the R word, you specify the size (radius) of the arc being machined.

Note that the value of *the R word must correspond to the path you are programming*. If programming the work surface path (using cutter radius compensation), the R word must reflect the *workpiece radius* being machined. If programming the cutter's centerline path (which we must demonstrate until we present cutter radius compensation), the R word must reflect *the radius of the cutter's centerline path*. Figure 3.9 shows what we mean by *the cutter's centerline tool path*.

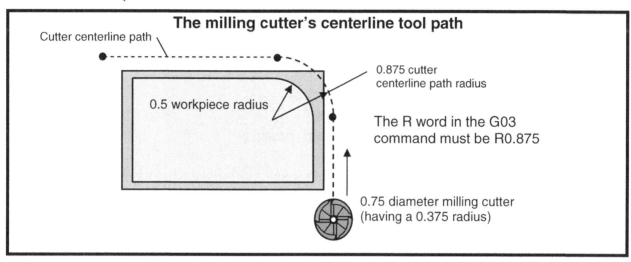

Figure 3.9 shows a drawing used for the example program to follow.

With an outside radius as shown in Figure 3.9, you must *add* the cutter radius to the workpiece radius to come up with the cutter's centerline path radius. If milling an inside radius (as would be the case when milling a circular pocket), you must *subtract* the milling cutter's radius from the workpiece radius to come up with the cutter's centerline path radius.

Figure 3.10 shows a full example of circular motion.

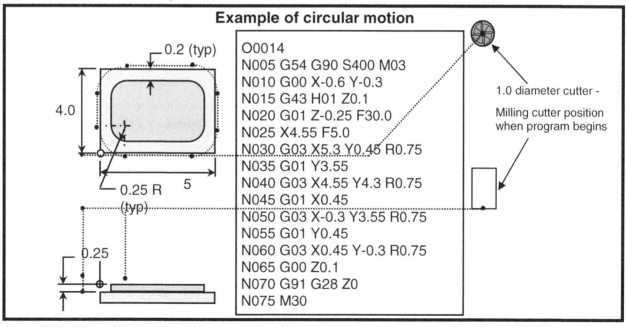

Figure 3.10 – Example program for circular motion

Program with comments:

```
O0014 (Program number)
N005 G54 G90 S400 M03 (Select fixture offset #1, absolute mode, start spindle fwd at 400 rpm)
N010 G00 X-0.6 Y-0.3 (Rapid to approach position in XY)
N015 G43 H01 Z0.1 (Instate tool length compensation, move to just above workpiece)
N020 G01 Z-0.25 F30.0 (Fast feed to work surface)
N025 X4.55 F5.0 (Mill lower surface)
N030 G03 X5.3 Y0.45 R0.75 (Mill lower-right radius)
N035 G01 Y3.55 (Mill right surface)
N040 G03 X4.55 Y4.3 R0.75 (Mill upper-right radius)
N045 G01 X0.45 (Mill upper surface)
N050 G03 X-0.3 Y3.55 R0.75 (Mill upper-left radius)
N055 G01 Y0.45 (Mill left surface)
N060 G03 X0.45 Y-0.3 R0.75 (Mill lower-left radius)
N065 G00 Z0.1 (Rapid to just above workpiece)
N070 G91 G28 Z0 (Rapid to the Z axis zero return position)
N075 M30 (End of program)
```

We're using a 1.0 diameter end mill. Notice that each of the four radii being machined is 0.25 R. This means the cutter's centerline path radius will be R0.75 (0.25 workpiece radius plus 0.5 cutter radius). Also notice that we're milling in a counter-clockwise direction around this workpiece, so all of the circular motions are in this program are commanded with G03.

In line N010, the cutter is being positioned (at rapid) clear of the left side of the workpiece in X (by 0.1) and flush with the lower surface to mill in Y (0.2 minus 0.5 cutter radius, Y-0.3).

In line N015, the cutter is brought to within 0.1 of the top of the workpiece (again, at rapid) and tool length compensation is being instated.

In line N020, the cutter fast feeds to the work surface (Z-0.25) at the fast feedrate of 30.0 ipm.

Machining begins in line N025, as the lower surface is milled. The tool ends this motion in a position ready to machine the first radius (the lower-right radius). That is, the end point for this straight-line motion is the *start point* for the lower-right radius.

In line N030, the first radius is milled. G03 specifies counter-clockwise motion, and the end point in XY brings the cutter to a position to begin the milling of the right side of the workpiece. And the R0.75 specifies the size of the radius to be machined (again, this is the cutter's centerline path radius: R0.75).

The motions toggle between G01 and G03 from this point as the cutter moves around the balance of the contour. In line N065 the cutter retracts from the work surface back to a clearance position above the workpiece in Z (Z0.1).

The R word is not modal

Notice again that each circular command follows a straight-line motion command. The G01 in each straight line motion command *is required*. Since circular commands are modal, if the subsequent G01 commands are left out, the control will retain the circular motion mode. Also note that *the R word is not modal*. Even though all of the radii in this program are 0.75" in size, *every* circular command requires the radius word.

More details about clockwise versus counter clockwise circular motion

Almost all of the circular motions you command will occur in the *XY plane* (as the example in Figure 3.10 shows). However, circular motions can also be performed in the two other planes (XZ and YZ). In these somewhat rare occurrences, the method by which you determine clockwise versus counter clockwise will change. Additionally, three plane selection G codes specify which plane is being used for the circular motion. G17 selects the XY plane. G17 is *initialized* at power-up, meaning if you are working exclusively in the XY plane (as you almost always will), there will be no need to specify G17 in your program. Notice that we're making this assumption in the program shown in Figure 3.10. There is no G17 in this program. G18 selects the XZ plane and G19 selects the YZ plane.

A better way of determining whether a motion is clockwise or counter clockwise is to *view the motion from the plus side of the uninvolved perpendicular axis*. When making an XY circular motion, as we have been, you view the motion from the plus side of the Z axis (again from above the print). But when making an XZ circular motion, you must view the motion from the plus side of the Y axis (this is the column side of a vertical machining center). When making a YZ circular motion, you must view the motion from the plus side of the Y axis (the right side of a vertical machining center).

Planning your own tool paths

Planning tool paths for hole-machining tools, like drills, is pretty simple. To drill a hole, you will first rapid the drill to the hole location in X and Y. Next you'll rapid the drill down to just above the work surface in Z. You'll machine the hole with G01 and then rapid the drill out of the hole in Z. If there are more holes to drill, you'll rapid the drill to the next hole-location (clearing any obstructions along the way) then machine the hole with G01, and retract with G00. You'll then repeat this for all other holes to be drilled.

While there are other types of holes that must be machined (peck-drilling, tapping, reaming, for example), their tool paths are still pretty easy to plan. And as you'll see in Lesson Sixteen, there are several hole-machining canned cycles that dramatically simplify the machining of holes.

But the tool paths for milling operations can be more difficult to plan – even for experienced machinists. You've already seen one complication based upon whether or not you'll be using cutter radius compensation (work surface tool path or cutter centerline tool path).

Milling tool paths can be difficult to plan. In Figure 3.10, the contour must be milled. You must plan an appropriate approach position. We chose a position left of the workpiece by a clearance distance in X and flush with the lower surface to mill in Y. This approach position, must of course, be based upon the kind of milling you will be doing (climb or conventional) and must be clear of any clamps or obstructions in the setup. From this point, the tool path for the workpiece in Figure 3.10 may be pretty easy to understand – but do you see anything wrong with it?

Key points for Lesson Nine:

- CNC control manufacturers provide interpolation types for the motions that their machines must make.
- You must know the point on the cutting tool that is being programmed.
- The three most basic motion types are rapid motion – G00, linear motion – G01 (also called straight-line motion), and circular motion – G02 (clockwise) and G03 (counter-clockwise).
- All motion types are modal.
- All motion types require that you specify the end point for the motion in the motion command.
- Use rapid motion whenever the tool is not cutting to minimize cycle time.
- Linear and circular motions additionally require that you specify a feedrate for the motion – and feedrate is also modal.
- Circular motion additionally requires that you specify the arc size for the motion – and the R word is the simplest way to do so.

Answer program is shown on the page just before Lesson Nine begins (p. 78).

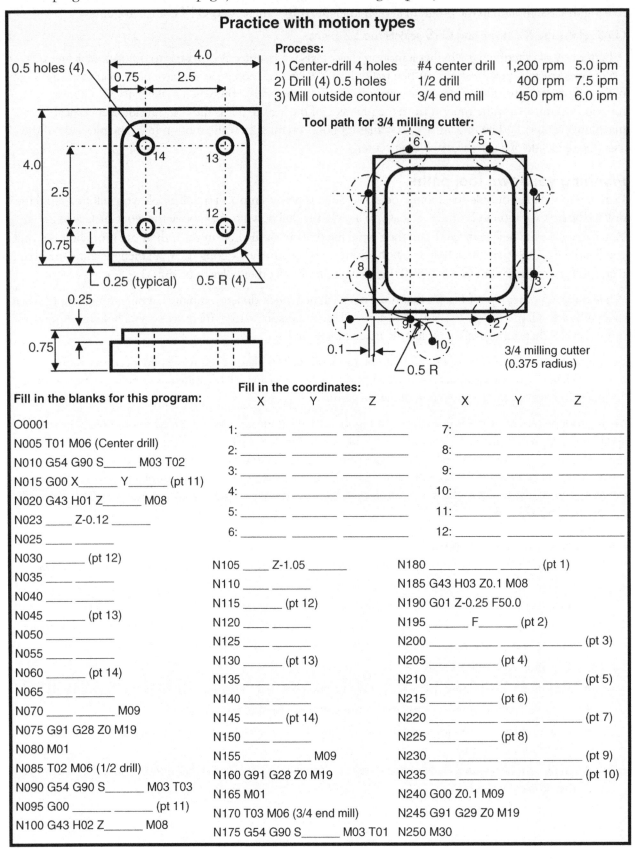

Practice with motion types

Process:

1) Center-drill 4 holes	#4 center drill	1,200 rpm	5.0 ipm
2) Drill (4) 0.5 holes	1/2 drill	400 rpm	7.5 ipm
3) Mill outside contour	3/4 end mill	450 rpm	6.0 ipm

Tool path for 3/4 milling cutter:

3/4 milling cutter (0.375 radius)

Fill in the blanks for this program:

O0001

N005 T01 M06 (Center drill)

N010 G54 G90 S_____ M03 T02

N015 G00 X_____ Y_____ (pt 11)

N020 G43 H01 Z_____ M08

N023 _____ Z-0.12 _____

N025 _____ _____

N030 _____ (pt 12)

N035 _____ _____

N040 _____ _____

N045 _____ (pt 13)

N050 _____ _____

N055 _____ _____

N060 _____ (pt 14)

N065 _____ _____

N070 _____ _____ M09

N075 G91 G28 Z0 M19

N080 M01

N085 T02 M06 (1/2 drill)

N090 G54 G90 S_____ M03 T03

N095 G00 _____ _____ (pt 11)

N100 G43 H02 Z_____ M08

Fill in the coordinates:

	X	Y	Z		X	Y	Z
1:				7:			
2:				8:			
3:				9:			
4:				10:			
5:				11:			
6:				12:			

N105 _____ Z-1.05 _____

N110 _____ _____

N115 _____ (pt 12)

N120 _____ _____

N125 _____ _____

N130 _____ (pt 13)

N135 _____ _____

N140 _____ _____

N145 _____ (pt 14)

N150 _____ _____

N155 _____ _____ M09

N160 G91 G28 Z0 M19

N165 M01

N170 T03 M06 (3/4 end mill)

N175 G54 G90 S_____ M03 T01

N180 _____ _____ _____ (pt 1)

N185 G43 H03 Z0.1 M08

N190 G01 Z-0.25 F50.0

N195 _____ F_____ (pt 2)

N200 _____ _____ _____ _____ (pt 3)

N205 _____ _____ (pt 4)

N210 _____ _____ _____ _____ (pt 5)

N215 _____ _____ (pt 6)

N220 _____ _____ _____ _____ (pt 7)

N225 _____ _____ (pt 8)

N230 _____ _____ _____ _____ (pt 9)

N235 _____ _____ _____ _____ (pt 10)

N240 G00 Z0.1 M09

N245 G91 G29 Z0 M19

N250 M30

Key Concept

4

Know The Compensation Types

CNC machining centers provide three kinds of compensation to help you deal with tooling related problems. In essence, each compensation type allows you to create your CNC program without having to know every detail about your tooling. The setup person will be entering certain tooling information into the machine separately from the program.

Key Concept Number Four is made up of four lessons:

> 10: Introduction to compensation
> 11: Tool length compensation
> 12: Cutter radius compensation
> 13: Fixture offsets

The fourth Key Concept is that you must know the three kinds of compensation designed to let you ignore certain tooling problems as you develop CNC programs. In Lesson Nine, we'll introduce you to compensation, showing the reasons why compensation is needed on CNC machining centers.

Two of the compensation types are related to cutting tools: tool length compensation and cutter radius compensation. We'll discuss them in Lessons Eleven and Twelve. The third compensation type, fixture offsets, is related to work holding devices – and we'll discuss fixture offsets in Lesson Thirteen. While we do show the use of fixture offsets in Key Concept Number One, we'll show more applications for them in Lesson Twelve.

Frankly speaking, not all of the material presented in Key Concept Number Four may be of immediate importance to you. If you do don't do any contour milling on your machines (maybe your machines only perform hole-machining operations), then you won't have much need for cutter radius compensation. In like fashion, you may not need to know any more about fixture offsets than what is shown in Lesson Six. Depending upon your needs, you may want to simply skim these lessons to gain an understanding of when and why the related features are needed – then move on. You can always come back and dig in should your needs change.

On the other hand, tool length compensation is a very important feature that is used in *every tool of every program you write*. You'll need to completely master its use.

Fundamentals of CNC

Lesson 10
Introduction To Compensation

An airplane pilot must compensate for wind direction and velocity when setting a heading. A race-car driver must compensate for track conditions as they negotiate a turn. A marksman must compensate for the distance to the target when firing a rifle. And a CNC programmer must compensate for certain tooling-related problems as programs are written.

What is compensation and why is it needed?

When you compensate for something, you are allowing for some unpredictable (or nearly unpredictable) variation. A *race car driver* must compensate for the condition of the race track before a curve can be negotiated. In this case, the unpredictable variation is the condition of the track. An *airplane pilot* must compensate for the wind direction and velocity before a heading can be set. For them, wind direction and velocity are the unpredictable variations. A *marksman* must compensate for the distance to the target before a shot can be fired – and the distance to the target is the unpredictable variation. The marksman analogy is remarkably similar to what happens with most forms of CNC compensation. Let's take it further...

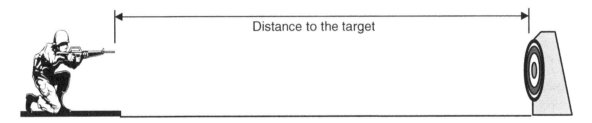

Distance to the target

Before a marksman can fire a rifle, they must judge the distance to the target. If the target is judged to be fifty yards away, the sight on the rifle will be adjusted accordingly. When the marksman adjusts the sight, they are *compensating for the distance to the target*. But even after this preliminary adjustment and before the first shot is fired, the marksman cannot be *absolutely* sure that the sight is adjusted perfectly. If they've incorrectly judged the distance – or if some other variation (like wind) affects the sight adjustment – the first shot will not be perfectly in the center of the target.

After the first shot is fired, the marksman will know more. If the shot is not perfectly centered, another adjustment will be needed. And the second shot will be closer to the center of the target than the first. Depending upon the skill of the marksman, it might be necessary to repeat this process until the sight is perfectly adjusted.

With *all* forms of CNC compensation, the setup person will do their best to determine the compensation values needed to perfectly machine the workpiece (just as the marksman does their best in judging the distance to the target and adjusting the sight). But until machining actually occurs, the setup person cannot be sure that their initial compensation values are correct. After machining, they may find that another variation (like tool pressure) is causing the initial adjustment to be incorrect. Depending upon the tolerances for the surfaces being machined, a second adjustment may be required. After this second adjustment, machining will be more precise.

There is even a way to make an initial adjustment (prior to machining) that ensures excess material will remain on the machined surface after the first machining attempt (this technique is called *trial machining*). This guarantees that the workpiece will not be scrapped when the cutting tool machines for the first time – and is especially important for very tight (small) tolerances. With tight tolerances, even a small machining imperfection will cause a scrap workpiece.

Once the cutting tool has machined for the first time, the setup person will stop the cycle and measure the surface. If they have used the trial machining technique, there will be more material yet to remove. They will then make the appropriate adjustment and re-run the cutting tool. The second time the cutting tool machines, the surface will be within its tolerance band, probably right at the *target* dimension (the dimension they're shooting for).

More on tolerances

Again, all dimensions have tolerances. And from Lesson Four, you know that you must *program the mean value of the tolerance band* for every coordinate you include in your programs. The mean value, of course, is right in the middle of the tolerance band.

Companies vary when it comes to how tight (small) the tolerances are that they machine on their CNC machining centers. Generally speaking, overall tolerances over about 0.010 inch (about 0.25 mm) are considered pretty open (easy to hold with today's CNC machining centers). Tolerances between 0.002 and 0.010 inch (0.050 – 0.25 mm) are common, and still not considered to be very tight. But tolerances under 0.002 inches can be more difficult to hold. And under 0.0005 inch (0.0013 mm) – which many companies do regularly hold on their CNC machining centers – can be quite challenging – especially when many workpieces must be produced.

The initial setting for compensation

Again, the setup person will do their best to assemble and measure certain cutting tool attributes (like length and diameter). They will then enter their measured values into the machine (into something called *tool offsets*). But even if they *perfectly* measure and enter tooling values, and even if the programmer specifies the mean value for every tolerance in the program, there is no guarantee that every cutting tool will perfectly machine each dimension to the mean value of its tolerance band.

Tool pressure, which is the tendency for a cutting tool to deflect from the workpiece during machining, will always affect the way a cutting tool machines. It usually has the tendency of pushing the cutting tool away from the surface being machined. While the surface being machined should be *close* to its mean value (again, assuming the setup person perfectly measures and enters tooling information), it may not be perfectly at the mean value.

How tight is the tolerance? The tighter the tolerance, the more likely it will be that the deflection caused by tool pressure will cause the machined surface to be outside the tolerance band when the cutting tool machines for the first time. This could cause a scrap workpiece. And again, this is the reason why trial machining is required – to ensure that the first workpiece gets machined correctly.

And by the way, we've assumed that the setup person has *perfectly* measured and entered tooling information. Any mistakes will, of course, increase the potential for problems holding size on the first workpiece to be machined.

When is trial machining required?

First let's talk about when trial machining is *not* required. Trial machining is not required for most roughing operations – like rough milling and rough boring. While a measurement should still be taken right after the roughing operation to confirm that the appropriate amount of *finishing stock* is being left for the finishing tool (and an adjustment must be made if not), trial machining is not necessary.

There are certain cutting tools that do not allow adjustments, and trial machining cannot be done for these tools. With drills, for example, you cannot control the diameter that the drill machines. Hole-diameter is based upon the diameter of the drill. The same goes for taps, reamers and counter-boring tools. So for machining operations that cannot be adjusted, you will not be trial machining.

Also, whenever tolerances are greater than about 0.003 inch (0.075 mm) or so, it is not unreasonable to expect the setup person (or tool-setter) to assemble and measure cutting tools accurately enough so that the cutting tool will machine within the tolerance band (even considering tool pressure) on the first try. While an adjustment may be necessary *after machining* to bring the dimension precisely to the target value, trial machining should not be necessary. Examples include the depth of many (blind) holes and a certain milling operations.

As stated, trial machining is only required when machining tight tolerances. Though the actual cut-off point for when trial machining is required varies based upon the skill of the setup person and the operation being performed, it is not uncommon to trial machine for dimensions that have tolerances under about 0.002 inch (0.050 mm). This includes finish boring operations and many milling operations.

What happens as tools begin to wear?

So say the setup person uses trial machining techniques for critical dimensions and the first workpiece passes inspection. Now the production run begins. As cutting tools continue to machine workpieces, of course, they will begin to show signs of wear. Certain tools, like drills, will continue to machine properly for their entire lives without having an impact on the dimensions they machine (hole size and depth).

But as certain tools begin to wear, like milling cutters and boring bars, a small amount of material will wear away from the tool's cutting edge (a boring bar or an end mill will actually get a little smaller in diameter). Depending upon the tolerance band for the dimension being machined by the cutting tool, it may be necessary to make additional sizing adjustments *during the cutting tool's life* to ensure that the cutting tool continues to machine acceptable workpieces.

What do you shoot for?

The value for each dimension that you are aiming for as you make adjustments is called the *target value*. In some companies, CNC setup people and operators are told to always target the mean value of the tolerance band (the same value the programmers includes in the program). But to prolong the time between needed adjustments, some companies ask their CNC people to target a value that is closer to the high or low limit of the tolerance band, whichever will prolong the time between adjustments. In any event, CNC people must know the target value for each dimension they must machine (this information should be in the production run documentation).

Why do programmers have to know this?

Admittedly, much of this discussion is more related to setup and operation than it is to programming. But again, a programmer must understand enough about machine usage to be able to direct people that run the machine. And these discussions go to the heart of one of the most important reasons why compensation is required on CNC machining centers: *to allow sizing adjustments without needing to modify the program.*

Understanding offsets

All three compensation types use *offsets*. Offsets are storage locations for values. They are very much like memories in an electronic calculator. With a calculator, if a value is needed several times during your calculations, you can store the value in one of the memories. When the value is needed, you simply type one or two keys and the value returns. In similar fashion, the setup person or operator can enter important tooling-related values into offsets. When they are needed by the program, a command within the program will *invoke* the value of the offset. And by the way, just as a calculator's memory value has no meaning to the calculator until it is invoked during a calculation, neither does a CNC offset have any meaning to the CNC control until it is invoked by a CNC program.

Like the memories of most calculators, offsets are designated with *offset numbers*. Offset number one may have a value of 6.5439. Offset number two may have a value of 6.2957. For cutting tool related offsets, offset numbers are made to correspond in some way to *tool station numbers*. For example, the tool length compensation value for the cutting tool placed in station number one is commonly entered in offset number one.

Unlike the memories of an electronic calculator that will be lost when the calculator's power is turned off, CNC offsets are more permanent. They will be retained even after the machine's power is turned off – and until an operator or setup person changes them.

Offsets are used with each compensation type to tell the control important information about tooling. From the marksman analogy, you can think of offset values as being like the amount of sight adjustment a marksman must make prior to firing a shot. Tooling related information entered into offsets includes each cutting tool's length, each milling cutter's radius, and program zero assignment values for work holding devices.

Offset organization

Machining center controls vary with regard to how many offsets are available and how they are organized. First of all, rest assured that you will always have enough offsets to handle your applications. Most machine tool builders supply many more offsets than are required even in the most elaborate applications (an exception to this statement may sometimes be with regard to fixture offsets). Most machining centers have two distinct sets of offsets, one for cutting tools (tool offsets) and another for program zero assignment (fixture offsets).

Offsets related to cutting tools

All cutting tools will require a tool length compensation offset. Milling cutters that perform contour milling operations will additionally require a cutter radius compensation offset. Control manufacturers vary when it comes to how offsets are organized and displayed.

One common method they use is to provide one offset *register* per offset. This means the same table of offsets will be used for both tool length and cutter radius compensation values. Figure 4.1 shows the display screen for this kind of offset organization.

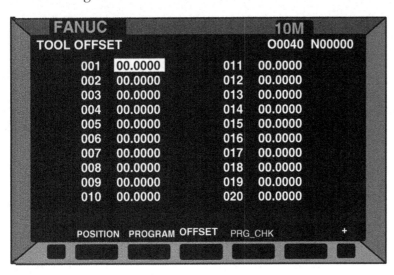

Figure 4.1 – Display screen having but one value per offset number

With this kind of offset organization, tool length and cutter radius compensation share the same group of offsets.

A second and more logical method of offset organization provides two values per offset number – one for tool length compensation and the other for cutter radius compensation. Figure 4.2 shows it.

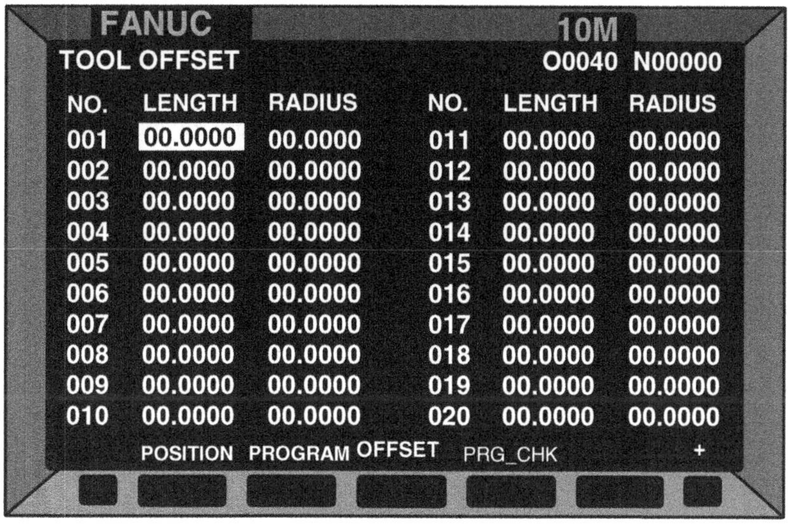

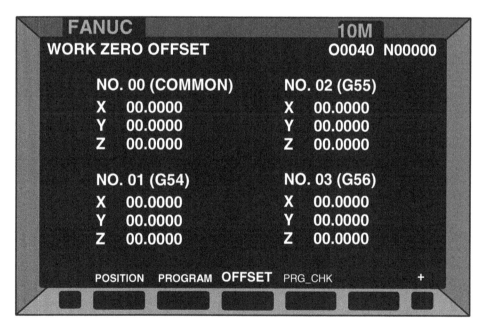

Figure 4.2 – Display screen having two values per offset number

Offsets related to program zero assignment

Program zero assignment values are placed in fixture offsets, as introduced in Lesson Six. Each offset has at least three registers, one for X, one for Y, and one for Z. If the machining center has any rotary axes, there will also be a register for each rotary axis. Figure 4.3 shows the first page of fixture offset display screen for a popular control.

Figure 4.3 – Fixture offset display screen

How offsets are instated

Again, offsets have no meaning to the control until they are *instated* by a program command. For offsets used with tool length compensation, an *H word* is used to instate the offset (we'll discuss tool length compensation in Lesson Eleven). **H01**, for example, instates the value in offset number one.

For cutter radius compensation, most controls use a D word to instate the related offset (though some controls do use the H word for this as well). **D31** instates the value in offset number thirty-one. We'll discuss cutter radius compensation in Lesson Twelve.

For controls that have but one register per offset, there is a bit of a problem. Again, this requires the same table of offsets to be shared for both tool length and cutter radius compensation. You cannot use the same offset number for both tool length compensation and cutter radius compensation (as you can if there are two offset registers per tool offset). Say, for example, tool number one is an end mill that will be using cutter radius compensation. You pick offset number one in which to store the tool length compensation value for this tool. This means, of course, that you must use *another* offset number in which to store the cutter radius compensation value. We'll provide some suggestions for handling this problem in Lessons Eleven and Twelve.

Key points for Lesson Ten:

- Compensation types allow you to ignore certain tooling-related problems as you write programs.

- Compensation types allow sizing and trial machining for critical workpiece attributes.

- Setup people and operators must know the *target value* for each dimension they must machine.

- Critical dimensions require trial machining for the first workpiece being machined.

- Tool wear can affect the dimensions machined by a tool. In some cases, adjustments must be made during the tool's life.

- Offsets are used in which to store compensation values.

- The offset tables vary from one machining center to another.

Talk to experienced people in your company...
...to learn more about how tight tolerances are held.

1) What are the tightest tolerances held on your CNC machining centers?

2) Are any of the tolerances you hold so tight that trial machining is required? If so, ask to see the trial machining techniques being used.

3) Once as setup is made and production is run, do CNC operators have to make sizing adjustments to deal with tool wear?

4) How do you determine the target value for dimensions being machined?

5) Ask to see the offset display screen pages on your CNC machining centers.

Lesson 11
Tool Length Compensation

Tool length compensation allows a programmer to ignore the precise length of each tool as a program is written. It is used for every tool in every program you write – so you must understand this important CNC feature.

You know that program zero assignment values for the X and Y axes are entered into fixture offsets from the *spindle center* to the program zero point in X and Y (while the machine is at its zero return position). So when you specify a position of **X1.0 Y1.0** in a program, the machine will be able to send the *spindle center* (and tool center) to this position – relative to program zero.

In the Z axis, you know that the program zero assignment value is entered into the fixture offset from the *spindle nose* to the Z axis program zero surface (again, while the machine is at its zero return position). But for cutting tool positioning, you don't want to specify Z axis positions from the *spindle nose*. This would be very cumbersome – and it would require that you know the precise length of each tool before you could even write the program. Instead, you specify Z axis positions to the *tip* of each cutting tool. That is, when you specify a position of **Z0.1**, the *tool tip* will move to this position. In order to be able to program the tool tip in the Z axis, a feature called *tool length compensation* must be used. Mastering tool length compensation is the focus of lesson number eleven.

The reasons why tool length compensation is needed
Cutting tools used on machining centers differ from one another. For one thing, there are a variety of cutting tool *types* that are used on machining centers, including center drills, spot drills, drills, taps, reamers, boring bars, end mills, and face mills (among many others). Each type of tool requires a different way of gripping the actual cutting tool in its holder. Some tools (like some straight shank tools) use a collet system. Others (like end mills) use a set-screw to hold the cutting tool in place. Yet others (like face mills and taps) require a very special style of tool holder – designed especially for the cutting tool.

No two tools will have exactly the same length
Given the vast assortment of cutting tools available for use on CNC machining centers, it is unlikely that any two tools used in a program will be exactly the same length. And you will not know precisely how long each tool will be when you write the program. Figure 4.4 shows five different types of cutting tools to illustrate this point.

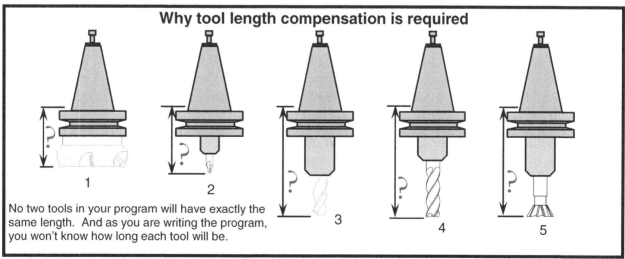

Why tool length compensation is required

No two tools in your program will have exactly the same length. And as you are writing the program, you won't know how long each tool will be.

1   2   3   4   5

Figure 4.4 – Five cutting tools that might be used in by a CNC machining center program

Tool length compensation will allow you to write programs even though you don't know how long the cutting tools will be.

A given tool's length will vary from one time it is assembled to the next
When a cutting tool is assembled more than once (even with the same components), its length will usually vary. Consider, for examples, straight shank tools that are placed in collet holders. Each time you assemble the tool, it will be of a different length. Tool length compensation will allow you to use the same program over and over again, even though each tool's length changes from one time the job is run to the next.

Tool data is entered separately from the program
The same program will work regardless of how long each cutting tool is. The program tells the control where to look for the length of each tool. During setup, the setup person (or someone) assembles and measures each cutting tool. The length of each tool is then placed in the appropriate location (a tool offset register).

Sizing and trial machining must often be done
In Lesson Ten, we discuss the importance of being able to trial machine in order to machine the first workpiece correctly. And during a given tool's life you know that tool wear may cause the surface being machined to change. Tool length compensation allows the setup person and operator to easily hold size for Z axis related dimensions (pocket depths, hole-depths, etc.). The program need not be changed when workpiece dimensions must be adjusted.

What about interference and reach?
As is discussed in Lesson Eight, you must be concerned with whether or not your cutting tools will reach the surfaces to be machined – without over-traveling – and without interfering with the fixture of other obstructions on the workholding device. Tool length compensation will *not help* with these concerns.

Programming tool length compensation
There are actually two popular methods for using tool length compensation. As long as you use fixture offsets to assign program zero, programming remains *exactly* the same regardless of which method you choose to use. The differences between the two methods have to do with program zero assignment (the Z value in your fixture offset) and tool length compensation values (the values actually entered into the tool length compensation offset registers). Again, these differences are not related to programming.

Tool length compensation is instated with a **G43** word. Included within the **G43** command is an *H word* that specifies the offset number in which the tool length compensation value is stored. You must also include a Z word in the **G43** command, telling the machine where you want the *tool tip* to be positioned.

*The **G43** command will always be the cutting tool's first Z axis motion.* Said another way, you instate tool length compensation during every tool's first Z axis motion. This, of course, is the tool's approach movement to the workpiece in the Z axis.

Once tool length compensation is instated, it remains in effect until the next tool. All Z axis motions you need to the tool to make will be relative to the tool tip.

Since you will instate tool length compensation (again) during the *next tool's* first Z axis motion – using the appropriate offset of course – and since offsets are *not* accumulative – you need not cancel tool length compensation. There is a G code labeled *tool length compensation cancel* (it happens to be **G49**), but if you the techniques shown in this text, you need not use **G49** in your programs.

Choosing the offset number to be used with each tool
As you know, offsets are storage registers for values. Each offset to be used with tool length compensation will contain a tool length compensation value for one tool. To keep from entering the tool length compensation value in the wrong offset, the programmer must use a logical approach for selecting offset numbers.

Keep it simple. Use the offset number that corresponds to the cutting tool's *magazine station number.* For example, use offset number one for the tool in tool station number one. Use offset number two for the tool in tool station number two. And so on.

This tool length compensation offset will be *instated* in the program during each tool's first Z axis motion. An H word is used to specify the offset number. And again, we recommend that you make the H word number match the tool station number (the T word number for each tool).

An example program

Figure 4.5 shows the drawing to be used for this example.

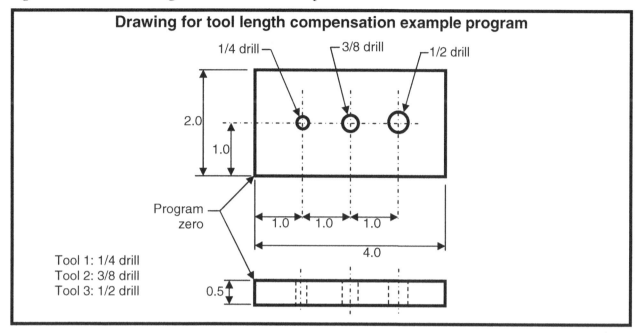

Figure 4.5 – Drawing for example program

Program with comments:

    O0003 (Program number)

    N005 T01 M06 (Load tool number one in spindle)

    (1/4 drill)

    N010 G54 G90 S1200 M03 T02 (Select fixture offset #1, absolute mode, start spindle fwd at 1200 rpm, get tool number two ready)

    N015 G00 X1.0 Y1.0 (Rapid to hole location in X and Y)

    N020 **G43 H01 Z0.1** (Instate tool length compensation for tool one, approach in Z to just above work surface)

    N025 M08 (Turn on the coolant)

    N030 G01 Z-0.65 F4.0 (Drill hole)

    N035 G00 Z0.1 M09 (Rapid out of hole, turn off coolant)

    N040 G91 G28 Z0 M19 (Rapid to tool change position, orient spindle)

    N045 M01 (Optional stop)

    N050 T02 M06 (Load tool number two in spindle)

    (3/8 drill)

    N055 G54 G90 S1000 M03 T03 (Select fixture offset #1, absolute mode, start spindle fwd at 1000 RPM, get tool number three ready)

    N060 G00 X2.0 Y1.0 (Rapid to hole position in X and Y)

    N065 **G43 H02 Z0.1** (Instate tool length compensation for tool two, approach in Z to just above work surface)

    N070 M08 (Turn on coolant)

    N075 G01 Z-0.7 F5.0 (Drill hole)

N080 G00 Z0.1 M09 (Rapid out of hole, turn off coolant)

N085 G91 G28 Z0 M19 (Rapid to tool change position, orient spindle)

N090 M01 (Optional stop)

N095 T03 M06 (Place tool number three in spindle)

(1/2 Drill)

N100 G54 G90 S800 M03 T01 (Select fixture offset #1, absolute mode, start spindle fwd at 800 RPM, get tool number one ready)

N105 G00 X3.0 Y1.0 (Rapid to hole in X and Y)

N110 **G43 H03 Z0.1** (Instate tool length compensation for tool three, approach in Z to just above work surface)

N115 M08 (Turn on coolant)

N120 G01 Z-0.75 F6.0 (Drill hole)

N125 G00 Z0.1 M09 (Rapid out of hole, turn off coolant)

N130 G91 G28 Z0 M19 (Rapid to tool change position, orient spindle)

N135 M30 (End of program)

Lines N020, N065, and N110 instate tool length compensation for each of the three tools. Notice that each of these commands is the first Z axis movement for the tool (its approach movement in Z). Each instating command includes the G43 word, the appropriate H word (that matches the tool station number that is currently in the spindle), and a Z word. Once tool length compensation is instated, it will remain in effect until the next tool. Again, it never has to be canceled using this style of programming.

The setup person's responsibilities with tool length compensation

Before this program can be run, of course, the setup person must perform several tasks. You know, for example, that they must mount the workholding device (probably a vise in this example), measure the program zero assignment values (techniques shown in Lesson Five) and enter them into fixture offsets (Lesson Six). They must also load the program.

Cutting tools must be assembled and loaded into the machine's automatic tool changer magazine – tool station one for the 1/4 drill, tool station two for the 3/8 drill, and tool station three for the 1/2 drill. Though we haven't shown how yet, tool length compensation values must also be determined and entered into the appropriate offsets. Let's now discuss how tool length compensation values are determined.

Again, there are two popular ways to use tool length compensation. We'll first show our recommended method.

Recommended method: Using the tool's length as the tool length compensation (offset) value

With our recommended method, the Z axis program zero assignment value (fixture offset Z value) is the distance from the spindle nose at the Z axis zero return position to the Z axis program zero point – a large negative value– just as is shown in Lessons Five and Six. With our recommended method, *the length of each cutting tool will be its tool length compensation value.*

The length of the cutting tool is the *distance from the tool tip to the spindle nose* and will always be a positive value. This will be the value that is entered into the tool's tool length compensation offset register. Figure 4.6 shows a typical tool's length.

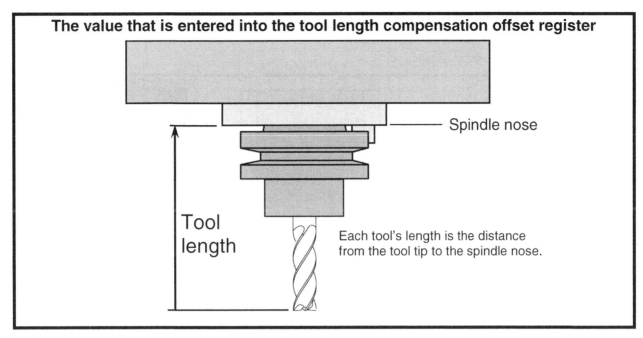

The value that is entered into the tool length compensation offset register

Spindle nose

Tool length

Each tool's length is the distance from the tool tip to the spindle nose.

Figure 4.6 – Tool length is the distance from the tool tip to the spindle nose – always a positive value

Most machining centers use tool holders that have a tapered shank (CAT-40, CAT-50, BT-40, and BT-50 tool holders are very common examples of tapered shank tool holders). Figure 4.4 shows some examples. The shank taper matches the taper in the spindle itself – and the tool locates in the spindle against this taper. From the illustration in Figure 4.6, notice that there is a small gap between the spindle nose and the flange of the tool holder. This gap is about 0.125 inch.

We're pointing this out because you might *incorrectly assume* that tool length is the distance from the tool tip to the end of the flange. When holding a tool holder in your hands, this might appear to be logical. But again, the tool's length is the distance from the tool tip to the *spindle nose*, not to the end of the flange.

Determining tool length compensation values
The tool length for each tool can be measured right on the machine during setup, or it can be measured *off line* using a tool length measuring device. Measuring on the machine during setup takes time – and companies that are highly concerned with reducing setup time prefer to measure cutting tools off line. Let's look at both methods.

Measuring tool lengths right on the machine
Even if your company does measure tool lengths off line, there will be times when you must still measure tool lengths on the machine (maybe after a dull tool is replaced) – so it is quite important that *all* setup people and operators know how to measure tool lengths on the machine. Techniques to do so are quite similar to those used to measure program zero assignment values (shown in Lesson Five).

1) Make the workholding setup – or place some kind of flat block on the machine table
It is important to have a nice, flat surface on which to work. The top of a vise works nicely.

2) Place a gauge block on the flat surface
The three inch side of a 1-2-3 block works nicely.

3) Without a tool in the spindle, make the spindle nose touch the block.

Use the joystick and handwheel to carefully touch the spindle nose to the block. At this point, reset (set to zero) the Z axis relative position display.

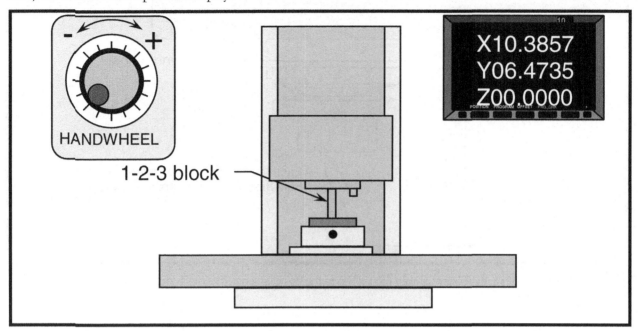

4) Retract the Z axis and load a cutting tool to be measured. Bring the tool tip to the same block.

Use the joystick and handwheel, and cautiously move the spindle nose away from the block. Load a cutting tool to be measured. Now, carefully move the *tool tip* to the same block that was just touching the spindle nose. The Z axis display will follow along. When the tool tip is touching the block, the Z axis display will be showing you the tool's length. This is the value that must be entered into the tool length compensation offset register for this tool. Write it down.

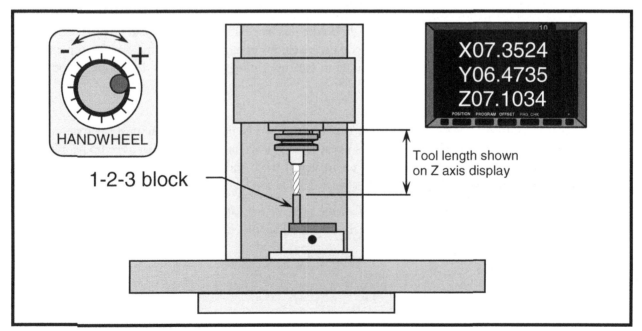

5) Repeat step four for all cutting tools that must be measured.

Measuring tool lengths off line – with a tool length measuring gauge

There are many suppliers that can provide devices that are specially designed for tool length measuring. These devices make it quite easy to measure tool lengths on a bench – away from the CNC machining center. And of course, the machine can be running production while a person assembles cutting tools and measures tool lengths in preparation for up-coming jobs. In this manner, downtime between production runs can be reduced.

But you don't have to buy a fancy tool length measuring gauge. With a little effort, just about any height gauge can be used. Figure 4.7 shows an example.

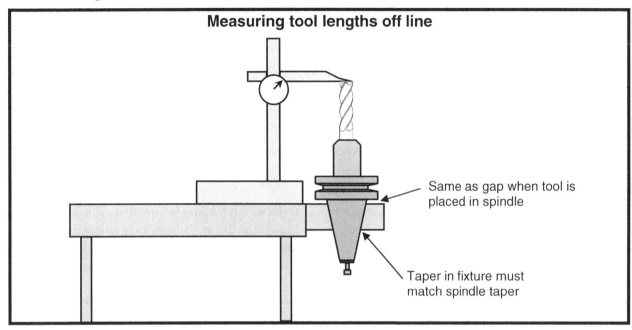

Figure 4.7 – A simple and inexpensive height gauge can be used to measure tool lengths.

Entering tool length compensation offsets

Regardless of which method is used to determine tool length values (measuring on the machine or off line), the tool length compensation values must be entered into tool offsets before the program can be run. One benefit of measuring on the machine is that most machining centers provide a way to *transfer* the Z axis position display value right to the tool length compensation offset. This eliminates the possibility for entry mistakes.

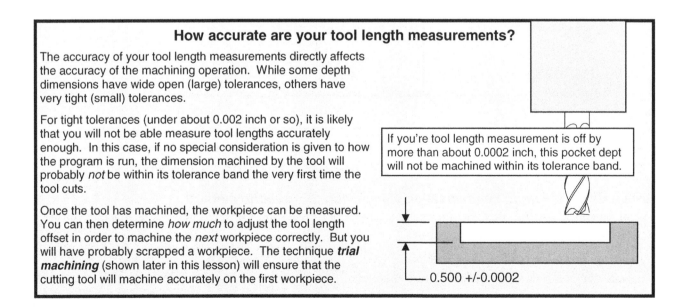

Using the distance from the tool tip to program zero as the tool length compensation (offset) value

While we don't recommend using this method, it is quite popular with *vertical* machining centers (it is quite difficult to apply this method to horizontal machining centers when rotary devices are used). If your company uses this method, you'll probably have to conform. This technique does not (feasibly) allow tool length compensation values to be determined off line, which is one of the reasons we don't recommend it. But there are many situations in which tools *cannot* be measured off line:

- One person is responsible for all CNC tasks – this person doesn't have time to measure tool lengths off line.
- Lot sizes are very small and cycle times are very short – and no one can keep up with the number of tools that must be measured off line.
- Short lead times – no one knows what cutting tools will be needed in upcoming jobs.
- There are not enough cutting tool components – tools needed in upcoming jobs cannot be assembled.

In these situations, tool length compensation values *must* be measured on the machine during setup. But do keep in mind that our recommended method can still be used. If you are new to CNC machining centers and will have control of how things are done, we urge you to use our recommended method. But again, many companies are using this second method – and in this case – you'll probably have to adapt.

With this method, the Z axis program zero assignment value is *zero*. That is, the fixture offset Z register will be set to a value of zero. The tool length compensation value (that is entered into the tool length compensation offset register) is the distance from the tool tip (at the Z axis zero return position) to program zero. The polarity is negative – all of your tool length compensation offsets will be very large negative values. Figure 4.8 shows the offset value.

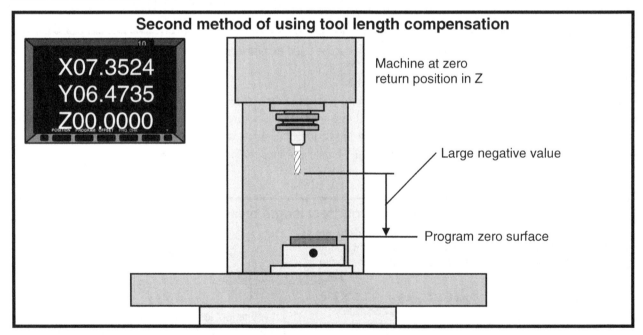

Figure 4.8 – Tool length compensation value when tool tip to program zero is used as the offset

Measuring tool length compensation values with this second method

Again, when you use this method, tool length compensation values must be measured on the machine during setup. Here's how.

1) Make the workholding setup and load a workpiece
Program zero will be the top surface of the workpiece.

2) Send the machine to the Z axis zero return position and reset (set to zero) the Z axis relative display.

This sets the point of reference for your measurement.

3) Load a tool to be measured and manually bring the tip of the tool to the Z axis program zero surface.
Using the joystick and handwheel, cautiously bring the tool tip to the program zero surface. The Z axis display follows along. With the tool tip touching the program zero surface in Z, the Z axis display will be showing you the tool length compensation value for this tool. This is the value that must be entered into the tool length compensation offset register.

Reasons for using our recommended method
Here we list a few reasons to use the recommended method. Note that if your company uses the second method shown, these reasons might help you convince people in your company to change.

Tool length compensation values can be measured off line - We've already mentioned this. Minimizing down time between production runs is a very important goal in many companies.

Cutting tools can be used from job to job without being re-measured – This is a very important reason. The length of a cutting tool will not change (unless it is disassembled) between production runs – but the Z axis program zero surface probably will. If using our recommended method, offset values will remain the same for tools used from job to job. If you use the second method, the tool length compensation value for *all* tools must be re-measured in every setup. Consider how many tools get used in consecutive jobs (center drills, spot drills, common drill-and-tap combinations, and common milling cutters). Indeed, many companies assign standard tool stations to most often-used cutting tools. Using the second method can result in a great deal of duplicated effort.

Cutting tools can be used from machine to machine – In similar fashion, a cutting tool's length does not change when it is placed in a different machine. If a tool is not disassembled after a production run (and if it is not dull), write down the tool's length and keep it with the tool. The next time the tool is needed, you won't have to re-measure it – regardless of which machining center uses it.

The work holding setup doesn't have to be made before tool lengths can be measured – You can measure tool lengths at any time – even on the machine.

Multiple identical tools can be setup and kept ready for action – For those tools that are most prone to wearing out – or for those tools that you use the most, you can keep several identical tools ready to go. When a tool dulls, simply replace it in the machine and enter the new tool length compensation value (that has been measured off line).

Working with multiple program zero points is much easier – Though we haven't shown any applications yet, if your company uses a rotary device to expose several surfaces to the spindle for machining during the CNC cycle, you'll find it much easier to do so with our recommended method. This is the case with most horizontal machining centers.

Typical mistakes with tool length compensation
Mistakes with tool length compensation can have some pretty severe consequences – so we want to prepare you for what can happen when mistakes are made.

Forgetting to instate tool length compensation
As you know, you must instate tool length compensation during every tool's first Z axis movement, including a G43 and H word along with the Z axis departure. If you forget to do so, here's what can happen. Remember that tool length compensation is modal. If you forget to instate tool length compensation for the fourth tool in your program, the control will use the tool length compensation value from the third tool. If the fourth tool is shorter than the third, at least the machine will not try to crash the tool into the workpiece – but it still won't send the tool to the correct Z axis position.

If you forget to instate for the first tool in the program (and if tool length compensation has not been instated since power-up), tool length compensation is in its *cancelled state*. If using our recommended method (tool length is offset value), the machine will think that the nose of the spindle is the tool tip. The machine thinks you have a tool with a zero length. The machine will bring the spindle nose to the programmed Z surface, crashing the tool into the workpiece along the way.

Forgetting to enter the tool length compensation value

When going from job to job, the setup person must remember to enter all tool length compensation values for the up-coming job. If they forget to do so, and if there is a value in the offset from the previous job, the machine will (incorrectly) use this offset value for the new job. If the current value of the offset is zero, the machine will think your tool has zero length – and again – it will try to bring the spindle nose to the programmed Z surface.

Mismatching offsets

Remember, the H word must correspond to the tool station number. As you write the H word in your program, get in the habit of looking back up into the program for the T word in the most recent tool change (**M06**) command. The H word must match the tool station number.

Trial machining with tool length compensation

Tool length compensation will allow trial machining and sizing for *depth* (Z) dimensions that cutting tools machine. Again, *trial machining* is required on the first workpiece when a dimension machined by a cutting tool has a very tight tolerance. This technique ensures that the tool will not machine too much material on its very first try. If done for each tool that has a tight tolerance, the first workpiece will pass inspection.

Trial machining for depth dimensions involves five steps:

1: Recognition of a tight tolerance that worries you – If you're worried that a dimension's tolerance is so small that your initial tool length offset measurement is not accurate enough to make the tool machine the dimension within the tolerance band, then trial machining must be done. For example, say you notice a 0.500 deep pocket with a very tight depth tolerance of plus or minus 0.0002 inch. You're worried that your initial offset setting is off by more than 0.0002 inch – or that tool pressure will cause the tool to machine improperly.

2: Increase the value of the tool length compensation offset by about 0.010 inch (0.25 mm) – Again, this is done *after* you have measured and entered the tool length compensation offset value. In our pocket milling example, say you have measured the milling cutter's length and found it to be 5.2376 inch long – and you have initially placed this value (5.2376) in the tool length compensation offset register. You'll increase this value to *5.2476* to perform the trial machining operation.

3: Let the tool machine under the influence of the trial machining offset and stop the cycle after the tool is finished – In our example, you will allow the milling cutter to machine the pocket. With the increased offset, the machine will keep the milling cutter 0.010 inch further away from the workpiece during machining (forcing our pocket to come out too shallow).

4: Measure the current dimension and reduce the tool length compensation offset accordingly – In our example, say you measure the pocket depth and find it to be 0.492 inch deep. It is currently 0.008 inches undersize. You must reduce the offset by 0.008 inch, making it 5.2396 (5.2476 minus 0.008). (By the way, if you had *not* used trial machining techniques and just let the tool cut, it would have machined the pocket about 0.002 inches too deep. And since the tolerance is only plus or minus 0.0002 inch, this workpiece would have been scrap.)

5: Re-run the tool under the influence of the adjusted offset – This time, the cutting tool will machine the pocket depth in our example properly – very close to your target dimension (0.500 deep in our case). Any tiny deviation you notice will be caused by the difference in tool pressure from the first time the cutter machines (the normal amount of stock is being removed) to the smaller depth-of-cut after trial machining.

When trial machining is not required

Again, only tight depth tolerances require trial machining with tool length compensation offsets. You typically do *not* need trial machining for drilling, tapping, reaming, and most other hole-machining operations.

But just because a cutting tool does not require trial machining techniques doesn't mean you don't have to measure and adjust after the tool has machined for the first time. If the depth dimension has a large tolerance, the tool will machine *somewhere* within the tolerance band on its first try (the workpiece will be acceptable). But it is still important to adjust the offset in such a way that the *next* workpiece machined will have the dimension come out to its target value.

In our pocket example, say the pocket depth dimension (0.500 inch) has a tolerance of plus or minus 0.005 inch – 0.010 inch overall. In this case trial machining is not required – so you let the milling cutter cut with the initial offset setting. Once the tool is finished you measure the pocket depth and find it to be 0.498 inch deep. This depth is well within the tolerance band, but not right at its target value. In this case, you should decrease the offset by 0.002 inch to make the milling cutter machine 0.002 inch deeper on the next workpiece – making the pocket depth come out right to its target dimension.

Sizing with tool length compensation

Sizing is required when the wear a cutting tool experiences during its life affects the surfaces it machines. Frankly speaking, this rarely occurs with depth dimensions (controlled with tool length compensation). If you do notice changes in depth dimensions during a cutting tool's life, remember that you can make the tool machine deeper by *reducing* the tool length compensation value by the amount of the deviation caused by tool wear.

A tip for remembering which way to adjust the offset

A common mistake made by beginning setup people and operators is adjusting the tool length compensation offset in the wrong direction. Here's a way to remember which way is which.

With our recommended method, think about what will happen if a tool length compensation offset is set to *zero*. In this case, the machine will think the cutting edge is at the spindle nose, and bring the spindle nose to the programmed surface. So, *reducing the offset value will make the tool go deeper.*

What if I use the second method shown for tool length compensation?

If you use the distance from the tool tip to program zero as the offset (large negative values in the offset registers), believe it or not, techniques used for adjusting offsets are exactly the same – if you understand polarity.

Again, if you want a tool to go deeper *you must reduce the tool length compensation offset value.* If you already have a negative offset value in the offset, as is the case with the second method shown, this means you must make the already negative value *more negative* by the amount of your desired adjustment. If for example, the tool length compensation offset is currently -12.2726 when you measure the pocket depth and find it to be too shallow by 0.002 inch, you must *reduce* -12.2726 by 0.002 inch, making it *-12.2746.*

Do I have to make all these calculations when adjusting offsets?

Most CNC controls allow you to modify offset values *incrementally.* Fanuc calls this feature *input plus* (actually INPUT+ on the display screen). With this feature, you need only know the amount of needed offset adjustment. If you need an offset to be reduced by 0.002 inch, you simply type -0.002 and press the INPUT + key. The control will automatically calculate the new value for the offset and enter it.

Why can't I just change the Z coordinate/s in the program to make sizing adjustments?

There are three reasons to use offsets to make all sizing adjustments. *Never change programmed coordinates to make sizing adjustments.*

All programmed coordinates must specify mean values

First, and maybe most importantly, it is important that all programmed coordinates specify the mean value of the tolerance band they machine. If milling a 0.500 deep pocket with a tolerance of plus or minus 0.002, the mean value is 0.500, and a position of Z-0.5 must be specified in the program (assuming program zero in Z is at the top of the pocket). Programming mean values is important because it provides some consistency through out the program – and consistency from one time the job is run to the next.

Say you machine this pocket and find it to be 0.002 inch too shallow (0.498 deep). If you change the programmed Z coordinate at the pocket bottom from Z-0.500 to Z-0.502, admittedly, this pocket *will be* machined 0.002 inch deeper.

But let's go a little further. What caused the 0.002 inch deviation? In this case, you must not have correctly measured the cutting tool's length (to be off by this much). Additionally, tool pressure may be affecting the way the tool machines. But the program is correct. Changing the program to deal with tooling problems doesn't make sense.

Also, consider what will happen when this tool eventually dulls and gets replaced. It's likely that you will correctly measure its new length. But when it machines, the incorrect program coordinate (Z-0.502) will cause this tool to machine the pocket too deep.

The same thing goes for the next time the job is run. If the program still has the Z-0.502 coordinate from the last time it was run, the setup person will have no idea how the cutting tool will machine on its first try.

Programs cannot be changed while machine is running.

Many sizing adjustments, especially during a production run, can be done while the machine is running. Tool offsets can be modified while the machine is in cycle. Most machining centers do not allow you to modify the program while the machine is running.

Lots of program modifications may be required

In our example, there is only one pocket to machine. But say you must machine *fifty* pockets. This means there will be fifty Z-0.500 words in the program to change. With offset adjustments, only one value must be changed to affect how all fifty pockets will be machined.

Again, never modify programs to make sizing adjustments!

Key points for Lesson Eleven:

- Tool length compensation lets the programmer ignore the precise length of each tool prior to writing the program.
- Tool length compensation is used for every tool in every program and is instated on each tool's first Z axis approach movement to the work surface.
- Tool length compensation is instated with a G43 word that includes an H word and a Z word. The H word specifies the offset number to use.
- Make the H word number (offset number) the same number as the tool station number.
- With our recommended method, the tool offset will contain a value equal to the tool's length (from tool tip to spindle nose).
- With our recommended method, the fixture offset Z value must be the distance from the spindle nose to the Z axis program zero surface (a large negative value).

Lesson 12
Cutter Radius Compensation

Cutter radius compensation is only used for milling cutters. Just as tool length compensation lets you ignore the precise length of cutting tools as you write programs, cutter radius compensation allows you to ignore the precise diameter of milling cutters used for contour milling.

You know from Lesson Nine that milling cutters (like end mills) can be used for contour milling operations. Motions can be linear (straight-line) or circular. To this point, we have shown the contour milling tool path based upon a milling cutter's *centerline* path. As you know, calculating coordinates for a milling cutter's centerline path requires that you consider the milling cutter's radius for *every* coordinate – which can make calculating coordinates quite difficult.

In Lesson Twelve, we're going to show you how to minimize the calculations that must be done when determining coordinates needed for contour milling. Instead of using the milling cutter's centerline coordinates, you will be using coordinates that are right on the work surface to be milled. And again, these coordinates are much easier to calculate. Mastering the use of cutter radius compensation will be the focus of Lesson Twelve.

Will you need to learn this feature?
Unlike tool length compensation – which is used for every cutting tool in every program – cutter radius compensation is only used for milling cutters, and only when contour milling (milling on the periphery of the cutter). If your company doesn't perform any contour milling operations, you won't be needing cutter radius compensation.

Even if this is the case, you will still want to know the reasons *why* cutter radius compensation is required for contour milling. So at the very least, read on until we start discussing how cutter radius compensation is programmed. You may then want to skip to Lesson Thirteen. If the need ever arises, you can always come back to this lesson and dig in.

Reasons why cutter radius compensation is required
Let's begin by discussing why you must master cutter radius compensation. Some of the reasons are quite similar to the ones given for tool length compensation.

Calculations are simplified for manual programmers
When performing contour milling operations *without* cutter radius compensation (as shown in Lesson Nine), you must specify coordinates for the milling cutter's centerline path. This requires that you consider the size of the milling cutter in *every* calculation – complicating each calculation. With cutter radius compensation, and as a manual programmer (not using a computer aided manufacturing [CAM] system), you will specify coordinates that are right on the work surface, ignoring the size of the milling cutter. Figure 4.8 shows the two different tool paths.

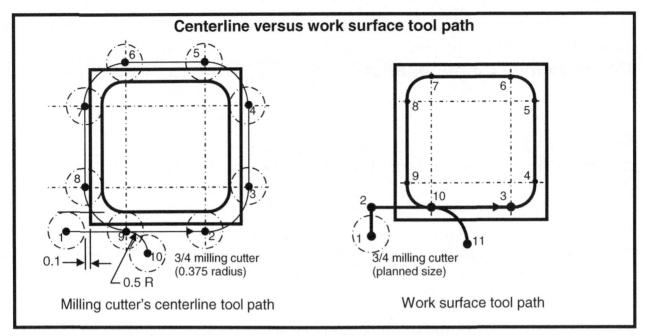

Figure 4.8 – Difference between the milling cutter's centerline path and the work surface tool path

When programming the work surface path, the only time you must consider the size of the milling cutter is during its approach movement. The end point of this approach movement is called the *prior position*. It is point number one in the drawing to the right side of Figure 4.8. The prior position must be at least the milling cutter's radius away from the first surface to mill (starting at point number two in the work surface tool path in Figure 4.8).

Even with the drawing to the left in figure 4.8, calculating centerline coordinates is not terribly difficult. It involves simply adding or subtracting the cutter radius to or from each work surface position. This is because every milled surface is parallel to an axis – and all circular movements are full ninety degree arcs. Consider how much more difficult it will be to calculate the cutter's centerline path for the drawing shown in Figure 4.9.

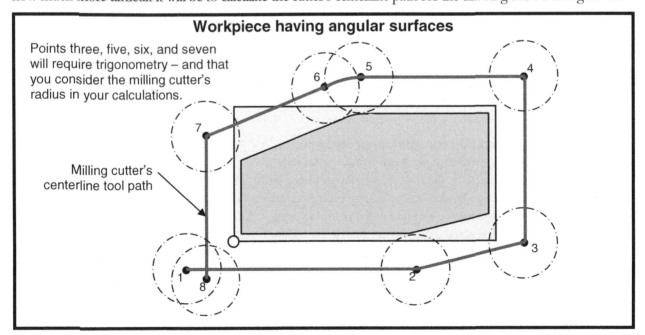

Figure 4.9 – Milled contour involving angular surfaces

While cutter radius compensation will not eliminate the need for trigonometry, it will simplify the related calculations. Look at Figure 4.10.

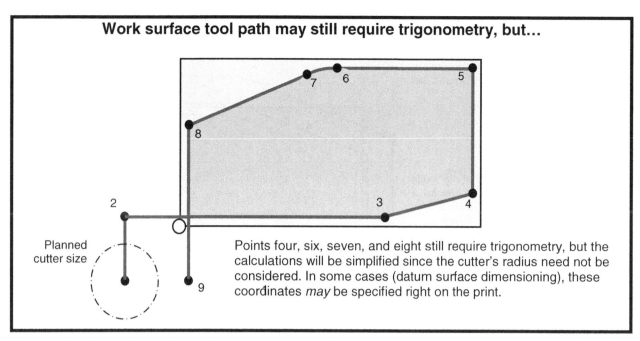

Figure 4.10 – Work surface path coordinates are always easier to calculate than the milling cutter's centerline path coordinates

Do you have a CAM system?

Simplifying calculations is a benefit only for *manual programmers*. If you will be using a computer aided manufacturing (CAM) system to prepare CNC programs, simplifying calculations will not be of concern to you. Your CAM system will be able calculate the milling cutter's center line path just as easily as it can calculate the work surface path. We'll discuss how cutter radius compensation is used with CAM systems later in this lesson.

Range of cutter sizes

This reason for using cutter radius compensation applies to *everyone*, regardless of programming methods (manual programming or CAM system programming). And it applies just as much to setup people and operators (maybe more so) as it does to programmers. Just as tool *length* compensation allows the length of the cutting tool to vary without requiring a program change, so does cutter radius compensation allow the milling cutter's *diameter* (and of course, its radius) to vary without requiring a program change.

Consider the program shown during Lesson Nine in Figure 3.10. It uses a 1.0 diameter end mill. In order for this program to machine the contour properly, the milling cutter *must* be precisely 1.000 inch in diameter (a new 1.000 inch end mill). If it is not, the contour will not be machined to its intended size. Figure 4.11 shows what will happen if you program a milling cutter's center line path (without cutter radius compensation), but don't use the intended milling cutter size.

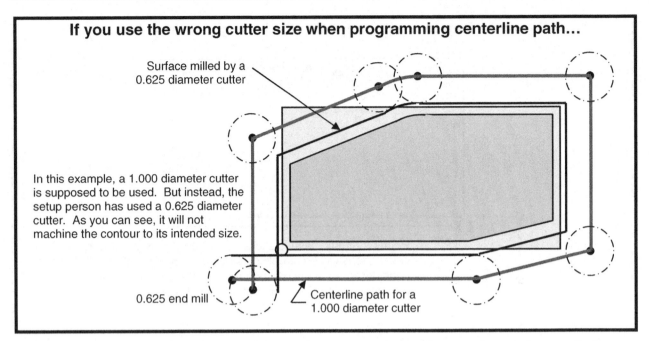

If you use the wrong cutter size when programming centerline path...

Surface milled by a
0.625 diameter cutter

In this example, a 1.000 diameter cutter
is supposed to be used. But instead, the
setup person has used a 0.625 diameter
cutter. As you can see, it will not
machine the contour to its intended size.

0.625 end mill

Centerline path for a
1.000 diameter cutter

Figure 4.11 – If you don't use the intended cutter size when programming a milling cutter's centerline coordinates, (without cutter radius compensation) the contour will not be machined to its correct size

When cutter radius compensation is used, the machine will *automatically adjust all motions* based upon the size of the milling cutter *currently* being used. The setup person enters the size of the current milling cutter (its radius) in the cutter radius compensation offset.

While we don't want to get too far ahead, the range of cutter sizes allowed by cutter radius compensation is much smaller than it is with tool length compensation. With tool length compensation, a cutting tool can be just about any length and the program will still work. The only limitation (from the program's standpoint) is the Z axis range of motion (over-travel limits). A cutting tool can be well under three inches long or well over fifteen inches long and the program will still function properly (as long as the Z axis does not over-travel).

With cutter radius compensation, the range of cutter sizes is much smaller. When planning the diameter of a milling cutter to use, you'll need to keep the range within about an inch or so (about 25.0 mm). The milling cutter's approach position in the X and Y axes (how far it is from the first work surface to mill) – its prior position – determines how large the milling cutter can be.

Do you use re-sharpened (re-ground) cutters?

Many companies re-sharpen their dull milling cutters and use them again. When a milling cutter is re-sharpened, its diameter will become smaller by about 0.020 inch or so (0.50 mm). Figure 4.12 shows some re-sharpened milling cutters. The milling cutter's tool path must, of course, be modified whenever a re-sharpened cutter is used – due to its smaller diameter. Again, cutter radius compensation will *automatically* modify the milling cutter's tool path based upon the current size of the re-sharpened cutter. If you *do not* use cutter radius compensation, you cannot use sharpened cutters without changing all of the cutter's motions in the program.

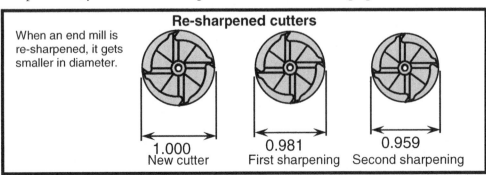

Re-sharpened cutters

When an end mill is
re-sharpened, it gets
smaller in diameter.

1.000	0.981	0.959
New cutter	First sharpening	Second sharpening

Figure 4.12 – Many companies use re-sharpened cutters

Trial machining and sizing

Just as tool length compensation allows trial machining and sizing for *depth* surfaces, so does cutter radius allow trial machining and sizing for *XY* surfaces (surfaces milled with the periphery of the milling cutter). Like depth surfaces, many XY surfaces have close tolerances. If you do not use cutter radius compensation, there will be no way to make sizing adjustments without actually changing the tool path coordinates in the program (which is usually difficult to the point of being infeasible – and we *never* recommend changing programmed coordinates to make sizing adjustments).

Also as with machining depth dimensions, *tool pressure* will affect how XY surfaces are milled. Again, just because a milling cutter's coordinates have been programmed perfectly (to the mean value for each tolerance) – and just because a milling cutter of exactly the intended size is being used – it is no guarantee that the milling cutter will machine the XY surface/s perfectly to size. The tighter the tolerance/s to be held, the more likely the milling cutter will *not* correctly machine the surface/s.

There is also a *tooling-related problem* that affects the accuracy of contour milling operations. Milling cutters have a tendency to *run-out* in their holders. That is, the milling cutter will not be perfectly concentric with its tool holder. Run-out will cause the milling cuter to machine more material than it should – having the same effect as using a slightly larger milling cutter.

Milling cutters also have a tendency to *wear* during their lives. As the cutter wears, a small amount of material will be removed from its outside diameter. In effect, the milling cutter becomes smaller in diameter. With tight tolerances, this small amount of variance will cause problems. *Sizing* must be done during the tools life to keep the milling cutter machining surfaces to size.

For these reasons, you must have the ability to perform trial machining and sizing with XY milling operations. Whether you are simply milling one straight surface (like the right end of the workpiece), or machining an elaborate contour, cutter radius compensation will allow you to do so.

Rough and finish milling with the same set of coordinates

This reason applies only to *manual programmers*. Rough milling an XY contour involves machining the contour while leaving a small amount of finishing stock. Though the amount of finishing stock will vary based upon workpiece material and the style of milling cutter used, it is usually between 0.010 and 0.050 inch. This, of course, requires two milling passes – and *two sets of coordinates*. One pass is for the roughing operation and anther is for the finishing operation. Consider, for example, the drawing shown in Figure 4.12.

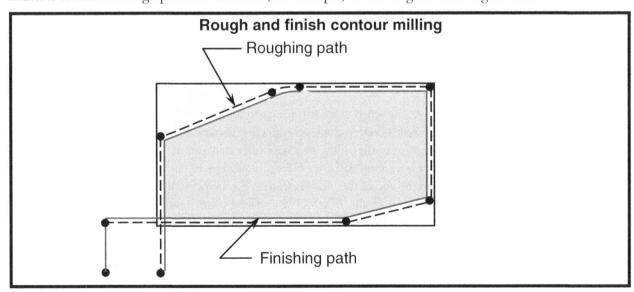

Figure 4.12 – The rough and finish pass required for a contour milling operation.

It can be difficult enough to calculate the coordinates required for the *finishing* path – even when cutter radius compensation is used and these coordinates are on the work surface. It will be more difficult to calculate *another* set of coordinates that is a small distance away from the finished work surface path. With cutter radius compensation, the *same set of coordinates* used to perform the finish milling operation can be used to perform the rough milling operation. If you do not use cutter radius compensation (you program the milling cutter's centerline path), you must calculate the two sets of coordinates – and both must consider the size of the milling cutter you will be using.

Do you have a CAM system?

Again, this is only important to manual programmers. If you use a computer aided manufacturing (CAM) system to create programs, roughing passes will not be of concern. The CAM system can generate the set of roughing coordinates just as easily as it can generate the set of finishing coordinates – on the work surface or based upon the planned milling cutter's centerline path.

How cutter radius compensation works

In the program, you will be telling the machine the relationship between the milling cutter and the *path* of the milling cutter. To do so, you specify whether the milling cutter is on the *right or left side of the path*. The setup person will tell the machine (in the cutter radius compensation offset) the size of the milling cutter being used (its radius). With this information, the machine will keep the milling cutter away from all programmed surfaces by cutter's radius. Figure 4.13 shows an example.

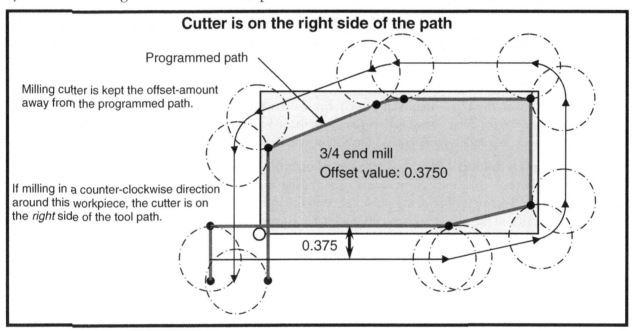

Figure 4.13 – How the machine keeps the milling cutter on the right or left side of the programmed path

In Figure 4.13, the programmer has programmed work surface coordinates (work surface tool path). Since the milling cutter will be moving in a generally counter-clockwise direction around this path, they will have specified in the program that the milling cutter is on the *right side* of this programmed path. (If this is confusing, look in the direction the milling cutter is moving and ask *which side of the work surface the milling cutter is on?* In this example, the milling cutter is on the right side of the work surface.) At the present time, a 3/4 end mill is being used. So the setup person has placed a value of 0.375 (the milling cutter's radius) in the appropriate offset. The machine will automatically keep this milling cutter precisely 0.375 from all surfaces as it moves around this path.

Now here's the beauty of cutter radius compensation. Say the 3/4 milling cutter wears out and the setup person can't find another one. All that is available is a 7/8 (0.875) end mill. They will simply change the offset to 0.4375 (the radius of the 0.875 end mill). When the program is run, the machine will keep *this* end mill 0.4375 away from the programmed path – and mill the contour properly.

Steps to programming cutter radius compensation

As with tool *length* compensation, there are two ways to use cutter radius compensation. But unlike tool length compensation, how programs are written is directly related to the method you choose. With the method we recommend, which is especially intended for manual programmers, you *program the work surface path and the cutter radius compensation offset value is the milling cutter's radius.* With the other method, preferred by many computer aided manufacturing (CAM) system programmers, you program the milling cutter's centerline path and the cutter radius compensation offset value is the difference (in radius) between the planned milling cutter size and the actual size of the milling cutter being used. We'll spend most of this lesson discussing our recommended method.

Here are the three basic steps you must master in order to program cutter radius compensation:

- Instate cutter radius compensation

- Program the tool path to be machined

- Cancel cutter radius compensation

Admittedly, if you are studying cutter radius compensation for the first time, this is going to get a little complicated. Cutter radius compensation tends to be the most difficult CNC feature to fully understand and master. But if your company does any contour milling, stick with it. Based upon the reasons just shown, it will be well worth the time it takes to master.

Step one: Instate cutter radius compensation

Machining centers allow a great deal of flexibility when it comes to instating and using cutter radius compensation. We're going to be showing a method that will work in almost all cases.

Instating cutter radius compensation involves at least three positioning commands that engage the milling cutter to the first surface to be milled. These motions include:

- An XY motion to the *prior position*

- One or more Z motions to the Z axis work surface

- A command instating cutter radius compensation that brings the milling cutter (in X and/or Y) to the first surface to mill

Let's discuss these commands one at a time.

The XY motion to the prior position

This positioning movement brings the center of the milling cutter to a position in X and Y that makes it possible to instate cutter radius compensation. It is usually done at rapid, while the milling cutter is well clear of the workpiece. Indeed, this motion is usually done while the milling cutter is above the workpiece in the Z axis.

We call the end point for this motion the *prior position* because it is the XY position just prior to the instating command. The *next* XY motion will instate cutter radius compensation.

The prior position must be *at least* the milling cutter's radius (the value stored in the cutter radius compensation offset) away from the surface to mill. And it is important to choose a prior position that allows the milling cutter to form a ninety degree angle (right angle) with the first surface to be milled.

That is a lot to remember, so let's look at an example. Figure 4.14 shows the prior position.

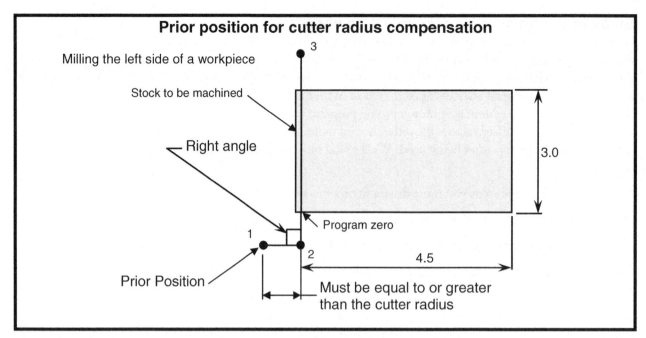

Figure 4.14 – The prior position for cutter radius compensation

In Figure 4.14, we're milling the left side of a workpiece that will be held in a vise. The prior position is point number one. This position must be the at least the milling cutter's radius away from the first surface to mill (beginning at point number two). And notice that we've chosen this prior position in such a way that as we instate cutter radius compensation (this will be during the movement to point number two) and begin machining (to point number three), a ninety degree angle is formed.

While cutter radius compensation allows you to use a *range* of milling cutter sizes, it does not allow you to completely ignore the milling cutter size. You must, for example, specify an intended cutter size in the setup documentation. And again, your chosen prior position must be far enough away from the first milled surface that the *largest* intended cutter will clear. The prior position sets the limit for *maximum cutter size.*

For the example in Figure 4.14, say you intend to use a 1.000 inch diameter end mill. The prior position (again, point one) must be *at least* 0.5 inch (the radius of the intended cutter size) away from the first surface to mill (starting at point number two). Notice that the program zero point is the lower-left corner of the *finished* workpiece. The milling cutter is currently well above the workpiece in Z. You program a rapid motion to the prior position with

O0001 (Program number)

N005 T01 M06 (1.0 end mill)

N010 G54 G90 S350 M03 T02 (Select fixture offset #1, absolute mode, start spindle fwd at 350 rpm, and get tool station one ready)

N015 **G00 X-0.5 Y-0.6** (Rapid to prior position)

When a 1.000 inch cutter moves to the position specified in line **N015**, it will be perfectly flush with the first surface to mill. While this will work as a prior position for a 1.000 end mill, it does not allow for *any* variation (like trial machining and sizing) during the production run. *You have set the maximum cutter size to a 1.000 diameter end mill.* If the setup person enters *anything* larger than 0.500 inch in the cutter radius compensation offset, the machine will generate an alarm. (The machine will think the cutter is already violating the first surface to mill.)

Let's revise the prior positioning movement to

N015 G00 X-0.6 Y-0.6

Now there is plenty of clearance between our planned 1.000 end mill and the first surface to machine. In fact, an end mill up to 1.200 inches in diameter can be used without generating an alarm. So, *one way to determine maximum cutter size is to double the distance from the prior position to the first surface to mill.* (Note that there are some other limiters for maximum cutter size. You'll see them as we continue.)

Don't forget to document the maximum cutter size – Since your prior position sets the maximum cutter size (again, anything larger and an alarm will be sounded), be sure to specify the maximum cutter size in the setup documentation. While you're at it, also specify the smallest cutter that can be used (based upon the strength of the cutter versus the amount of material being machined). In our example, if there is about 0.100 inch material to machine and the workpiece is 1.0 inch thick, we may not want the setup person using an end mill smaller than about 3/4 inch (0.75) in diameter.

The Z motion/s to the Z axis work surface

With the milling cutter at the prior position in the X and Y axes, it's time to move it to the Z axis work surface. In most cases, when the milling cutter is at the prior position, it will be well clear of the workpiece in the X and Y axes, meaning you can rapid the tool to the Z axis work surface. For our example in Figure 4.14, we'll say the workpiece is 1.00 thick. We'll want to move the tip of the end mill a little further below the bottom surface of the workpiece. We'll do so by first positioning the milling cutter just above the workpiece, then sending it to the Z axis work surface.

O0001 (Program number)

N005 T01 M06 (1.0 end mill)

N010 G54 G90 S350 M03 T02 (Select fixture offset #1, absolute mode, start spindle fwd at 350 rpm, and get tool station one ready)

N015 G00 X-0.6 Y-0.6 (Rapid to prior position)

N020 **G43 H01 Z0.1** (Instate tool length compensation)

N025 **Z-1.1** M08 (End mill is now at Z axis work surface and coolant is on)

Line N015 is the movement to the prior position in the X and Y axes. Line N020 instates tool length compensation and moves the end mill to just above the work surface. Line N025 moves it to the Z axis work surface (still at rapid). Note that you could include Z-1.1 in line N020 instead of Z0.1 to bring the end mill directly to the Z axis work surface.

The command instating cutter radius compensation that positions the cutting tool to the first surface to mill

At this point, you're ready to instate cutter radius compensation. The instating command will include three things: a G41 or G42 specifying the relationship between the milling cutter and the tool path, a D *word* specifying the offset number containing the cutter radius compensation value, and a motion to the first work surface to be machined.

G41 or G42?

One of two G codes will be used in the instating command.

- G41 – milling cutter is to the left of the its path

- G42 – milling cutter is to the right of its path

One way to determine which of G41 or G42 you must use is to look in the direction the milling cutter will be moving as it machines the work surface/s and ask *"Which side of the work surface is the milling cutter on?"* If the milling cutter is on the left side of the work surface, you will use G41 in the instating command. If the cutter is on the right side of the work surface, you will use G42. Figure 4.15 shows some examples.

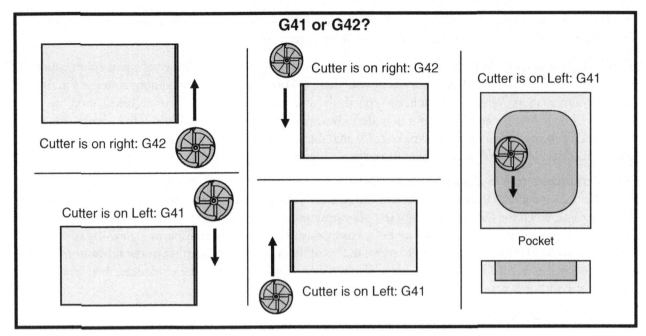

Figure 4.15 ~ Deciding whether to use G41 or G42

If you have basic machining experience, there is an easier way to determine whether to use **G41** or **G42** in the instating command. As long as you're using a right-hand milling cutter (almost all milling cutters are of the right-hand type), *climb milling requires* **G41** *to instate* cutter radius compensation and *conventional milling requires* **G42** *to instate*. Figure 4.16 shows the same examples – but this time relative to climb versus conventional milling.

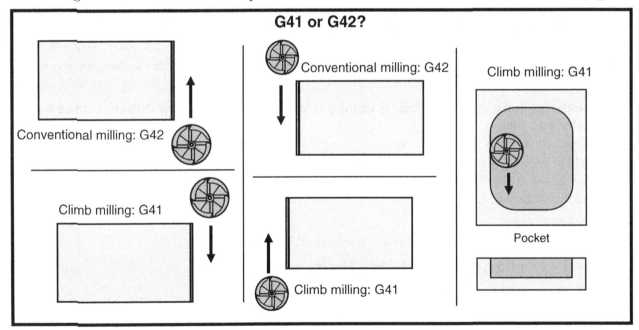

Figure 4.16 ~ Deciding whether to use G41 or G42 based upon milling style

The offset used with cutter radius compensation

The command that instates cutter radius compensation also requires that you specify the *offset number* in which the setup person will place the cutter radius compensation value. While instating, the machine will use this value to bring the cutter flush with the first surface to machine. It will also use this value to keep the cutter flush with all surfaces as the contour is being machined.

From Lesson Ten, remember that machining centers vary when it comes to offset organization. Again, some have but one register per offset and others have two registers per offset. If you have but one register per offset, they must be shared between tool length and cutter radius compensation. And you will use the offset number that corresponds to the tool station number in which to store the tool *length* compensation value. It cannot be used again for cutter radius compensation – you must choose another.

For machines that have but one register per offset, we recommend adding a constant number to the tool station number to come up with the offset number to use with cutter radius compensation. This constant number must be equal to or larger than the number of cutting tools the machine's automatic tool changer can hold.

For example, if your machining center's automatic tool changer magazine can hold thirty tools, add *thirty* to the tool station number to come up with the cutter radius compensation offset number. For this machine, if tool number five is a milling cutter, you will use offset number five in which to store its tool length compensation value and offset number thirty-five in which to store its cutter radius compensation value. Since this style of offset table is very popular, we use this technique in this text (adding thirty to the tool station number to determine the offset number used with cutter radius compensation).

If you have a machining center that has two registers per offset, choosing the cutter radius compensation offset number is much easier. Simply make it correspond to the milling cutter's tool station number. For this kind of machine, if station five is a milling cutter, you will use offset number five in which to store its tool length compensation value *and* its cutter radius compensation value (again, with this machine, there are two registers per offset – one for length and another for radius).

In the program, a *D word* is used to specify the offset number used with cutter radius compensation. The D word must be included in the instating command.

SPECIAL NOTE! There are some control models that use the H word to specify the offset number to be used with both tool length compensation *and* cutter radius compensation. The Fanuc 3M, and some 0M models, for example, use the H word in this fashion. These models do not have the letter address D on their control panels.

The motion to the first work surface
Also included in the instating command is a motion that brings the milling cutter to the first work surface to be machined. With this motion, you are no longer programming the milling cutter's centerline position. The end point for position is right on the work surface. Here is Figure 1.14 again.

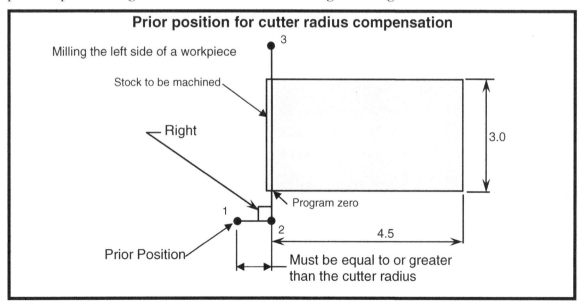

Prior position for cutter radius compensation

Milling the left side of a workpiece

Stock to be machined

Right

Program zero

3.0

Prior Position

1

2

4.5

Must be equal to or greater than the cutter radius

The instating command moves the milling cutter from the prior position (point number one) to the first work surface to machine (point number two). Again, notice that point number two is right on the work surface. Also, the instating command can be done at rapid (G00) if the milling cutter clears the workpiece during the motion – or in a straight line motion (G01) with a feedrate if it does not. *You are not allowed to instate cutter radius compensation during a circular motion command (G02 or G03).*

Since points one and two in our example are well clear of the workpiece in the Y axis (with our 1.00 planned cutter size), we'll make the instating command at rapid. Here are the commands so far:

O0001 (Program number)

N005 T01 M06 (1.0 end mill)

N010 G54 G90 S350 M03 T02 (Select fixture offset #1, absolute mode, start spindle fwd at 350 rpm, and get tool station one ready)

N015 G00 X-0.6 Y-0.6 (Rapid to prior position)

N020 G43 H01 Z0.1 (Instate tool length compensation)

N025 Z-1.1 M08 (End mill is now at Z axis work surface and coolant is on)

N030 **G41 D31 X0** (Instate cutter radius compensation)

In line N030, G41 specifies that the milling cutter will be on the *left side* of the surface during machining. Also, notice that a right-hand end mill will be climb milling – again, requiring G41 in the instating command. D31 tells the machine to look in offset number thirty-one to find the cutter radius compensation value (0.500 if a 1.000 diameter is currently being used). X0 is the first (and in our example, the only) surface to mill. At the completion of line N030, the milling cutter's diameter will be flush with the X0 surface (a 1.000 inch end mill's center position will be at X-0.5 Y-0.6 at the end of line N030).

Finally, cutter radius compensation is instated. As you might agree, instating compensation is the most difficult part of using it.

Step two: Program the tool path to be machined

Once cutter radius compensation is instated, you must program the work surface path for the contour to be machined. It is usually important that you end the contour with the milling cutter well clear of the last machined surface. If you leave the milling cutter in contact with the last milled surface, it's likely that the milling cutter will leave a witness mark on the surface when it is retracted in the Z axis.

The tool path in Figure 1.14 is quite simple. There is only one surface to machine:

O0001 (Program number)

N005 T01 M06 (1.0 end mill)

N010 G54 G90 S350 M03 T02 (Select fixture offset #1, absolute mode, start spindle fwd at 350 rpm, and get tool station one ready)

N015 G00 X-0.6 Y-0.6 (Rapid to prior position)

N020 G43 H01 Z0.1 (Instate tool length compensation)

N025 Z-1.1 M08 (End mill is now at Z axis work surface and coolant is on)

N030 G41 D31 X0 (Instate cutter radius compensation)

N035 **G01 Y3.6 F5.0** (Mill left side)

Line N035 contains a G01 (very important here) to switch to straight-line cutting motion. Y3.6 will cause the end mill to machine the left side of the workpiece, leaving it well clear of the workpiece.

This work surface in this example is so simple that it doesn't allow us to stress an important point. Let's look at another example, shown in Figure 4.17.

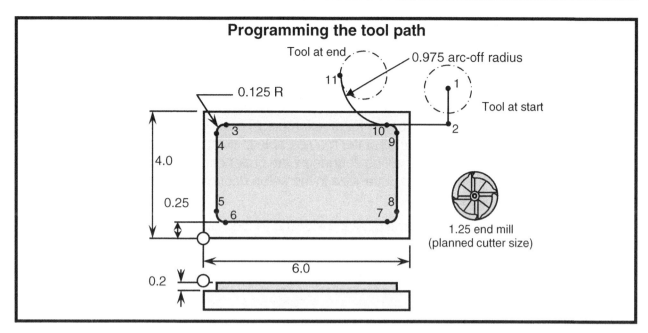

Figure 4.17 – More elaborate example of a work surface tool path using cutter radius compensation

While we won't show the program yet (it's coming up as one of the examples later in this lesson), we want to make a point about the work surface path. To instate cutter radius compensation for this example, you send the milling cutter to the prior position in XY (point one), move to the work surface in Z, and then, using **G42** and a D word, instate cutter radius compensation during a motion to the first milled surface (point number two).

At this point, the milling cutter is flush with the first surface to mill (its center is above point number two in Y by the value placed in the cutter radius compensation offset). You're ready to program the work surface tool path.

The work surface tool path takes the milling cutter from point two to point three (straight-line motion) – from point three to point four (circular motion) – from point four to point five (straight-line motion) – and so on through point number ten. Remember that since you are programming the work surface tool path, the radius of all circular motions (R word) will be the actual workpiece radius, **R0.125** for each circular motion in our example.

When the milling cutter reaches point number ten, it is finished milling the contour, but if you allow it to retract in Z at point ten, it will leave a nasty witness mark on the workpiece (a gouge-line on the side of the milled surface). Most programmers will come up with *one more* work surface tool path motion to bring the milling cutter away from the last surface to mill. This is best done in a circular motion (commonly called an *arc-out motion*). The size of this arc must, of course, be larger than the radius of the largest planned cutter. It must also include enough clearance to allow the largest planned cutter to move away from the last surface being milled.

For our example, when using the planned cutter size (1.25 inch in diameter), the 0.975 arc-out radius will provide 0.350 inch motion away from the last surface milled (0.975 arc-out radius minus 0.625 milling cutter radius). This distance provides clearance for the 0.25 inch step around the contour. When the cutter reaches point number eleven, it will be completely clear of the workpiece (by 0.1 inch) when using a 1.25 diameter cutter.

Here are a few rules that affect the way a work surface tool path must be programmed:

- The cutter radius compensation offset value must be equal to or smaller than the smallest inside radius in the tool path or an alarm will be sounded – In our example, the offset must be equal to or smaller than 0.975 – the motion from point ten to point eleven. You must be especially concerned with this rule when machining pockets. And, by the way, this is also a limiter for maximum cutter size.

- Be careful with non-motion commands – Once cutter radius compensation is instated, the machine will be constantly *looking ahead* to see the motions coming up in a program. Some machines cannot look very far. If it cannot determine the direction of the next XY motion, the machine will generate an alarm.

- Be careful when machining recesses in X and Y – The cutter must be able to fit into the recess. (See the drawing below for an example.) This is yet another limiter for maximum cutter size. A small cutter may be able to fit into a recess, but a larger one may not. Again, an alarm will be sounded if the cutter cannot fit into the recess. Be sure to specify a maximum cutter size in the setup documentation that considers this potential problem.

- Two movements in the same direction – Some (especially older) machines do not allow this. If two motions in the same direction are commanded, the machine will generate an alarm.

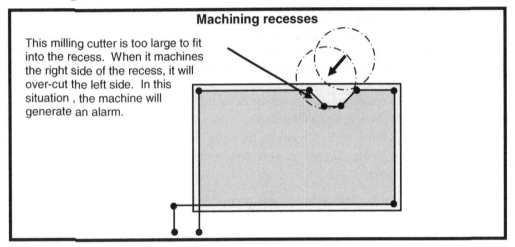

Machining recesses

This milling cutter is too large to fit into the recess. When it machines the right side of the recess, it will over-cut the left side. In this situation , the machine will generate an alarm.

Step three: Cancel cutter radius compensation

Like many CNC features, cutter radius compensation is *modal*. The machine will continue to keep the milling cutter on the right or left side of all programmed motions until cutter radius compensation is cancelled. A **G40** word is used to cancel cutter radius compensation. The very next X and/or Y coordinate (either within the **G40** command or after it) will be a taken as a centerline position.

Though it is not always possible to do so, we recommend retracting the milling cutter in the Z axis prior to canceling cutter radius compensation. This way, the cutter will be clear of the workpiece should an unexpected X and/or Y movement take place during cancellation. Once again, here is Figure 4.14.

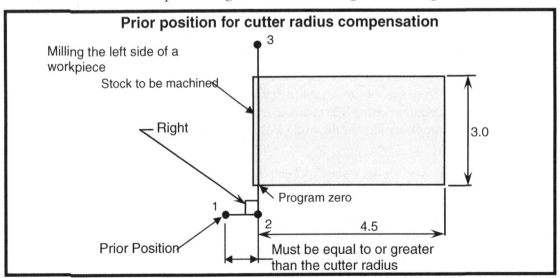

Prior position for cutter radius compensation

Milling the left side of a workpiece

Stock to be machined

Right

Program zero

Prior Position

Must be equal to or greater than the cutter radius

3.0

4.5

Here is the entire milling operation, including the cancellation of cutter radius compensation:

O0001 (Program number)

N005 T01 M06 (1.0 end mill)

N010 G54 G90 S350 M03 T02 (Select fixture offset #1, absolute mode, start spindle fwd at 350 rpm, and get tool station one ready)

N015 G00 X-0.6 Y-0.6 (Rapid to prior position)

N020 G43 H01 Z0.1 (Instate tool length compensation)

N025 Z-1.1 M08 (End mill is now at Z axis work surface and coolant is on)

N030 G41 D31 X0 (Instate cutter radius compensation)

N035 G01 Y3.6 F5.0 (Mill left side)

N040 **G00 Z0.1** M09 (Retract milling cutter to above work surface, turn off coolant)

N045 **G40** (Cancel cutter radius compensation)

N050 G91 G28 Z0 M19 (Move to tool change position, orient spindle)

N055 M01 (Optional stop)

.

.

.

In line **N040**, we retract the tool to above the workpiece – then in Line **N045**, we cancel cutter radius compensation. From this point on, any XY position specified in the program will be to the spindle centerline.

Don't forget to cancel!

You *must* remember to cancel cutter radius compensation. If you do not, the next tool will remain under the influence of cutter radius compensation – and will move in an unusual manner. Consider, for example, a drilling operation that follows a contour milling operation. If you forget to cancel cutter radius compensation after milling, the center drill will move to XY positions that are incorrect by the amount of the cutter radius compensation offset value, even though specified hole-centerline coordinates in the program are correct. This problem can be quite difficult to diagnose.

What if I have more than one contour to mill?

If you have multiple contours to mill, you must handle them separately – Consider, for example, having to mill the right side of a workpiece *and then the left side*. This requires two separate work surfaces. First, instate, machine, and cancel for the right side. Then move on to the left side. Instate, machine, and cancel. If you must rapid the tool in XY at any time during a milling cutter's motions, you probably have more than one contour to machine – and again – you must handle them separately.

Example

We offer a complete example to help solidify your understanding of cutter radius compensation.

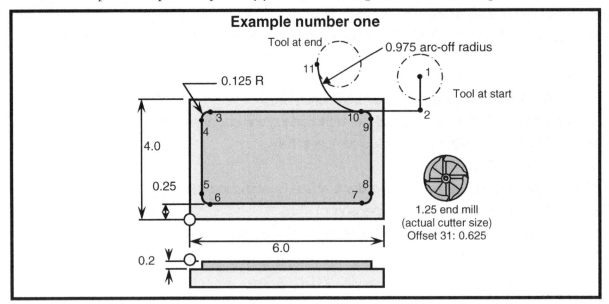

Example number one

The milling cutter will be milling the 0.25 inch step around the outside of this workpiece. The planned cutter size is a 1.25 diameter end mill (having a 0.625 radius). Based upon the distance from point one to point two in the program, the maximum cutter size for this program is 1.45 inches in diameter.

Program with comments:

O0026 (Program number)

(1.25 end mill in station 1)

N005 G54 G90 S400 M03 (Select fixture offset #1, absolute mode, start spindle fwd at 400 rpm)

N010 **G00 X6.725 Y4.475** (Rapid to prior position pt 1)

N015 G43 H01 Z0.1 (Instate tool length compensation, move to just above workpiece)

N020 G01 Z-0.2 F40.0 (Fast feed to work surface)

N025 **G42 D31 Y3.75** (Instate cutter radius compensation to pt 2)

N030 **X0.375 F5.0** (Mill to point 3)

N035 G03 X.25 Y3.625 R.125 (Circular mill to point 4)

N040 G01 Y0.375 (Mill to point 5)

N045 G03 X.375 Y0.25 R.125 (Circular mill to point 6)

N050 G01 X5.625 (Mill to point 7)

N055 G03 X5.75 Y.375 R.125 (Circular mill to point 8)

N060 G01 Y3.625 (Mill to point 9)

N065 G03 X5.625 Y3.75 R.125 (Circular mill to point 10)

N070 **G02 X4.65 Y4.725 R.975** (Arc-off the work surface to point 11)

N075 **G00 Z0.1** (Retract to above the work surface in Z)

N080 **G40** (Cancel cutter radius compensation)

N085 G91 G28 Z0 (Rapid to Z axis zero return position)

N090 M30 (End of program)

Since the cutter is 0.725 away from the right side of the workpiece in line N010, we're allowing the fast-feed approach in line N025 (at 40.0 ipm). In line N030, when milling begins, we include the cutting feedrate.

Lesson 13
Fixture Offsets

Tool length and cutter radius compensation allow you to ignore certain attributes of cutting tools as you writer your CNC programs. In like fashion, fixture offsets allow you to ignore the precise location of the work holding device/s on the machine table as you write programs.

You know from Lesson Six that fixture offsets are used to assign program zero. Based upon our discussions so far, you know that the distances *from* the zero return position *to* the program zero point must be determined (either measured or calculated). The polarity of these values is almost always negative since the zero return position is usually close to the plus over-travel limit of each axis. These values are entered into fixture offset number one to assign program zero. (You might want to review Lesson Six before continuing with Lesson Thirteen.)

We've only shown how to assign *one* program zero point, using fixture offset number one. In Lesson Thirteen, we'll be showing when *multiple* program zero points are required as well as how they are assigned and programmed.

Also, with the method shown in Lesson Six, the *point-of-reference* for fixture offset entries is the zero return position. There are situations when the zero return position doesn't make the best point of reference for fixture offset entries. In Lesson Thirteen, we'll show how the point of reference can be shifted – and when you can benefit from shifting it.

And finally, there are a few tricks you can use with fixture offsets to enhance your machining center's performance. We'll include them in Lesson Thirteen.

Do you need to learn more about fixture offsets?
There are many companies that don't need any more from fixture offsets than we show in Lesson Six. They have but one program zero point per program. The setup person measures program zero assignment values during setup and enters them into fixture offset number one. They're satisfied using fixture offsets in this manner.

However, when it comes to becoming more efficient with program zero assignment (and in turn – reducing setup time), many companies can benefit by improving their methods. You cannot become more efficient, of course, unless you know something better is possible.

While the material we present in Lesson Thirteen may not be of immediate need to you, you will definitely want to understand its content. If you do decide to skip ahead, be sure to come back to this material once you've acquainted yourself with your machining center and its usage.

Assigning multiple program zero points
Most programs for vertical machining centers require but one program zero point per program – but there are exceptions. Consider, for example, the setup shown in Figure 4.19.

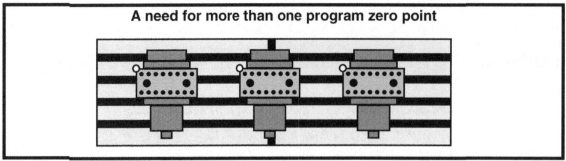

Figure 4.19 ~ Three setups made on the table of a vertical machining center.

In this example, three are three identical workpieces being run on the table at the same time. With this application, the company will save on tool changing time (one tool will machine all three workpieces before the next tool).

If only one program zero point is assigned for *all three* workpieces, the programmer must, of course, know the *exact* distance between the workpieces. And of course, the setup person must position the workpieces precisely in the setup. This can be very cumbersome, especially for vise setups.

Fixture offsets will allow you to assign up to *six* program zero points for use by a single program. (Actually, there is an option that allows up to *forty-eight* fixture offsets, but most machine tool builders supply only six.) With multiple fixture offsets, you can run multiple identical workpieces – as shown in Figure 4.19, different sides of the same workpiece (possibly utilizing an indexing device), or completely different workpieces. We show a rotary device application for multiple fixture offsets in Lesson Nineteen.

With multiple program zero points comes the need to determine multiples sets of program zero assignment values. And one way to *assign* multiple program zero points is simply an extension of what you already know. This method of program zero assignment works very well when there are no relationships among the program zero points to be assigned – as is the case when workpieces are held in vises.

With this method, the distances from the zero return position to each program zero point in all axes is determined and entered into each fixture offset. Figure 4.20 shows the program zero assignment values using this method.

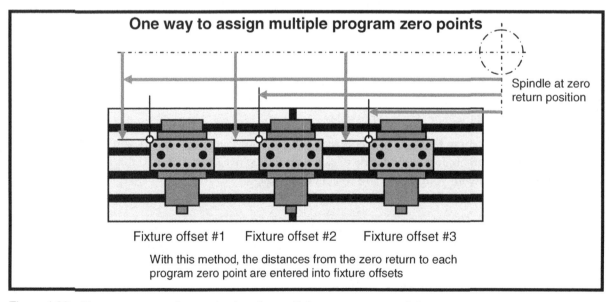

Figure 4.20 ~ Program zero assignment values for multiple program zero points

When using this method, all program zero assignment values will be negative. For the example in Figure 4.20, the program zero assignment values for the left-most workpiece will be placed in fixture offset number one.

For the middle workpiece, they will be placed in fixture offset number two. And for the right-most workpiece, they will be placed in fixture offset number three.

The Z axis program zero assignment values for each fixture offset must also be determined and entered into the appropriate fixture offset – as is shown in Figure 4.21.

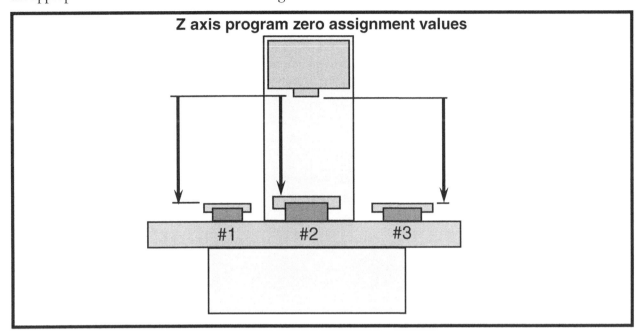

Figure 4.21 – The program zero assignment values for the Z axis

Here is an example of the fixture offset page of the display screen with fixture offsets one through three filled in. Again, notice the large negative values. These are the distances from the zero return position to the program zero point in each axis (and for each fixture offset).

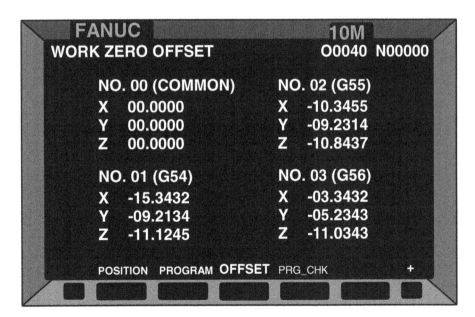

Key points for Lesson Thirteen:

- Fixture offsets allow you to ignore the position/s of your work holding tool/s on the machine table as you write programs.

- You can assign up to six program zero point to be used by a program (up to forty-eight if you have purchased the additional fixture offset option).

- One way to assign multiple fixture offsets is to determine (measure) the distances from the zero return position to each individual program zero point in each axis and enter these program zero assignment values into the related fixture offsets.

- You can shift the point of reference for fixture offset entries from the zero return position to a more logical position. The common fixture offset is used for this purpose. This method works well when your workholding setups are very predictable.

- The common offset can be used to deal with variations among machines, to help re-align after a crash, and to help with program verification.

Talk to experienced people in your company...

... to learn more about fixture offsets.

1) Ask if any of the setups made on your machining centers use multiple program zero points.

2) Ask if repeated setups are qualified, and if they are, whether G10 is being used to program fixture offset settings.

3) Ask whether the point of reference for fixture offset entries has been shifted to a more local position. Does your company use sub-plates?

　　　Fundamentals of CNC

You Must Provide Structure To Your CNC Programs

While there are many ways to write programs, you must ensure that your programs are safe and easy to use, yet as efficient as possible. This can be a real challenge since safety and efficiency usually conflict with one another.

Key Concept Number Five contains two lessons:

> 14: Introduction to program structure
> 15: Four types of program format

The fifth Key Concept is you must structure your CNC programs using a strict format – while incorporating a design that accomplishes the objectives you intend. Though it is important to create *efficient* programs, *safety and ease-of-use* must take priority – at least until you gain proficiency. In Key Concept Number Five, we'll be showing techniques that stress *safety as the top priority*.

To this point in the text, we have been presenting the building blocks of CNC programming – providing the needed individual tools. Machine components, axes of motion, program zero, and programmable functions are presented in Key Concept Number One. Preparation steps in Key Concept Number Two. Motion types in Key Concept Number Three. Compensation types in Key Concept Number Four. In Key Concept Number Five, we're going to draw all of these topics together, showing you what it takes to write CNC machining center programs completely on your own – making you a *self-sufficient programmer*.

Lesson Fourteen will introduce you to a CNC program's structure, showing you the reasons *why* programs must be strictly formatted. We'll also review some the program-structure-related points we've made to this point – and present a few new ones. And we'll address some variations related to how certain machine functions are handled.

In Lesson Fifteen, we'll show the four types of program format as they are applied to both vertical and horizontal machining centers. These formats can be used as a crutch to help you write your first few programs.

You may be surprised at how much you already know about a program's structure, especially if you have been doing the exercises included in this text and/or those in the workbook. We have introduced many of the CNC words used in programming, and we have been following the structure-related suggestions that we will be recommending here in Key Concept Number Five.

For example, do you recognize any of these words and commands? If so, write their meanings in the space provided.

O0001: _____

T01 M06: _____

G54 G90 S1000 M03 T02: _____

G91 G28 Z0 M19: _____

M01: _____

G43 H01 Z0.1 M08: _____

M30: _____

If you can't remember one or more of these CNC words or commands, don't worry. We'll be explaining them in detail during Key Concept Number Five. But if you do recognize most of them, you're already well on your way to understanding program structure.

Lesson 14
Introduction To Program Structure

Structuring a CNC program is the act of writing a program in a way that the CNC machine can recognize and execute safely, efficiently, and with a high degree of operator-friendliness.

You know that CNC programs are made up of commands, that each command is made up of words, and that each word is made up of a letter address and a numerical value.

You also know that programs are executed sequentially – command by command. The machine will read, interpret, and execute the first command in the program. Then it will move on to the next command. Read – interpret – execute. It will continue doing so until the entire program has been executed.

And you have seen several complete programs – you have even worked on a few if you have done the exercises in this text and in the workbook. You have probably noticed that there is quite bit of consistency and *structure* in CNC programs.

Our focus in Lesson Fourteen will be to help you understand more about the structure that is used in CNC programming.

Objectives of your chosen program structure

CNC machines have come a long way. In the early days of NC (before computers), a program had to be written *just so*. If *anything* was out-of-place, the machine will go into an alarm state – failing to execute the program. While today's CNC machines are *much* more forgiving, you must still write CNC programs in a rather strict manner.

There are many ways to write a workable program – and the methods you use in structuring your programs will have an impact on three important objectives:

- Safety
- Efficiency
- Ease-of-use (operator friendliness)

It may be impossible to come up with a perfect balance among these objectives. Generally speaking, what you do to improve one objective will negatively affect the other two. When faced with a choice, a beginning programmer's priorities should always lean toward *safety* and *ease-of-use*. Our recommended programming structure stresses these two objectives. We will, however, show some of the efficiency-related short-comings of our recommended methods – so you can improve efficiency as you gain proficiency.

We're going to be assuming that *you have control* of the structure you use to write programs. Your company may, however, already have a programmer that is writing programs with a different structure. As long as these programs are working – and satisfying the company's objectives – you're going to have to adapt to the established structure. If you understand the reasons for formatting, and if you understand one successful method for structuring programs, it shouldn't be too difficult to adapt.

Reasons for structuring programs with a strict and consistent format

Let's begin by discussing the reasons *why* you must write your programs using a strict structure.

Familiarization

You must have some way to get familiar with CNC programming. You'll need some help writing your first few programs. The formats we show in Lesson Fifteen will provide you with this help. You'll be able to use our given formats as a crutch until you (eventually) have them memorized.

Actually, the formats we show in Lesson Fifteen will keep you from having to memorize anything. Instead, you will look at a word or command in the format and then you must remember its use. Compare this to recognizing the road signs you see as you drive an automobile. It is unlikely that you could recite every road sign from memory – but when you see one – you immediately know its meaning. Think of our given formats as like a set of road signs designed to help you write CNC programs.

As you'll see in Lesson Fifteen, you'll have road signs to help you *when you begin writing a program* (program start-up format), *when you're finished with a tool* (tool ending format), *when you begin a new tool* (tool start-up format), and *when you end a program* (program ending format).

Consistency

If you have been doing the exercises in this text and in the workbook, you've already worked on a few actual programs, filling in the blanks with needed CNC words. You have also seen several complete example programs in this text. You probably noticed that these programs are written in a very consistent manner. And the commands within each tool of each program are consistent with the other tools in the program.

Consistency within programs is important for three reasons. First, consistency helps you to become familiar with programming. Repeated commands soon become memorized.

Second, and more importantly, consistency among programs will help everyone that must work with your programs, including other programmers, setup people, and operators. Your programs will be easier to work with if you always structure them in the same manner.

Third, it is important to be able to repeat your past successes. If a program is running properly – achieving all of the objectives you intend – using its structure in your *next* program will ensure continued success.

Re-running tools in the program

This is the most important reason for structuring your programs using a strict format. As you know, *trial machining* involves five steps: recognizing a close tolerance, adjusting an offset to leave additional stock, machining under the influence of the trial machining offset, measuring the machined dimension and adjusting the offset accordingly, *and re-running the tool.*

Trial machining is but one time when tools must be re-run. Say you're verifying a long program. You are fifteen tools into a twenty-tool program when you find a mistake. You must stop the cycle to correct the mistake. With the mistake corrected, you'll want to pick up where you left off – at the beginning of the fifteenth tool. You wouldn't want to re-run the entire program (from the beginning) just to get to tool number fifteen (doing so would be a waste of time).

Your ability to re-run tools is directly related to the structure you use to program. *If you don't structure your programs properly, it will not be possible to re-run tools.*

Here is a specific example that stresses why program formatting is so very important for re-running tools. Say you are programming two tools that run in sequence (say, tools one and two). Both tools happen to run at 400 rpm. Close to the beginning of the first tool, you have an **S400** word in the command that starts the spindle. As you continue writing the commands for the second tool, you decide to leave out the **S400** word, since the spindle speed is *modal.*

Everything will work just fine as long as the program runs in sequence – from beginning to end (the second tool immediately follows the first tool). But say the operator runs the entire program before they discover that the second tool has done something wrong (maybe it's a drill that has not gone deep enough). After correcting the problems, if the operator attempts to re-run the second tool by itself (as they should), it will run at the same speed as the last tool in the program — *probably not 400 rpm!* This is but one of many times when you must include redundant words (words that are already instated) in each tool – just to gain the ability to re-run tools.

In essence, you must make each tool in the program *independent from the rest of the program.* Think of the programming commands for each tool in a program as making a *mini-program*, self sufficient and capable of activating all necessary machine functions.

You will often be tempted to leave out (necessary) redundant words because they are modal and remain in effect from tool to tool. To fight this temptation, think of each tool in the program as if it is the first tool. Everything you include in the program for the first tool (except the program number) will be needed in the program for each successive tool. This includes the tool change command (like **T01 M06**), the fixture offset designator (like **G54**), the absolute mode selection (**G90**), the spindle starting words (like **S1000 M03**), the next tool (like **T02**), movement to the first XY position in both axes (like **G00 X2.0 Y2.0**), the tool length compensation command (like **G43 H01 Z0.1**), the coolant activation (**M08**), and the feedrate word in the tool's first cutting command (like **F5.0**).

In Lesson Fifteen, we will be providing two complete sets of program format – one for vertical machining centers and the other for horizontal machining centers. These formats include all of the words needed in each tool – allowing any tool in the program to be re-run.

Efficiency limitations

Again, the highest priority for our given formats is safety. Even so, beginners must exercise extreme caution when verifying their first few programs. Next we emphasize *ease-of-use*. Programs written with our recommended formats will be relatively easy to work with.

Our given formats do *not* place an emphasis on efficiency. While programs written with these formats will not be wasteful, they will not be as efficient as they could be. But remember, when you start emphasizing efficiency, safety and ease-of use will probably suffer. Here are some of the efficiency-related limitations of our given formats:

One activity per command – For the most part, our given formats will cause the machine to do *one thing at a time*. In reality, the machine can sometimes be doing more. For example, our recommended format for tool start-up will have the spindle to start in one command and the movement to the first XY position in the next. While these two functions can be done together, it is helpful for beginners to concentrate on but one thing at a time.

Approach motions – During each approach to the workpiece, we first move the cutting tool to its first XY position. Then we move it to its first Z position. While these commands can be combined to minimize program execution time, a three-axis approach movement tends to be scary for entry-level setup people and operators.

Rapid approach distance – In our given formats, we recommend using an approach distance of 0.100 inch (about 2.5 mm) when approaching qualified (known, consistent) surfaces. While it is more efficient to reduce rapid approach distance, it is not as safe. As you gain confidence and proficiency, you should consider reducing rapid approach distance to improve efficiency.

Machine variations that affect program structure

In Lesson Fifteen, we're providing two sets of format – one for vertical machining centers and another for horizontal machining centers. But even within a given machine type, there are lots variations among machine tool builders.

Our given formats are aimed at basic machining centers – those without a lot of bells and whistles. About the only programmable functions we address are automatic tool changer, spindle, coolant, and feedrate. If your machine has other accessories, you must consult your machine tool builders programming manual to learn how they are programmed. As is discussed in Lesson One, most accessories are handled with M codes – so look for the machine's list of M codes.

Even within our limited coverage of accessories, there are variations. Here we list them.

M code differences

M code number selection is left completely to the discretion of your machine tool builder/s, and no two builders seem to be able to agree on how all M codes should be numbered. For *very* common machine functions like spindle on/off (**M03, M04, M05**) and flood coolant on/off (**M08, M09**), machine tool builders have standardized. But for less common functions like indexers, pallet changers, automatic clamping, chip conveyers, high pressure coolant systems, and through-the-tool coolant systems, you must find the list of M codes in your machine tool builder's programming manual.

Here is a list of very common M codes you will need for our program formats. Fortunately, most machine tool builders utilize these M code numbers just as we show below.

> M00 - Program stop (halts the program's execution until the operator reactivates the cycle)
>
> M01 - Optional stop (used to cause the machine to stop between tools during program verification)
>
> M03 - Spindle on CW (for right hand tools)
>
> M04 - Spindle on CCW (for left hand tools)
>
> M05 - Spindle off (not normally needed with our formats since M19, M06, M30 will stop the spindle)
>
> M06 - Tool change (places the tool in the ready station into the spindle)
>
> M08 - Flood coolant on (cools and lubricates the machining operation)
>
> M09 - Coolant off (used to turn off coolant before each tool change)
>
> M19 - Spindle orientation (used to save a little time at each tool change – spindle begins orienting for a tool change on the way to the tool change position)
>
> M30 - End of program (on many machines, M02 can also be used)

Automatic tool changer commands

All true CNC machining centers have automatic tool changers to automatically load tools into the machine's spindle. Automatic tool changers incorporate some kind of tool storage magazine to hold tools that are not in use. They also incorporate some kind of tool-change-arm that grabs tools and exchanges them between the magazine and the spindle. Machining centers vary with regard to how many tools they can hold.

Even with the wide variety of automatic tool changing systems in use today, programming remains remarkably similar. There are but two basic programming styles.

T word brings a tool to the ready station, M06 commands the tool change

The vast majority of automatic tool changers are programmed in this manner. We have been showing this method in all of the example programs in this text. With this method, a T word is used to specify the tool to be brought to the *ready* or *waiting* station. The T word by itself does *not* make the tool change. It simply rotates the magazine, bringing the specified tool to the ready position (the position that allows it to be placed into the spindle). An **M06** word commands the tool change, placing the tool that is currently in the ready station into the spindle. **M06** also places the tool that is currently in the spindle back into the magazine. The command

> T07 M06

will first rotate the magazine, bringing tool number seven to the ready station. Then it will place tool number seven in the spindle. The tool that is currently in the spindle when this command is executed will be placed back into the magazine, probably in its original tool stations (though tool changer designs vary in this regard).

Almost all current model machining centers have *random access* tool changers. This means you use tools in any order, regardless of tool station number. Though our example programs have been using cutting tools in sequential order (tool one, then tool two, then tool three, and so on), you can use tools in any order in your programs.

Do you have a double-arm tool changer?

If the automatic tool changer has a *double-arm* changing arm (as many do), you'll have an added benefit. As the tool in the spindle is machining a workpiece, you can command that the *next* tool be placed in the ready station – so one tool can be machining while the next is getting ready. This will save tool changing time. We've been assuming this style of automatic tool changer is being used in all example programs shown in this text. Consider these commands from a previous example program:

```
O0003 (Program number)
N005 T01 M06 (Load tool number one in spindle)
(1/4 drill)
N010 G54 G90 S1200 M03 T02 (Select fixture offset #1, absolute mode, start spindle fwd at 1200
    rpm, get tool number two ready)
N015 G00 X1.0 Y1.0 (Rapid to hole location in X and Y)
N020 G43 H01 Z0.1 (Instate tool length compensation for tool one, approach in Z to just above
    work surface)
N025 M08 (Turn on the coolant)
N030 G01 Z-0.65 F4.0 (Drill hole)
N035 G00 Z0.1 M09 (Rapid out of hole, turn off coolant)
N040 G91 G28 Z0 M19 (Rapid to the tool change position, orient spindle)
N045 M01 (Optional stop)
N050 T02 M06 (Load tool number two in spindle)

(3/8 drill)
N055 G54 G90 S1000 M03 T03 (Select fixture offset #1, absolute mode, start spindle fwd at 1000
    rpm, get tool number three ready)
N060 G00 X2.0 Y1.0 (Rapid to hole position in X and Y).
    .

    .
```

In line **N005**, tool number one is being placed in the spindle. While tool number one begins machining, tool number two is rotated to the ready station (in line **N010**). Later, in line **N050**, when the tool change is commanded, the magazine will probably have completed its rotation to station number two, so the tool change can take place immediately. Tool number two will be placed in the spindle (and tool number one will be placed back in the magazine). In line **N055**, the magazine will begin rotating to tool number three as tool number two begins machining. This process is repeated for all tools used in the program

Where is the tool change position?

With the vast majority of machining centers, the tool magazine remains stationary. That is, *the spindle must move to a special position before a tool change can occur.* This special position is called the *tool change position*. If the machine is not at the tool change position when an **M06** is commanded, most machines will generate an alarm (though there are machines with which the **M06** will cause a rapid motion to the tool change position and then make the tool change).

For most vertical machining centers, the tool change position is the Z axis zero return position (**G28**). For most horizontal machining centers, the tool change position is the *Y and Z* axis zero return position. In our example (for a vertical machining center), notice that line **N040** sends the Z axis to its zero return position prior to the tool change command in line **N050**.

What is the M19 doing in line N040?

M19 activates a machine function called a *spindle orient*. Before a tool change can occur, the spindle must rotate to align the *keyway* in the cutting tool holder with the *key* in the tool changer-arm. The spindle orient function performs this spindle rotation. While **M06** *will* automatically perform a spindle orient, including **M19** during the movement to the tool change position will save some time (about one-half to three seconds per tool change, depending upon the machine). The spindle will be oriented properly by the time the machine reaches the tool change position.

Why is the T word repeated in each M06 command?

You may be wondering why we repeat the tool station number in the **M06** command. In line **N050**, for example, there is a **T02** word, even though the **T02** word has already been specified in line **N010**. This is another example of a seemingly redundant word that must be repeated for the purpose of being able to re-run tools. If we must re-run tool number two and if tool number two is not currently in the spindle, the restart block will be the **M06** command (line **N050** in the example above). Without a T word in this command, how will the machine know which tool to place in the spindle when you re-run tool number two?

Do you have a single-arm tool changer?
A single arm-tool changer has but one change-arm. It must replace the tool from the spindle back to the magazine before the next tool can be loaded into the spindle. For most machines with a single-arm tool changer, programming is quite similar. A T word still rotates the magazine to the ready station and an **M06** still makes the tool change.

But with a single-arm tool changer, you are *not* allowed to get the next tool ready while the tool in the spindle is machining the workpiece. Here is the same program shown above, modified for a single arm-tool changer. Notice that the only difference is the T word has been removed from the command after the tool change (lines **N010** and **N055**)

> O0003 (Program number)
> N005 **T01 M06** (Load tool number one in spindle)
> (1/4 drill)
> N010 G54 G90 S1200 M03 (Select fixture offset #1, absolute mode, start spindle fwd at 1200 rpm)
> N015 G00 X1.0 Y1.0 (Rapid to hole location in X and Y)
> N020 G43 H01 Z0.1 (Instate tool length compensation for tool one, approach in Z to just above work surface)
> N025 M08 (Turn on the coolant)
> N030 G01 Z-0.65 F4.0 (Drill hole)
> N035 G00 Z0.1 M09 (Rapid out of hole, turn off coolant)
> N040 **G91 G28 Z0 M19** (Rapid to the tool change position, orient spindle)
> N045 M01 (Optional stop)
> N050 **T02 M06** (Load tool number two in spindle)
> (3/8 drill)
> N055 G54 G90 S1000 M03 (Select fixture offset #1, absolute mode, start spindle fwd at 1000 rpm
> N060 G00 X2.0 Y1.0 (Rapid to hole position in X and Y)
> .
> .

Tool change at beginning or end?
A bit of a controversy exists regarding whether it is best to make the first tool change at very beginning of the program – or to assume the first tool is in the spindle at the beginning of the program, making a tool change at the end of the program that places the first tool in the spindle.

We *strongly* recommend making the first tool change at the beginning of the program – as our example programs – and our given formats – show. With this method, you can rest assured that the correct tool will be in the spindle when machining begins.

If you assume that the first tool is in the spindle when the program begins, the results could be disastrous. Any number of things could cause the wrong tool to be the spindle when the setup person or operator activates the program.

Understanding the G28 command
Again, the tool change position on most machining centers is the zero return position in at least one axis. Before a tool change can occur, the machine must be sent to this position. The most *universal way* to command the machine to move to the zero return position is to use **G28**. This method will work on all Fanuc and Fanuc-compatible machining centers.

G28 is a *two-step* command. Two things will happen whenever a G28 is commanded. First the machine will move (at rapid) the axes included in the G28 command to an *intermediate position*. Then the machine will rapid the axes to the zero return position. When an axis reaches its zero return position, a corresponding *axis-origin-light* comes on – indicating that the axis is at its zero return position.

The *intermediate position* confuses most people. In absolute mode, which most programmers prefer for general purpose programming, the intermediate position is specified relative to program zero. In incremental mode, it is specified relative to the tool's current position. The best way to gain an understanding is to give a few examples, explaining after each.

Consider this command, which we use in all example programs in this text.

    G91 G28 Z0

In *step one* of this G28 command, the tool will move to an intermediate position that is *incrementally nothing* (zero) from its current position in the Z axis. In *step two*, it will go to the zero return position (in Z only). The X and Y axes will not move. For all intents and purposes, we're telling the machine to move *straight to its zero return position in the Z axis*.

    G91 G28 X0 Y0 Z0

In *step one* of this G28 command, the tool will move to an intermediate position that is incrementally nothing (zero) from its current position in X, Y, and Z. In *step two*, it will go to the zero return position in X, Y, and Z (simultaneously). We're telling the machine to move *straight to its zero return position in X, Y, and Z*.

    G91 G28 X0 Y0 Z3.0

In *step one* of this G28 command, the tool will move to an intermediate position that is incrementally nothing (zero) from its current position in X, Y. But in Z, it will move up three inches. Maybe the tool is in a pocket you need to clear before moving in X and Y. In *step two*, it will go to the zero return position in X, Y, and Z (simultaneously).

Watch out! Here's what can happen in *absolute* mode. Consider this command.

    G28 X0 Y0 Z0

Assuming the machine is currently in absolute mode (G90) when this G28 command is executed, *step one* tells the machine to move to the program zero point (probably a crash) in X, Y, and Z. Then, in *step two*, the machine will move to the zero return position in X, Y, and Z (if it still can).

This is the reason why we recommend temporarily switching to the incremental mode with G28. We're not really programming any coordinates incrementally, we're simply trying to send the machine straight to the zero return position.

A possible problem with initialized modes

You know that many modal CNC words are *initialized* (automatically instated at power-up). If you work in the inch mode, for example, G20 is initialized. And if you work exclusively in the inch mode, you shouldn't need to include a G20 word in any of your programs. You may have noticed that none of the example programs shown in this text include G20.

We're *assuming* that the inch mode is currently instated when these programs are run. As long as no one has changed the measurement system mode to metric mode (G21), these programs will run properly.

(By the way, what do you think will happen if the machine is somehow in the *metric mode* when a program written for the inch mode is run? The machine will interpret all measurement-system-related words (like coordinates, feedrate, program zero assignment values, and tool length compensation values) as being in *millimeters*. An X position of X1.0 [which is supposed to be 1.0 inch from program zero in X] will be interpreted as 1.0 millimeter from the program zero position. This has the effect of scaling down the program's motions by a scale factor of 25.4 [the conversion factor from inch to metric]. Obviously, this would not machine the workpiece properly – and would probably be very confusing to the setup person and operator.)

Most programmers don't like to assume that the machine is still in all of its initialized modes when their programs are run. So they include a series of commands in the program (commonly called *safety commands*) to re-instate all of the initialized modes – even though the related words may be redundant. Consider these commands:

N005 G17 G20 G23 (Select XY plane, inch mode, cancel stored stroke limit)

N010 G40 G50 G64 (Cancel cutter radius compensation, cancel scaling, select normal cutting mode)

N015 G67 G69 G80 (Cancel modal custom macro call, cancel rotation, cancel canned cycle)

Remember that most machines allow only three compatible G codes per command – so we need three commands to specify nine modes. While we have not discussed some of these G codes, rest assured that they involve modes that must be instated when your programs run.

How to use our given formats

In Lesson Fifteen, we show four types of program format:

1) Program-startup format

2) Tool-ending format

3) Tool-startup format

4) Program-ending format

Using the formats is easy. When you need to begin writing a new program, follow the program-startup format. Once the first tool is ready to machine, you're on your own to program its cutting motions. When you are finished programming the motion commands for the first tool, follow the tool-ending format. Then follow the tool-startup format to begin the second tool. You're on your own again to program the motion commands for the second tool. Continue toggling among tool-ending format, tool-startup format, and cutting motion commands until you are finished programming the motion commands for the last tool in the program. Then follow the program-ending format. Here is a flow chart of this process for a five-tool program.

Program startup
  Machining operations
Tool ending
Tool startup
  Machining operations
Tool ending
Tool startup
  Machining operations
Tool ending
Tool startup
  Machining operations
Tool ending
Tool startup
  Machining operations
Program ending

Again, the formats will work as a crutch until you have them memorized. While you'll have to plug in appropriate speeds, feeds, axis positions, and other pertinent information, the *basic structure* of the program will be easy – and it will remain the same for every program you write.

Lesson 15

Four Types Of Program Format

The formats shown in this lesson will keep you from having to memorize most of the words and commands needed in CNC programming. As you'll see, a large percentage of most programs is related to structure.

You know the reasons why programs must be formatted using a strict structure. In Lesson Fifteen, we're going to show the actual formats. We will show two sets of format, one for vertical machining centers and another for horizontal machining centers. We will also explain every word in each format in detail. We will also show an example program that stresses the format's use.

There are four types of program format:

>Program startup format
>
>Tool ending format
>
>Tool startup format
>
>Program ending format

Any time you begin writing a new program, follow the program start-up format. You can copy this *structure* to begin your program. The actual values of some words will change based on what you wish to do in your own program, but the structure will remain the same every time you begin writing a new program.

After writing the program start-up format, you write the motions for the first tool's machining operations. When finished with the first tool motions, you follow the tool ending format. You then follow the tool start-up format for the second tool and write the second tool's machining motion commands. From this point, you toggle among tool ending format, tool startup format, and machining motion commands until you are finished machining with the last tool. You then follow the program ending format.

One of the most important benefits of using these formats is that you *will not have to memorize anything.* You simply copy the structure of the format.

Our formats assume that you are using fixture offsets to assign program zero. Later in this lesson, we'll discuss the changes that must be made to these formats if you must use G92 in the program to assign program zero. (Again, you should only use G92 to assign program zero if your machining center does not have fixture offsets.)

Format for vertical machining centers

This format is used for vertical machining centers that have fixture offsets. This format assumes that the machine has a double-arm automatic tool changer, that the tool change position is the Z axis zero return position, and that the machine is resting at the tool change position in Z when the program begins. When this program ends, the machine is left at the Y and Z axis zero return position. This makes a convenient workpiece loading position (with the table out toward the operator in the Y axis).

Again, you are to use the strict *structure* of these given formats. But the values that are shown in **bold** will change from program to program and from tool to tool.

Program Start-Up Format:

O0001 (program number)

N001 G17 G20 G23 (Select XY plane, inch mode, cancel stored stroke limit)

N002 G40 G50 G64 (Cancel cutter radius compensation, cancel scaling, select normal cutting mode)

N003 G67 G69 G80 (Cancel modal custom macro call, cancel rotation, cancel canned cycle)

(TOOL NAME FOR FIRST TOOL)

N005 **T01** M06 (Load first tool in spindle)

N010 G**54** G90 S**300** M03 **T02** (Select fixture offset, absolute mode, turn spindle on fwd at desired RPM, get next tool ready)

N015 G00 X**5.0** Y**5.0** (Move to first XY position)

N020 G43 **H01** Z**0.1** (Instate tool length compensation, move to first Z position)

N025 M08 (Turn on the coolant)

N030 **G01** . . . . F**3.0** (In first cutting movement, be sure to include a feedrate)

Tool Ending Format:

N075 M09 (Turn the coolant off)

N080 G91 G28 Z0 M19 (Return to tool change position, orient the spindle during motion)

N085 M01 (Optional stop)

Tool Start-Up Format:

(TOOL NAME FOR THIS TOOL)

N135 **T02** M06 (Ensure that the next station is still ready and make the tool change)

N140 G**54** G90 S**450** M03 **T03** (Select fixture offset, select absolute mode, turn spindle on CW at desired RPM, get next tool ready)

N145 G00 X**4.0** Y**4.0** (Move to this tool's first XY position)

N150 G43 **H02** Z**0.1** (Instate tool length compensation, move to first Z position)

N155 M08 (Turn on the coolant)

N160 **G01**…. F**4.0** (In first cutting movement, be sure to include a feedrate)

Program Ending Format:

N310 M09 (Turn coolant off)

N315 G91 G28 Y0 Z0 M19 (Return to tool change position in Z, orient spindle during motion)

N320 G28 Y0 (Return to zero return position in Y)

N325 M30 (End of program)

Here are some notes to help you understand the format.

N001

Lines **N001** through **N003** are safety commands to confirm that the machine is still in its initialized states.

G17 selects the XY plane. If another plane (XZ or YZ) has been incorrectly instated with **G18** or **G19**, circular motions, cutter radius compensation, and several other CNC features will not behave as expected.

Our given formats assume you work in inch mode, and *G20* selects the inch mode. If you work in metric mode, you must instead specify a **G21**. If the incorrect measurement system mode is selected, the machine will not move as expected.

G23 cancels something called *stored stroke limit*. This feature allows the programmer to specify a zone that cutting tools are not allowed to enter (*G22* is the G code used to specify a stored stroke limit). Most programmers elect not to use this feature, which is why we ensure that it is cancelled in this safety command.

N002

G40 ensures that cutter radius compensation is cancelled as your program begins.

G50 ensures that a feature called *scaling* is cancelled. All programmed motions will be performed at full scale.

G64 instates the *normal cutting mode*. It cancels something called *exact stop check* (which makes the machine come to a stop after every motion) and *single direction positioning* (which ensures that every motion approaches its end point from the same direction). These features are discussed in Lesson Eighteen.

N003

G67 cancels a modal call to a *custom macro*. Custom macro is a very helpful feature, but is beyond the scope of this text.

G69 cancels something called *coordinate rotation*, which can be used to rotate a series of programmed coordinates and is discussed in Lesson Eighteen.

G80 cancels any of the hole-machining *canned cycles*, which are discussed in Lesson Sixteen.

N005

Our program startup format shows tool station number one (T01) being placed in the spindle. If you wish to start with another tool, of course, you must specify the tool station number for your first tool.

N010

Our program startup format shows a specific fixture offset number, speed, and next tool station number. You will be specifying the fixture offset, speed, and next tool station number needed by your first tool.

N015

This is the first rapid positioning movement in XY made by your first tool. The actual position, of course, will be based upon your job.

N020

This is your first tool's first Z axis motion, so you'll instate tool length compensation in this command. The H word value must, of course, match your first tool's tool station number. And the Z position in this command must be appropriate to the work surface this tool is approaching. (If your machining center has a totally enclosed work area, you can include the M08 in this command to turn the coolant on as well.)

N025

Assuming your workpiece and cutting tool require coolant, turn it on in this command. Again, if your machining center has a totally enclosed work area, the M08 can be included in the previous command.

N030

Each cutting tool's feedrate is part of the program startup format. Though the format shows a specific feedrate value, you will of course, specify the feedrate for your first tool in its first cutting command. Also, your first cutting command may not be a G01 command.

N075

Though we're showing coolant being turned off in a command by itself, you can usually include the M09 word in the tool's last motion command.

N080

This command is sending the machine to the Z axis zero return position, which is the tool change position for most vertical machining centers. We provide a full description of the G28 command in Lesson Fourteen. This command also includes an M19 word to cause the spindle to begin rotating to its orient position on the way to the tool change position. While the up-coming tool change command (M06) *will* orient the spindle, placing the M19 in this command will save tool changing time, since the spindle rotation will be completed by the time the machine gets to its tool change position.

N085

An *optional stop* word (M01) will cause the control to look to the position of an on/off switch (commonly labeled optional stop) on the control panel. If this switch is off, the control will continue with the program, ignoring the M01. If the switch is on, the control will stop the machine at this point. For most machines, the spindle, coolant, and axes are stopped until the operator reactivates the cycle. We recommend placing an M01 at the end of each tool to allow the setup person or operator to stop the machine after each tool. This is especially helpful during the program's verification, when it is necessary to check what one tool has done before moving on to the next tool.

N135 - N160

Notice how similar the tool startup format is to the program startup format. The only exceptions are the program number and safety commands in the program startup format. You must, of course, specify the correct tool, speed, next tool station number, etc. (those values shown in bold) in each tool's startup information.

N310 - N320

These commands should be quite familiar. About the only new command is the second G28 command that sends the machine to a convenient workpiece loading position in the Y axis. As you have seen in the programming activities, this may not entirely necessary based upon the size of the machine.

N325

Most machines use an M30 as the *end-of-program* word. This command will stop the machine. It will also turn off anything that is still running (like spindle and coolant). And M30 will rewind the program back to the beginning so that when the operator activates the next cycle, the machine will start from the beginning of the program.

A note about documentation

As you know, you can include documenting messages in your program. They must be included in parentheses. Unlike we have been showing in the example programs shown in this text, most machines require that all characters in parentheses be upper case characters.

As we show in the tool startup and program startup format, you should include a message naming each tool just above the tool change command. Also, it helps to skip a line just before the message to make it stand out (this helps to nicely separate the tools in your program).

You should also include documenting messages whenever you think it might be helpful to the setup person or operator. For a series of milling passes, for example, you could include the messages (PASS ONE), (PASS TWO), (PASS THREE), and so on to clarify where each pass begins.

If you include any program stop commands in your program (M00), you should include a message that tells the operator what it is they are supposed to do at the program stop – like

N065 M00 (CLEAR CHIPS FROM POCKET)

While we do not show it in the example programs in this text, you should also include documenting messages at the very beginning of your program, to specify information like the part number, the last revision number, the process operation number, the date, the programmer's name, and any other general information that will help people understand what your program is doing.

Example program for vertical machining centers

Here is an example program that stresses the use of the format . Although the program is quite simple (it is actually the same program shown earlier during our discussion of tool length compensation), it shows all of the principles of program formatting. Pay primary attention to the strict structure followed for each tool. The program can be broken into *mini-programs*, each making up one tool. Again, each tool is independent of the rest of the program.

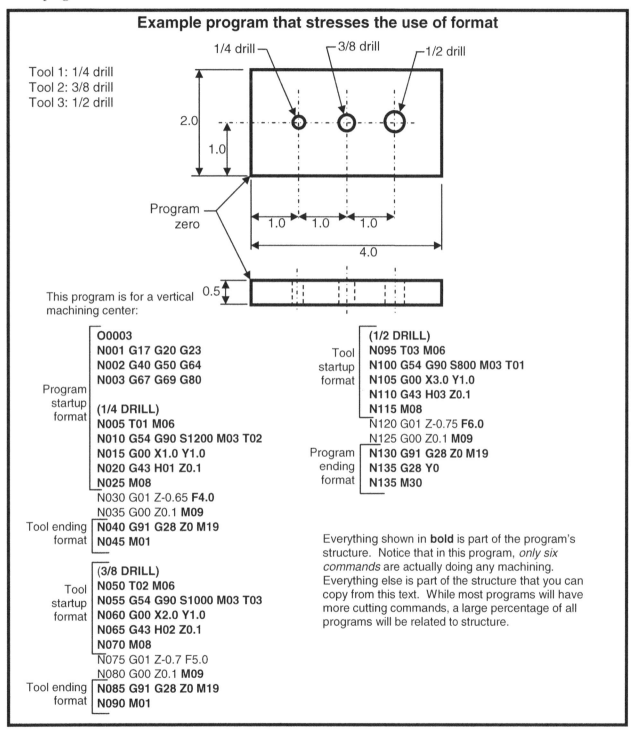

Example program that stresses the use of format

Tool 1: 1/4 drill
Tool 2: 3/8 drill
Tool 3: 1/2 drill

This program is for a vertical machining center:

Program startup format
```
O0003
N001 G17 G20 G23
N002 G40 G50 G64
N003 G67 G69 G80
```

Tool startup format
```
(1/4 DRILL)
N005 T01 M06
N010 G54 G90 S1200 M03 T02
N015 G00 X1.0 Y1.0
N020 G43 H01 Z0.1
N025 M08
```
```
N030 G01 Z-0.65 F4.0
N035 G00 Z0.1 M09
```
Tool ending format
```
N040 G91 G28 Z0 M19
N045 M01
```

Tool startup format
```
(3/8 DRILL)
N050 T02 M06
N055 G54 G90 S1000 M03 T03
N060 G00 X2.0 Y1.0
N065 G43 H02 Z0.1
N070 M08
```
```
N075 G01 Z-0.7 F5.0
N080 G00 Z0.1 M09
```
Tool ending format
```
N085 G91 G28 Z0 M19
N090 M01
```

Tool startup format
```
(1/2 DRILL)
N095 T03 M06
N100 G54 G90 S800 M03 T01
N105 G00 X3.0 Y1.0
N110 G43 H03 Z0.1
N115 M08
```
```
N120 G01 Z-0.75 F6.0
N125 G00 Z0.1 M09
```
Program ending format
```
N130 G91 G28 Z0 M19
N135 G28 Y0
N135 M30
```

Everything shown in **bold** is part of the program's structure. Notice that in this program, *only six commands* are actually doing any machining. Everything else is part of the structure that you can copy from this text. While most programs will have more cutting commands, a large percentage of all programs will be related to structure.

Figure 5.1 – Example program that stresses the use of the formats for a vertical machining center

Key points for Lesson Fifteen:

- There are four kinds of format to help you write programs: program startup format, tool ending format, tool startup format, and program ending format.

- You are only on your own to come up with the cutting motion commands for each tool – a large percentage of most programs is related to the program's structure.

- The startup formats include all CNC words to make each tool independent of the rest of the program – so tools can be re-run.

- The restart command for a tool is the tool change command (with the **M06**) if the tool is *not* currently in the spindle – it is the command following the tool change command if the tool *is* currently in the spindle.

- Tool ending format includes an optional stop word (**M01**) so setup people and operators can easily stop the machine after each tool to see what the tool has done (by turning on the optional stop switch).

Key Concept

6

Special Features That Help With Programming

You now have the basic tools you need to write CNC programs. However, writing programs with only the tools we have shown will be very tedious. In Key Concept Number Six, we'll be showing features that make programming easier, shorten the program's length, and in general, facilitate your ability to write programs.

Key Concept Number Six contains four lessons:

16: Hole-machining canned cycles
17: Sub-programming techniques
18: Other special programming features
19: Programming rotary devices

Key Concept Number Six will be of great interest to you. While you have learned how to write simple CNC programs, you may be wondering if there are easier ways to get things done. With hole-machining, for example, we have shown only drilling operations. As you know, there are other kinds of hole-machining operations (like tapping, reaming, boring, and counter-boring). And our programming methods have been very tedious – requiring three commands per hole (rapid to XY position, drill hole, retract from hole). You were probably thinking *there's got to be a better way!* Well there is. And we'll be showing it in Lesson Sixteen as we discuss hole-machining canned cycles.

In Lesson Seventeen, you'll learn how to use a program-shortening feature called *sub-programming*. In Lesson Eighteen, we'll present a number of special features that can help you develop programs. While they won't all be of immediate use, you'll surely find at least some to be very helpful. And in Lesson Nineteen, we'll show how rotary devices are programmed.

Almost every CNC function presented thus far is *essential* to your ability to program CNC machining centers. You will be using the features and techniques presented in Key Concepts one through five in every program you write (with the possible exception of cutter radius compensation). Think of what you have learned so far as the rudimentary tools of CNC programming.

Key Concept Number Six will dramatically simplify many programming tasks. Though not every feature and technique shown in this Key Concept will be of use to every CNC user, you will surely pick up several techniques that will streamline your daily programming activities. They make programming easier, they make programs shorter, they reduce the potential for mistakes, and in general, they simplify the task of writing CNC programs.

You may be wondering why we waited so long to show these features, since they do make programming *so* much easier. Do not underestimate the importance of what you have learned so far. Just as it helps to understand how arithmetic calculations are done manually before you use a calculator, so does it help to understand the long (hard) way of programming machining operations before you move on to the more advanced methods. At the very least, the programming activities you have done so far should give you a good appreciation for what is presented in Key Concept Number Six. More importantly, you should now have a *very* good foundation in CNC that will let you easily grasp the importance of the features we show in Key Concept Number Six.

Some of the features shown in Key Concept Number Six are considered by Fanuc to be *options*. But most machine tool builders will include them with the machines they sell. If you come across one of these features that your machine does not have, remember that all of them are *field-installable*, meaning if you have a need for them, they can be added to your control at any time (for an additional price).

Here is a list of the special features we will discuss during this Key Concept in the order we cover them:

- G73 - G89 canned cycles - To simplify hole-machining operations (Lesson Sixteen)

- M98 - M99 Sub-programming - To minimize redundant commands (Lesson Seventeen)

- / - Block delete techniques - To give your operator a choice (Lesson Eighteen)

- Sequence number techniques (Lesson Eighteen)

- G02-G03 thread milling - For outside threads and when holes are too large to tap (Lesson Eighteen)

- Other G codes not discussed to this point (Lesson Eighteen)

- M codes not addressed to this point (Lesson Eighteen)

- Rotary devices (Lesson Nineteen)

As stated, you will not have immediate need for some of these features. In fact, some you may never need. But we urge you to study this information if for no other reason than to gain a better understanding of what is possible in CNC programming. You can always come back and review a needed feature when the need arises, *if you know it is available.*

Lesson 16
Hole-Machining Canned Cycles

Armed only with what you know so far, programming hole-machining operations is very tedious. Hole-machining canned cycles will dramatically simplify the programming of these very common machining operations.

Almost all programs have at least *some* hole-machining operations. If you have been doing the exercises in this text and in the workbook, you have seen how tedious, time consuming, and error-prone it can be to program hole-machining operations with **G00** and **G01**. You know that with **G00** and **G01**, each hole will require *at least* three commands, making your program quite long. And we've only performed basic drilling operations. Peck drilling, tapping, boring, and counter-boring operations will require even more commands per hole.

Canned cycles will dramatically simplify the programming of hole-machining operations. *Only one command is required per hole*, regardless of the machining style (drill, peck drill, tap, ream, bore, counter-bore, etc.). Additionally, canned cycles are *modal*, meaning once you instate a canned cycle, you can continue machining holes by simply listing hole-positions. Again, this will dramatically shorten the program's length, make programming easier, less time-consuming, and less error-prone.

The meaning of "canned"
A *canned* cycle is a series of preset movements that the CNC machine will execute based upon a limited amount of program information. The *zero return command* (**G28**) is a simple kind of canned cycle. **G28** actually makes the machine do *two* things. First, the machine will move to the *intermediate position*. Second, it will move to the *zero return position*.

If you have the *single block* switch turned on (a function that makes the machine execute one command in the program at a time), you actually have to activate the cycle twice to make the control complete the **G28** command. Again, the first time you activate the cycle (by pressing the *cycle start* button), the machine executes the motion to the intermediate position. The second time, it moves to the zero return position.

Hole-machining canned cycles are much more elaborate. Even the standard drilling cycle will make the machine do at least three things per command. With the chip-breaking peck drilling cycle, one command can actually generate over one-hundred movements.

Here's how simple it is to use canned cycles: You *instate* the canned cycle for the first hole to machine. With a standard drilling cycle, for example, the instating command includes the cycle type (**G81**), the first hole position in XY, the hole's bottom position in Z (commonly its depth), and the machining feed rate. The instating command actually machines the first hole. To machine the rest of the holes, you simply list their XY coordinates, one set per command. After the last hole, you must *cancel* the cycle with a **G80** word.

Here is a list of the most common hole-machining canned cycles in approximate order of popularity:

> G80 – Cancel any of the canned cycles
>
> G81 – Standard drilling cycle
>
> G73 – Chip-breaking peck drilling cycle
>
> G83 – Deep hole peck drilling cycle
>
> G82 – Counter-boring cycle
>
> G84 – Right hand tapping cycle
>
> G74 – Left hand tapping cycle
>
> G86 – Standard boring cycle

G89 – Counter-boring with boring bar

G76 – Boring cycle without witness mark

G85 – Reaming cycle

Canned cycle commonalities

Canned cycles share two things in common. *First*, they are all *modal*. Once a canned cycle is instated, the machine will continue machining one hole per command until the canned cycle is cancelled. G80 is the word used to cancel canned cycles. Once you instate a canned cycle, you *must* remember to cancel it when you're finished machining holes. If you don't, the control will continue machining (unwanted) holes even after you have finished listing hole-coordinates.

Second, all canned cycles will cause these four basic activities:

1) A rapid positioning movement to the hole position in XY (if the tool is not already in this position).

2) A rapid positioning movement to the R plane (referred to as the *rapid plane*). This approach position is just above the surface to machine.

3) Using the canned cycle's machining style, the hole will be machined.

4) The cutting tool will retract from of the hole.

Description of each canned cycle

You choose how the hole will be machined using one of a series of G codes. For the most part, each hole-machining canned cycle is aptly named. Let's describe the machining operation performed by each cycle type in detail.

G80 – Cancel the canned cycle mode

This is not actually a canned cycle commanding word. It is the word used to cancel the canned cycle mode when you finish machining holes with any canned cycle. Again, you *must* remember to cancel the canned cycle when you're finished machining holes.

G81 –Standard drilling cycle

This canned cycle causes the tool to feed to the hole-bottom. Then it will retract at *rapid*. It is a the most popular canned cycle – commonly used for center drilling, spot drilling, drilling, reaming, and rough boring.

G73 – Chip-breaking peck drilling cycle

This canned cycle is used when drilling a material that has the tendency to produce long, stringy chips. If the chips do not break, they will have the tendency to gather around the drill (forming what machinists commonly call a *rat's-nest*). Eventually the bundle of chips will interfere with machining. This cycle is used to force the chips to break at regular intervals as the hole is drilled. The breaking of the chip is accomplished by a small retract movement (about 0.005 in). For example, if the peck amount (specified by a Q word in the canned cycle command) is set to 0.100 inch, the tool will plunge into the hole to a depth of 0.100 inch, then retract about .005 inch. The chip will break at this point. Then the tool will plunge another .100 inch and retract. It will continue with this process until the hole-bottom is reached.

The retract amount is adjustable by a *control parameter*, meaning it can be changed if you do not agree with its current setting. We have seen some machine tool builders that set the retract amount to an excessive value. If, on your machine, the tool appears to retract more than about 0.005 in per peck, you will be wasting cycle time whenever you use this cycle. (The parameter number for retract amount is specified in the control manufacturer's programming manual during the description of G73).

G83 – Deep-hole drilling cycle (full retract between pecks)

If drilling a hole having a depth that is greater than about three times the drill diameter, the chips machined by the drill will tend to bind up in the drill's flutes. If the drill continues into the hole, this binding will get progressively worse, and will eventually cause the drill to break. The deep-hole peck drilling cycle allows you to easily specify a series of pecks (peck depth is specified with a Q word in the **G83** command). After each peck, the machine will retract the drill (at rapid) all the way out of the hole to clear chips. The tool will then rapid back into the hole to within a small distance (set by another control parameter) of where it left off. It will then feed in another peck amount. The machine will continue this process until the hole-bottom is reached. The tool then retracts from the hole a final time.

Figure 6.1 provides a summary of the motions performed by the three drilling cycles.

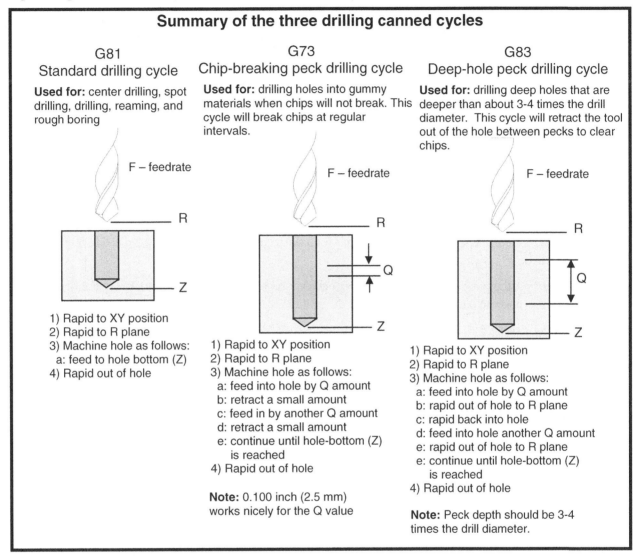

Figure 6.1 – The three drilling canned cycles

G84 – Right-hand tapping cycle

Since most tapped holes require right-hand threads, **G84** is the most common tapping cycle. The spindle must be running in the forward direction (**M03**) when this cycle is commanded. Once the tap feeds to the hole-bottom, the spindle direction will change (to reverse – **M04**) and the tap will feed back out of the hole. Then the spindle will change direction again (back to forward) to get ready to tap the next hole.

With many (especially older) machining centers, the machine cannot perfectly synchronize the spindle reversal at the hole bottom with the retract motion out of the hole. These machines require a special *tension/compression tap holder* that allows the tap to float (up and down) in its holder. Figure 6.2 shows one.

A tension/compression tap holder

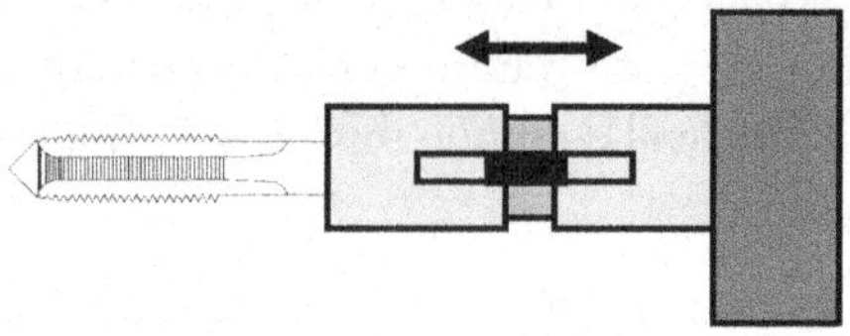

Tap holder can extend or contract by spring-loading – When released, it will snap back to its center position. This tap holder is required when the machining center does not have a feature called *rigid tapping*.

Figure 6.2 – A tension-compression tap holder

There is a feature called *rigid tapping* (also called *synchronous tapping*) that allows perfect synchronization between the spindle reversal at the hole-bottom and the retract motion. Machines that have rigid tapping can tap faster, provide better threads in the tapped hole, and eliminate the need for the special (and expensive) tension/compression tap holder.

Feedrate for tapping

As you know, most machining centers require that you program feedrate in per-minute fashion. To calculate the feedrate for tapping inch threads, you must multiply the *pitch* of the thread (pitch is one divided by the number of threads per inch) times the previously calculated spindle speed in rpm. For a 1/2-13 thread, for example, say your calculated speed is 230 rpm. Pitch for a thirteen-threads-per-inch thread is 0.0769 (1/13). The feedrate will be 17.68 ipm (230 times 0.0769).

With metric threads, the pitch is part of the thread specification. For an M16-1.5 thread, the major diameter is sixteen millimeters and the pitch is 1.5 millimeters.

Rapid plane for tapping

If using a tension-compression holder (the machine does not have rigid tapping), the tap can actually *float* in the tap holder. That is, it can extend and contract along the Z axis. If the tap holder extends as the hole is tapped, and if the rapid plane is only 0.100 inch (about 2.5 mm) above the work surface, it is possible that the tap will still be in the hole after the tap makes its retract motion out of the hole. If this happens, the tap will break during the machine's next XY movement. For this reason, most programmers will increase the approach distance to about 0.25 inch (about 6.0 mm) when tapping with tension-compression holders.

If using the feature rigid tapping, the tap does not float in its holder, so most programmers maintain the 0.1 inch approach distance (this is one of the reasons rigid tapping is faster – the tool doesn't need to move as far).

Tapping can be a little scary

Tapping canned cycles (**G84** and **G74**) will cause the machine to disable certain control-panel functions, making it a little difficult to safely verify tapping operations. Under normal operation, when the machine executes a tapping canned cycle, it will disable *feedrate override* and *feed hold*. Feedrate override is a multi-position switch (like a rheostat) that provides control of programmed feedrate. But when tapping, this function must be disabled. If tapping is not performed at the programmed feedrate, the tap will break as the hole is machined. Feed hold is a push-button that, when pressed, will cause axis motion to stop. If motion is stopped when the tap is in a hole, of course, the tap will break.

Without these two program-verification functions, tapping is somewhat dangerous. You must be sure that the tap is programmed properly before allowing a tap to enter a hole.

Coolant for tapping?

While some coolants work well for tapping certain materials, many companies use a special tapping compound. Some machining centers even have a special (automatic) function that applies tapping compound to the tap for each hole. Most machines do not. Instead, the program will stop (with an **M00**) command just before the tapping operation/s. During this program stop, the operator will clear the chips from holes to be tapped and apply the tapping compound to each hole.

When to tap

If the tapping compound must be manually applied, most programmers will save all tapping until the very end of the program. This way, the operator will only have to clear chips and apply tapping compound once.

G74 – Left-hand tapping cycle

This cycle is just like the **G84** tapping cycle, except it is used for left-hand taps. The spindle must be running in the *reverse* direction (**M04**) when this cycle is commanded. Left-hand tapping is not often required – and of course – requires a special left-hand tap.

Figure 6.3 shows the two tapping cycles.

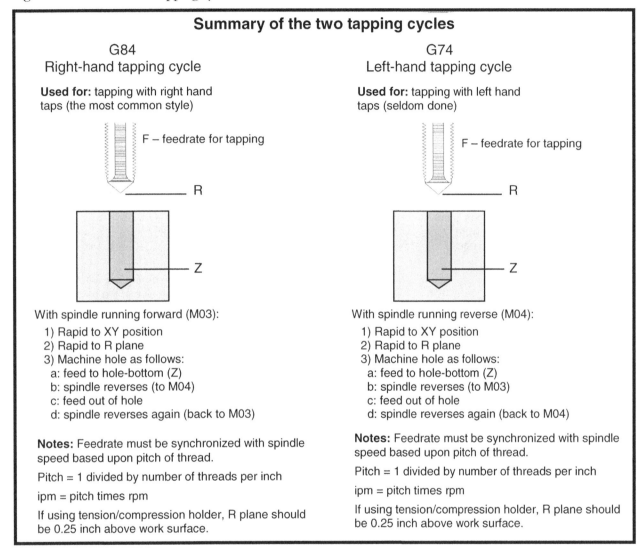

Summary of the two tapping cycles

G84 Right-hand tapping cycle	G74 Left-hand tapping cycle
Used for: tapping with right hand taps (the most common style)	**Used for:** tapping with left hand taps (seldom done)
F – feedrate for tapping	F – feedrate for tapping

With spindle running forward (M03):
1) Rapid to XY position
2) Rapid to R plane
3) Machine hole as follows:
  a: feed to hole-bottom (Z)
  b: spindle reverses (to M04)
  c: feed out of hole
  d: spindle reverses again (back to M03)
	With spindle running reverse (M04):
	1) Rapid to XY position
	2) Rapid to R plane
	3) Machine hole as follows:
	  a: feed to hole-bottom (Z)
	  b: spindle reverses (to M03)
	  c: feed out of hole
	  d: spindle reverses again (back to M04)

Notes: Feedrate must be synchronized with spindle speed based upon pitch of thread. Pitch = 1 divided by number of threads per inch ipm = pitch times rpm If using tension/compression holder, R plane should be 0.25 inch above work surface.	**Notes:** Feedrate must be synchronized with spindle speed based upon pitch of thread. Pitch = 1 divided by number of threads per inch ipm = pitch times rpm If using tension/compression holder, R plane should be 0.25 inch above work surface.

Figure 6.3 – Motions and programming words or tapping cycles

G82 – Counter-boring cycle

This cycle causes the tool to feed to the hole bottom, pause for a specified length of time (pause time is specified with a P word), and then retract from the hole at rapid. The pause at the hole-bottom allows the counter-boring tool (commonly an end mill) to *relieve tool pressure*, making the hole-bottom flat and precise.

G89 – Counter-boring cycle for a boring bar

This cycle is like a combination of the **G86** and **G82** commands. This cycle is used when a boring bar is machining a blind hole to a precise depth. The boring bar will first feed to the hole-bottom and then pause for a specified time to relieve tool pressure. The spindle will then stop. Finally, the tool will rapid out of the hole. As with **G86**, a *witness mark* will be left in the hole (see the discussion of **G86** for the reason why).

Figure 6.4 shows the two counter-boring cycles.

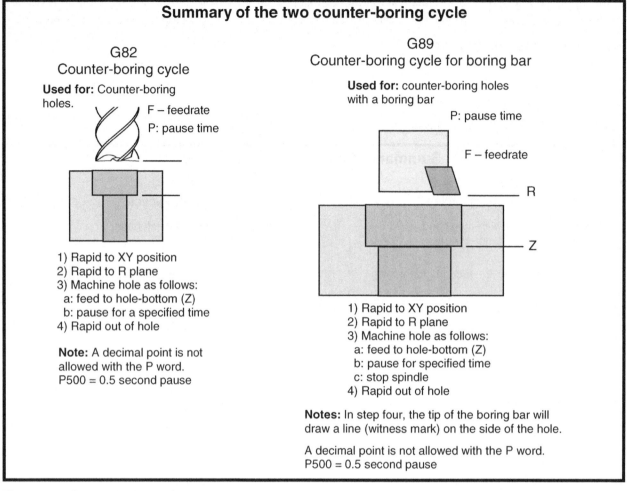

Figure 6.4 – Counter-boring cycles

G86 – Standard boring cycle (leaves drag line witness mark)

This cycle is used for rough or semi-finish boring, or for finish boring when it doesn't matter if a witness mark is left in the bored hole. The tool will feed to the hole-bottom, the spindle will stop, and the tool will retract from the hole (at rapid). Note that it is during the retract motion (with the tool tip still in contact with the hole surface) that a line will be drawn on the inside of the hole.

G76 – Fine boring cycle (leaves no witness mark)

This cycle is used for finish boring when no witness marks can be left in the hole after boring. For each hole, the boring bar will feed to the hole-bottom. The spindle will then stop. The spindle will then rotate to its orient position (the **M19** position). The tool's cutting edge will then be moved away from the hole-surface to allow some clearance between the tool tip and the hole (the direction and amount of the clearance motion is specified in the canned cycle command). The boring bar will then rapid out of the hole. Finally, the spindle will be restarted and the tool will move back to the hole center line.

Controlling move-over at hole-bottom

Two words are used to control the move-over amount (I and J). An *I word* in the **G76** command specifies the amount and direction of move-over along the X axis. A *J word* in the **G76** command specifies the amount and direction of move-over along the Y axis. Say, for example, the boring bar tip is pointing in the X plus direction when the spindle is at its orient position. In this case, an **I-0.001** can be included in the **G76** command to tell the machine to move over 0.001 inches the X minus direction at the hole-bottom and after the spindle is oriented. We recommend specifying a very small move-over amount (like 0.001 inch), just in case the setup person makes a mistake when loading the boring bar (having it pointing in the wrong direction).

In order to specify these words correctly, you must know (when programming) which way the tip of the boring bar will be pointing when the spindle is at its orient position. This may be impossible, since the boring bar may be held in a collet holder. Unless the setup documentation is very specific about how the boring bar must be held in the holder, you won't know which way the boring bar is pointing when the spindle is oriented.

For this reason, many programmers will include the words I0 and J0 in the G76 command. The setup person will modify the program during setup, once they know which way the boring bar will be pointing at the spindle's orient position.

A tip for boring bar tip pointing

Most tool holders used on machining centers have a dimple in one of the keyways held by the tool changing arm. And most machining centers only allow you to load the tool holder into the spindle in one position (the two keyways in the tool holder are of different widths). *Use the dimpled keyway as your target* for the boring bar tip. If you're using a collet holder, have the setup person rotate the boring bar in the holder until its tip is pointing at the dimpled keyway. Then tighten the collet. When the spindle is at its orient position, it will always be pointing in the same direction.

Figure 6.5 shows the two boring cycles.

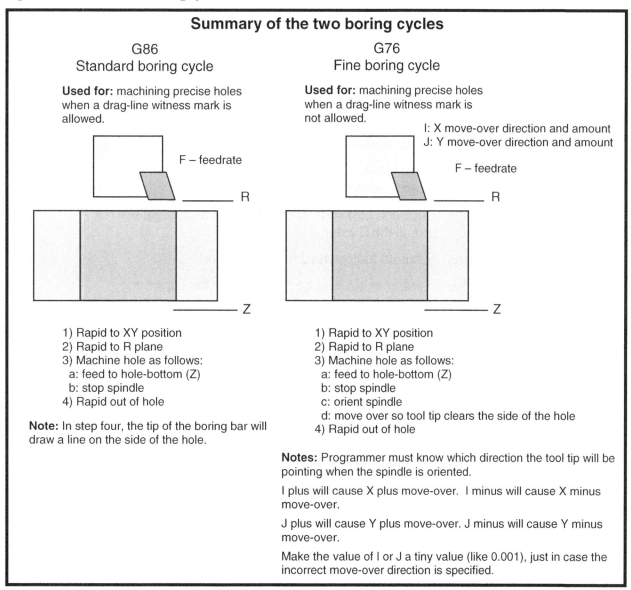

Summary of the two boring cycles

G86
Standard boring cycle

Used for: machining precise holes when a drag-line witness mark is allowed.

F – feedrate

R

Z

1) Rapid to XY position
2) Rapid to R plane
3) Machine hole as follows:
   a: feed to hole-bottom (Z)
   b: stop spindle
4) Rapid out of hole

Note: In step four, the tip of the boring bar will draw a line on the side of the hole.

G76
Fine boring cycle

Used for: machining precise holes when a drag-line witness mark is not allowed.

I: X move-over direction and amount
J: Y move-over direction and amount

F – feedrate

R

Z

1) Rapid to XY position
2) Rapid to R plane
3) Machine hole as follows:
   a: feed to hole-bottom (Z)
   b: stop spindle
   c: orient spindle
   d: move over so tool tip clears the side of the hole
4) Rapid out of hole

Notes: Programmer must know which direction the tool tip will be pointing when the spindle is oriented.

I plus will cause X plus move-over. I minus will cause X minus move-over.

J plus will cause Y plus move-over. J minus will cause Y minus move-over.

Make the value of I or J a tiny value (like 0.001), just in case the incorrect move-over direction is specified.

Figure 6.5 – Motions for the two boring cycles

G85 – Reaming cycle (most programmers use G81 for reaming)

This cycle causes the tool to feed to the hole-bottom and retract from the hole at the same feedrate. While this cycle is sometimes used for reaming, no machining will be done during the retract motion. For this reason, most programmers use the G81 standard drilling cycle for reaming.

G87 and G88 – Manual cycles (not recommended)

G87 and G88 are manual cycles that we do not recommend. They perform differently, based on the settings of certain control parameters. Each requires manual intervention on the operator's part. For the most part, the machining operations they perform (back-boring that requires a pilot to be placed on and removed from the tool during the cycle) are just as easily programmed in long-hand fashion using G00 and G01. With G00 and G01, you'll have complete control of the motions being made by the back-boring tool.

Words used in canned cycles

As you have probably noticed in Figures 6.1 through 6.5, canned cycles share many words in common. These words will have exactly the same meaning in all canned cycles. Although some canned cycles do not use some of these words, when you understand how the shared words work in one canned cycle, you will know how they work in all. Here is a list of all words used in canned cycles, which cycles they apply to, and a brief description of their use:

WORD:	STATUS:	DESCRIPTION:
N	All cycles	Sequence number
G73-G89	All cycles	Canned cycle type
X	All cycles	X coordinate of the hole-center
Y	All cycles	Y coordinate of the hole-center
R	All cycles	Rapid plane position above work surface (specified from program zero in Z)
Z	All cycles	Z position of hole-bottom (specified from program zero in the Z axis)
F	All cycles	Feedrate for the machining operation
L	All cycles	Number of holes to be machined in the command (for 0M and 3M controls, the a K word is used) (used with incremental mode only)
G98	All cycles	Retract out of the hole to the initial plane (initialized)
G99	All cycles	Retract out of the hole to the R plane
P	G82, G89	Pause time at hole bottom (P500 = .5 second)
Q	G73, G83	Peck drill amount per peck
I	G76	Amount and direction of move-over in X at bottom
J	G76	Amount and direction of move-over in Y at bottom

A simple example

Here is a simple example program that stresses the points made so far. Figure 6.6 shows the workpiece. We're simply drilling four holes using the standard drilling cycle, G81.

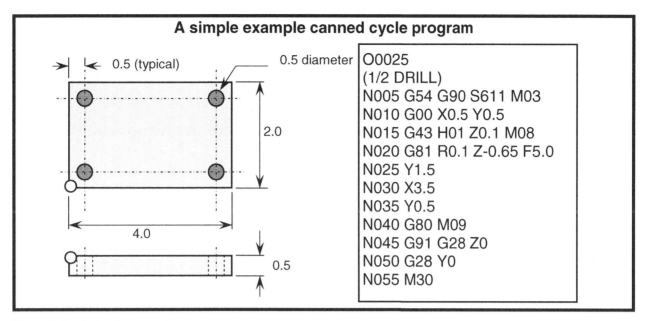

A simple example canned cycle program

```
O0025
(1/2 DRILL)
N005 G54 G90 S611 M03
N010 G00 X0.5 Y0.5
N015 G43 H01 Z0.1 M08
N020 G81 R0.1 Z-0.65 F5.0
N025 Y1.5
N030 X3.5
N035 Y0.5
N040 G80 M09
N045 G91 G28 Z0
N050 G28 Y0
N055 M30
```

Figure 6.6 – A simple example

Program with comments:

O0025 (Program number)

(1/2 DRILL)

N005 G54 G90 S611 M03 (Select fixture offset #1, absolute mode, start spindle fwd at 611 rpm)

N010 G00 **X0.5 Y0.5** (Rapid to first hole-location)

N015 G43 H01 Z0.1 M08 (Instate tool length compensation, position tool to just above work surface, start coolant)

N020 **G81 R0.1 Z-0.65 F5.0** (Drill lower-left hole)

N025 **Y1.5** (Drill upper-left hole)

N030 **X3.5** (Drill upper-right hole)

N035 **Y0.5** (Drill lower-right hole)

N040 **G80** M09 (Cancel canned cycle, turn off coolant)

N045 G91 G28 Z0 (Move to Z axis zero return position)

N050 G28 Y0 (Move to Y axis zero return position)

N055 M30 (End of program)

This program nicely illustrates the points made so far. In line **N010**, the tool is positioned over the first hole. In line **N015**, the tool is brought to just above the work surface.

Line **N020** instates the standard drilling cycle, telling the machine that the *R plane* is 0.1 inch above the workpiece (the tool is already in this position), the *hole bottom position* is -0.65 below the top of the workpiece (this happens to be the hole depth since the top surface is program zero), and the *feedrate* for drilling is 5.0 ipm. This command actually drills the first (lower-left) hole.

Now we simply list hole-locations – and only the moving axes must be included in the command. While you can include both X and Y coordinates in every command, it will lengthen the program and may result in a typing mistake. As with all motion commands, we recommend that you *only include moving axes* in each command. Line **N025** machines the upper-left hole. Line **N030** machines the upper-right hole. And line **N035** machines the lower-right hole.

Since there are no more holes to machine, the **G80** in line **N040** cancels the canned cycle.

This program uses **G81** – the standard drilling cycle. But say the material being machined is very gummy, and chips are stringing up around the drill. Our next example program machines the same workpiece, but uses **G73** (the chip breaking peck drilling cycle) instead of **G81**:

O0025 (Program number)

(1/2 DRILL)

N005 G54 G90 S611 M03 (Select fixture offset #1, absolute mode, start spindle fwd at 611 rpm)

N010 G00 X0.5 Y0.5 (Rapid to first hole-location)

N015 G43 H01 Z0.1 M08 (Instate tool length compensation, position tool to just above work surface, start coolant)

N020 **G73** R0.1 Z-0.65 **Q0.1** F5.0 (Drill lower-left hole)

N025 Y1.5 (Drill upper-left hole)

N030 X3.5 (Drill upper-right hole)

N035 Y0.5 (Drill lower-right hole)

N040 G80 M09 (Cancel canned cycle, turn off coolant)

N045 G91 G28 Z0 (Move to Z axis zero return position)

N050 G28 Y0 (Move to Y axis zero return position)

N055 M30 (End of program)

With only two changes (changing **G81** to **G73** and adding the **Q0.1** word), we've modified the way the holes will be machined.

Understanding G98 and G99

Fanuc and Fanuc-compatible machining centers provide the ability to easily clear clamps and other obstructions between holes when using canned cycles – moving over them in the Z axis.

Remember the four basic steps that all canned cycles will perform:

1) A rapid positioning movement to the hole position in XY (if the tool is not already in this position).

2) A rapid positioning movement to the R plane (referred to as the *rapid plane*). This approach position is just above the surface to machine).

3) Using the canned cycle's machining style, the hole will be machined.

4) The cutting tool will retract from of the hole.

In step number four, **G98** and **G99** specify to which of two possible Z positions the tool will be retracted. Your choices are:

- The R plane specified in the canned cycle (which is right above the work surface)
- The *initial plane* (the last programmed Z position prior to the canned cycle command) – this will be a position that is above the obstruction between holes

G98 specifies the *initial plane*. The initial plane is the cutting tool's last Z position prior to the canned cycle command. To make use of obstruction clearing, the initial plane must be specified above all obstructions.

G99 specifies the R plane, which is included in the canned cycle itself. The R plane is right above the work surface. **G98** and **G99** are modal, meaning if you have a series of holes that require the same retract position, **G98** or **G99** need only be specified in the first command of the series.

Though the word order of your canned cycle command is unimportant, we recommend placing the **G98** or **G99** at the *end* of the command. This will help you remember what the tool will do *after* the hole is machined.

G98 (initial plane) is initialized when the machine power is turned on – and it is reinstated whenever the canned cycle is canceled (with **G80**). So if you do not include a **G98** or **G99** in the canned cycle command for the first hole, the control will automatically retract the tool to the initial plane. This is the safer of the two retract positions.

If you position the tool to the R plane just prior to the canned cycle command (like we did in the example shown in Figure 6.6), the R plane and the initial plane will be at the same Z axis position. In this case, **G98** and **G99** will have no effect on the retract position – and you need not include either G code in the canned cycle command.

Figure 6.7 shows an example of obstruction clearing. Notice the two clamps between the holes.

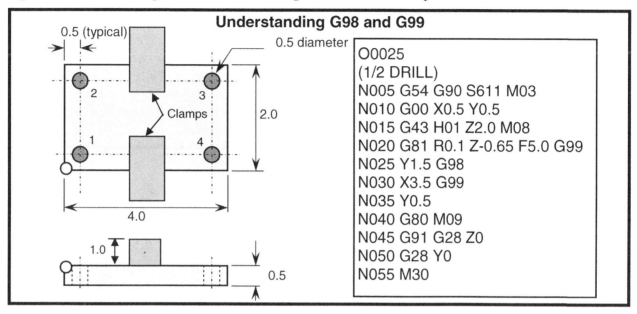

Figure 6.7 – Example showing obstruction clearing between holes

Program with comments:

```
O0025 (Program number)
(1/2 DRILL)
N005 G54 G90 S611 M03 (Select fixture offset #1, absolute mode, start spindle fwd at 611 rpm)
N010 G00 X0.5 Y0.5 (Rapid to first hole-location)
N015 G43 H01 Z2.0 M08 (Instate tool length compensation, position tool to initial plane 2.0 inches
    above work surface, start coolant)
N020 G81 R0.1 Z-0.65 F5.0 G99 (Drill lower-left hole, retract to R plane)
N025 Y1.5 G98 (Drill upper-left hole, retract to initial plane)
N030 X3.5 G99 (Drill upper-right hole, retract to R plane)
N035 Y0.5 (Drill lower-right hole, continue retracting to R plane)
N040 G80 M09 (Cancel canned cycle, turn off coolant)
N045 G91 G28 Z0 (Move to Z axis zero return position)
N050 G28 Y0 (Move to Y axis zero return position)
N055 M30 (End of program)
```

Again, there are two clamps between the holes. In line **N015**, the tool is brought to the *initial plane, which is two inches above the work surface.* This Z position is well above the clamps.

In line **N020**, the first hole is drilled. Since there is no clamp between this hole and the next one, a **G99** is placed in this command to allow the tool to retract from this hole to the R plane – saving a little time. Again, we recommend placing the **G98** or **G99** at the end of the command to help you remember what will happen *after* the hole is machined.

In line **N025**, the upper-left hole is machined. But there is a clamp between this hole and the next one, so a **G98** is included in this command to cause the tool to retract to the *initial plane*, two inches above the workpiece. When the XY positioning movement to the next hole occurs in line **N030**, the tool will be well above the clamp.

In line **N030**, the upper-left hole is machined, and since there is no clamp before the next hole, a **G99** is placed in this command. After the last hole is machined (in line **N035**) the tool will again retract to the R plane.

Canned cycles and the Z axis

The initial plane, R plane, and Z hole-bottom position values are *absolute coordinates*, taken from the Z axis program zero point. *The Z value in the canned cycle should not be thought of to as the hole-depth.* The Z word is the hole-depth only when the hole is being machined into the Z axis program zero surface (which is often the case). It is more correct to say that the Z value of the canned cycle command is the *hole-bottom position*. There will be times when you must machine holes into Z surfaces other than the program zero surface in Z. Figure 6.8 is an example showing a time when you will not be machining into the Z zero surface.

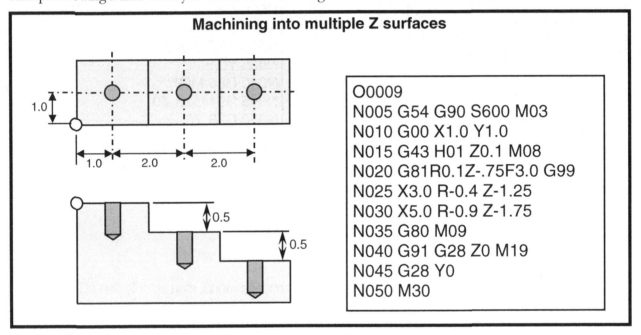

Figure 6.8 – Machining holes into multiple Z surfaces

As you can see, only one of the holes is being machined into the top (Z zero) surface of this workpiece. For the other two holes, you must manipulate the R plane and Z hole-bottom position. When you have multiple surfaces to machine with a tool, we recommend working from *top surface to bottom surface*.

Program with comments:

```
O0009 (Program number)
N005 G54 G90 S600 M03 (1/2 drill)
N010 G00 X1.0 Y1.0
N015 G43 H01 Z0.1 M08 (Above highest surface)
N020 G81 R0.1 Z-0.75 F3.0 G99 (Just like shown so far)
N025 X3.0 R-0.4 Z-1.25 (Note new R and Z values)
N030 X5.0 R-0.9 Z-1.75 (Note new R and Z values)
N035 G80 M09 (Cancel cycle)
N040 G91 G28 Z0
```

N045 G28 Y0

N050 M30

Notice how easy it is to manipulate Z axis positions by modifying R and Z values. You simply specify the *current* R plane and Z bottom-position for the holes that require surface changes. However, *you must be careful!* If you work from bottom to top, you may be in for a nasty surprise. Remember the basic sequence of any canned cycle:

> 1) A rapid positioning movement to the hole position in XY (if the tool is not already in this position).

> 2) A rapid positioning movement to the R plane (referred to as the *rapid plane*). This approach position is just above the surface to machine).

> 3) Using the canned cycle's machining style, the hole will be machined.

> 4) The cutting tool will retract from of the hole.

Consider this ***incorrect program***:

O0010 (Program number)

N005 G54 G90 S600 M03 (1/2 drill)

N010 G00 X5.0 Y1.0 (Right-most hole on lowest surface)

N015 G43 H01 Z0.1 (Above highest surface)

N020 M08

N025 G81 R-0.9 Z-1.75 F3.0 G99 (Okay so far)

N030 X3.0 R-0.4 Z-1.25 (This command causes a crash)

N035 X1.0 R0.1 Z-0.75 (This command causes another crash)

N040 G80 M09 (Cancel cycle)

N045 G91 G28 Z0

N050 G28 X0 Y0

N055 M30

This program causes a crash (actually two crashes). In line **N025** the first hole will be machined just fine. The tool will rapid in XY, then it will rapid to the R plane, plunge the first hole and rapid back out to the R plane (notice the **G99** in **N025**. But in line **N030**, the tool will *first* move in X to the new position (crash) before moving to the new R plane. If you must machine from bottom to top, you could, of course, simply change the **G99** in line **N025** to a **G98**. This will cause the tool to come up to the initial plane (0.1 above the upper-most surface) before moving in XY between holes. While this will cause a little wasted (rapid motion) time, the tool will not crash into the workpiece.

Extended example showing canned cycle usage

Here is a full example showing the use of several types of canned cycles (G81 – drilling, G84 - tapping, G82 – counter-boring, G73 – peck drilling, and G76 – fine boring).

Figure 6.9 shows a drawing of the workpiece to be machined. Notice the two clamps that must be avoided. We'll use an initial plane of two inches above the work surface to clear these clamps.

This program is going to be using the fine boring cycle (G76) so no witness mark drag line will be left in the hole. We must know which direction the boring bar tip will be pointing when the spindle is at its orient position. We'll say it is pointing in the X plus direction, meaning an X minus direction move (of 0.001 inch) must be specified at the hole-bottom after the spindle orient (this is specified with I-0.001).

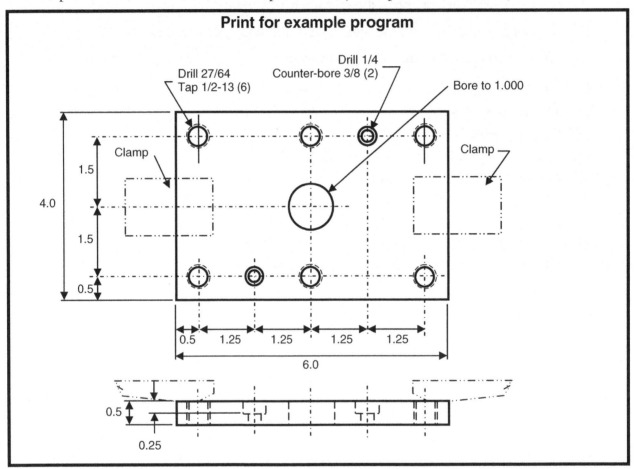

Figure 6.9 Print for canned cycle example program

Process to machine the workpiece:

Seq.	Description	Cycle	Tool	Feed	Speed
1	Center drill all holes	G81	#4 center drill	3.0 ipm	1200 rpm
2	Peck drill 27/64 holes (6)	G73	27/64 drill	5.0 ipm	611 rpm
3	Tap 1/2-13 holes (6)	G84	1/2-13 tap	17.6 ipm	230 rpm
4	Drill 1.000 hole to 31/32	G81	31/32 drill	5.0 ipm	340 rpm
5	Bore 1.000 diameter hole	G76	1.000 boring bar	2.0 ipm	500 rpm
6	Peck drill 1/4 holes (2)	G83	1/4 drill	2.5 ipm	1000 rpm
7	Counter bore 3/8 holes	G82	3/8 end mill	3.0ipm	600 rpm

The process specifies the intended G code type for the canned cycle to be used for each operation. This program stresses the use of as many different canned cycles as possible. You may not agree with the necessity for some of the canned cycle types chosen.

Program:

O0008
(CENTER DRILL)
N005 T01 M06
N010 G54 G90 S1200 M03 T02
N015 G00 X0.5 Y0.5
N020 G43 H01 **Z2.0** (Note two inch high initial plane)
N025 M08
N030 **G81 R0.1 Z-0.2 F3.0 G99** (Center drill hole number one)
N035 X1.75 (#2)
N040 X3.0 (#3)
N045 X5.5 **G98** (Note clamp after hole #4!)
N050 Y3.5 **G99** (Back to R plane! hole #5)
N055 X4.25 (#6)
N060 X3.0 (#7)
N065 X0.5 **G98** (Come up above clamp just in case #8)
N070 X3.0 Y2.0 (#9)
N075 **G80** M09 (Cancel cycle)
N080 G91 G28 Z0 M19
N085 M01

(27/64 DRILL)
N090 T02 M06
N095 G54 G90 S611 M03 T03
N100 G00 X0.5 Y0.5
N105 G43 H02 **Z2.0** (Note two inch high initial plane)
N110 M08
N115 **G73 R0.1 Z-0.75 Q0.1 F5.0 G99** (#1)
N120 X3.0 (#3)
N125 X5.5 **G98** (Clear clamp! #4)
N130 Y3.5 **G99** (#5)
N135 X3.0 (#7)
N140 X0.5 (#8)
N145 **G80** M09 (Cancel cycle)
N150 G91 G28 Z0 M19
N155 M01

(1/2-13 TAP)
N160 T03 M06
N165 G54 G90 S230 M03 T04
N170 G00 X0.5 Y0.5
N175 G43 H03 **Z2.0** (Note two inch high initial plane)
N180 M08

N185 **G84 R0.25 Z-0.65 F17.6 G99** (#1)

N190 X3.0 (#3)

N195 X5.5 **G98** (Note clamp! #4)

N200 Y3.5 **G99** (#5)

N205 X3.0 (#7)

N210 X.5 (#8)

N215 **G80** M09 (Cancel cycle)

N220 G91 G28 Z0 M19

N225 M01

(31/32 DRILL)

N230 T04 M06

N235 G54 G90 S340 M03 T05

N240 G00 X3.0 Y2.0 (#9)

N245 G43 H04 **Z0.1** (No need to clear clamps!)

N250 M08

N255 **G81 Z-0.85 R0.1 F5.0** (Since R plane and initial plane are the same, there is no need for either a G98 or G99)

N260 **G80** M09 (Cancel cycle)

N265 G91 G28 Z0 M19

N270 M01

(1.0000 BORING BAR)

N275 T05 M06

N280 G54 G90 S500 M03 T06

N285 G00 X3.0 Y2.0 (#9)

N290 G43 H05 **Z0.1** (No need to clear clamps!)

N295 M08

N300 **G76 Z-.6 R0.1 F2.0 I-0.001** (Since R plane and initial plane are the same, there is no need for a G98 or G99)

N305 **G80** M09 (Cancel cycle)

N310 G91 G28 Z0 M19

N315 M01

N320 T06 M06

(1/4 DRILL)

N325 G54 G90 S1000 M03 T07

N330 G00 X1.75 Y0.5 (#2)

N335 G43 H06 **Z0.1** (No need to clear clamps)

N340 M08

N345 **G83 Z-0.65 Q0.4 R0.1 F2.5** (Since the R plane and the initial plane are the same, there is no need for G98 or G99)

N350 X4.25 Y3.5 (#6)

N355 **G80** M09 (Cancel cycle)

N360 G91 G28 Z0 M19

N365 M01

```
(3/8 END MILL)
N370 T07 M06
N375 G54 G90 S600 M03 T01
N380 G00 X1.75 Y0.5 (#2)
N385 G43 H07 Z0.1 (No need to clear clamps)
N390 M08
N395 G82 Z-0.25 R0.1 F3.0 P500 (No need to clear clamps, so no need to use G98 or G99)
N400 X4.25 Y3.5 (#6)
N405 G80 M09 (Cancel cycle)
N410 G91 G28 Z0 M19
N415 G28 X0 Y0
N420 M01
N425 M30
```

Notes about the program:

N020, N105, N175

The first three tools need to clear the clamps. Notice that the **G43** command brings the tool up to the initial plane in Z (two inches above the work surface). This is the tool's last position just prior to the canned cycle command.

N030, N115, N185, N255, N300, N344, N395

Notice that the tool is already at the first hole-position when the canned cycle is instated (commanded in each tool's first XY motion). While the X and Y coordinates can be repeated in the canned cycle instating command, there is no need for them – and we recommend leaving them out.

N245, N290, N335

Since only one hole is being machined, there will be no need to clear obstructions, meaning the initial plane can be set the same as the R plane.

N255, N300, N345

Since the R plane and initial plane are the same for these tools, you can leave out the **G98/G99** completely. Though the initial plane (**G98**) will be in effect (it is reinstated at each **G80**), the tool will still come out of the hole to a Z position of **Z0.1**.

N045, N050, N125, N130, N195, N200

Again, **G98** and **G99** are modal. They are only required when you want to switch retract planes. Since they control what happens after the hole is machined, it helps to place them at the end of the command.

N075, N145, N215, N260, N305, N355, N405

G80 is used to cancel each canned cycle.

N255

You may be questioning why a standard drilling cycle (**G81**) is used when only one hole is being machined. You may feel that it is just as easy to program one hole with **G00** and **G01**. The need may arise during the program's verification for another machining method, like peck drilling. We recommend using canned cycles to machine all holes, even when there is only one hole to machine.

N300

We're using the **G76** fine boring cycle. We've said we know that the boring bar will be pointing in the X plus direction when the spindle is at its orient position, so we can include the **I-0.001** word in the **G76** command to make the boring bar move over in the X minus direction at the hole-bottom. If you don't know which way the boring bar will be pointing at the spindle orient position (maybe you are using a collet holder and a straight-shank boring bar), we recommend including I0 and J0 in the G76 command. Once the setup is made and the setup person knows which way the boring bar tip is pointing at the spindle orient position (by testing it), they must modify the G76 command.

N185

If you are using a tension/compression tap holder (your machine does not have rigid tapping), you must keep the rapid plane for tapping well above the work surface (0.25 in is sufficient). When the tap enters the hole, and especially during the spindle reversal at the hole-bottom, the tap will probably extend in the tension/compression holder. If it does, the tap may still be in the hole at the completion of the tapping cycle if a smaller R plane value is used.

Using canned cycles in the incremental positioning mode

You now know the function of canned cycles as they are used in the absolute positioning mode (G90). While we have been stressing the absolute positioning mode throughout this text, there is one benefit to working in the incremental mode (G91) with canned cycles. If you have a series of evenly spaced holes, as is the case with a grid pattern of holes, all of the evenly spaced holes along a line can be machined in one command. As you know, the absolute positioning mode allows but one hole per command to be machined. Consider the grid pattern of holes shown in figure 6.10. In the absolute mode, these one hundred holes require one hundred commands. In the incremental mode, this is reduced to about twenty.

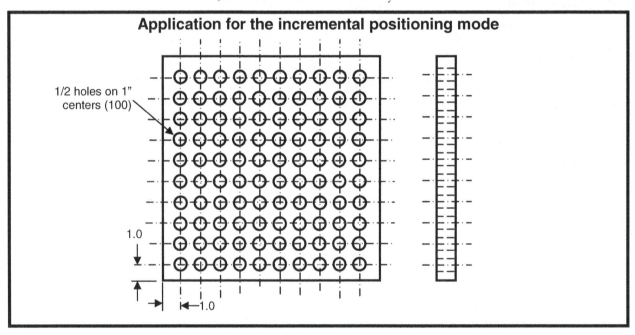

Figure 6.10 – Application for the incremental mode

An *L word* (K word for Fanuc OM and 3M controls) in the canned cycle command is used to specify the number of evenly spaced holes to be machined. If the L word is left out of the command (as is the case when you work in the absolute mode), the control will assume that *one hole* is to be machined.

In order to use canned cycles in the incremental mode, you must understand the meanings of canned cycle-related words in the incremental mode. Here are the canned cycle words that change in the incremental mode:

X is now the distance from the current X axis position to the center of the hole in X.

Y is now the distance from the current Y axis position to the center of the hole in Y.

R is now the distance from the current Z axis position of the tool to the R plane position.

Z is a little different. Z is the distance from the R plane to the bottom of the hole. If the R plane is 0.100 in above the workpiece, the Z value will be the depth of the hole plus 0.100 in. With incrementally commanded canned cycles, *Z is always minus.*

The rest of the canned cycle words (G98, G99, I, K, F, P, and Q) mean exactly the same thing as they do in the absolute mode.

Here is an example. You should be impressed with how few commands this grid pattern requires. Figure 6.10 shows the workpiece to be machined. The program will be drilling one hundred of 0.500 diameter holes through a 1.0 inch thick workpiece.

Program with comments:

```
O0001
N005 G54 G90 S600 M03 (1/2 drill)
N010 G00 X1.0 Y1.0 (Position of the lower-left hole)
N015 G43 H01 Z2.0 (Sets initial plane)
N020 M08
N025 G91 G81 R-1.9 Z-1.28 F5.0 G99 (Machine lower left hole in incremental mode– note R and
     Z values)
N030 X1.0 L9 (Machine first row of holes)
N035 Y1.0 (Machines first hole in second row)
N040 X-1.0 L9 (Machines nine holes in second row)
N045 Y1.0 (Machine first hole in third row)
N050 X1.0 L9 (Machine third row)
N055 Y1.0
N060 X-1.0 L9 (Fourth row)
N065 Y1.0
N070 X1.0 L9 (Fifth row)
N075 Y1.0
N080 X-1.0 L9 (Sixth row)
N085 Y1.0
N090 X1.0 L9 (Seventh row)
N095 Y1.0
N100 X-1.0 L9 (Eighth row)
N105 Y1.0
N110 X1.0 L9 (Ninth row)
N115 Y1.0
N120 X-1.0 L9 (Tenth row)
N125 G80 M09
N130 G91 G28 X0 Y0 Z0 M19
N135 M30
```

As you can see, this technique dramatically reduces the number of commands needed to machine the holes (from one hundred to about twenty). In line N010, we position the drill to the first hole to machine (the lower left hole). In line N025, the first hole is machined – notice the incremental specifications for R and Z. And as with the absolute mode, if the drill is already at the hole-position in XY, you need not include the XY specifications in this command (if you do, they will be X0 and Y0). Line N030 machines the rest of the holes in the first row (notice the **L9**). Line **N035** machines the first hole in the second row. Line **N040** machines the rest of the holes in the second row. This is repeated for all of the holes in the pattern.

Key points for Lesson Sixteen:

- Hole-machining canned cycles dramatically simplify the programming of hole-machining operations.
- With canned cycles, you instate the cycle in a command that machines the first hole. You then list the coordinates for the rest of the holes to be machined. Finally, you cancel the cycle.
- There are a variety of hole-machining styles that are selected by G codes.
- Canned cycles provide a way to clear obstructions between holes using two planes – the R plane (specified by **G99**) and the initial plane (specified by **G98**).
- The initial plane is the tool's last Z position prior to the canned cycle command.
- The R and Z words are absolute positions that can be changed for different Z surfaces.
- When you must machine a number of evenly spaced holes, you can use the incremental mode to minimize the number of required commands.

Write your program on a separate sheet of paper.

Practice with canned cycles

Process:

1) Mill (2) 0.375 radius slots	3/4 end mill	500 rpm	6.0 ipm	
2) Center drill all holes	#4 center drill	1,200 rpm	4.0 ipm	use G81
3) Drill (4) 1/4 holes	1/4 drill	1100 rpm	3.5 ipm	use G83
4) Counter-bore (4) 1/2 holes	1/2 end mill	800 rpm	4.5 ipm	use G82
5) Drill (2) 15/64 holes	15/64 drill	1,150 rpm	3.4 ipm	use G83
6) Ream (2) 0.25 holes	0.250 reamer	800 rpm	4.5 ipm	use G81
7) Drill (2) 5/16 holes	5/16 drill	700 rpm	5.0 ipm	use G73
8) Tap (2) 3/8-16 holes	3/8-16 tap	400 rpm	25.0 ipm	use G84

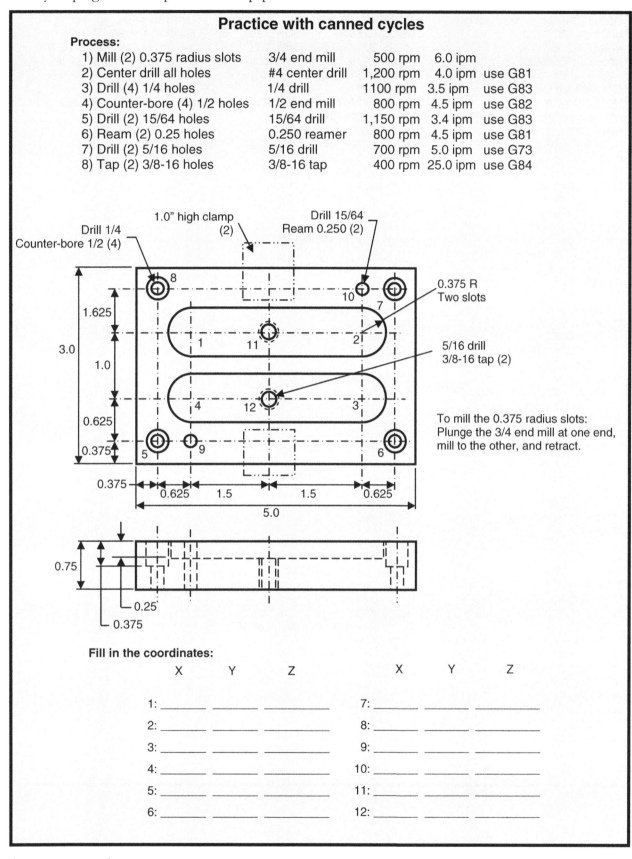

To mill the 0.375 radius slots: Plunge the 3/4 end mill at one end, mill to the other, and retract.

Fill in the coordinates:

	X	Y	Z		X	Y	Z
1:	_____	_____	_____	7:	_____	_____	_____
2:	_____	_____	_____	8:	_____	_____	_____
3:	_____	_____	_____	9:	_____	_____	_____
4:	_____	_____	_____	10:	_____	_____	_____
5:	_____	_____	_____	11:	_____	_____	_____
6:	_____	_____	_____	12:	_____	_____	_____

Answer program is on next page.

Answer program for canned cycles:

O0005 (Canned cycles practice)
N001 G17 G20 G23
N002 G40 G50 G64
N003 G67 G69 G80

(3/4 END MILL)
N005 T01 M06
N010 G54 G90 S500 M03 T02
N015 G00 X1.0 Y2.0 (pt 1)
N020 G43 H01 Z0.1
N025 M08
N030 G01 Z-0.25 F6.0
N035 X4.0 (pt 2)
N040 G00 Z0.1
N045 Y1.0 (pt 3)
N050 G01 Z-0.25
N055 X1.0 (pt 4)
N050 G00 Z0.1 M09
N055 G91 G28 Z0 M19
N060 M01

(CENTER DRILL)
N065 T02 M06
N070 G54 G90 S1200 M03 T03
N075 G00 X0.375 Y0.375 (pt 5)
N080 G43 H02 Z2.0
N085 M08
N090 G81 R0.1 Z-0.12 F4.0 G99
N095 X1.0 G98 (pt 9)
N100 X4.625 G99 (pt 6)
N105 Y2.625 (pt 7)
N110 X4.0 G98 (pt 10)
N115 X0.375 G99 (pt 8)
N120 X2.5 Y2.0 R-0.15 Z-0.37 G98 (11)
N125 Y1.0
N130 G80 M09
N135 G91 G28 Z0 M19
N140 M01

(1/4 DRILL)
N145 T03 M06
N150 G54 G90 S1100 M03 T04
N155 G00 X0.375 Y0.375 (pt 5)
N155 G43 H03 Z2.0
N160 M08
N165 G83 R0.1 Z-0.855 Q0.5 F3.5 G98
N170 X4.625 G99 (pt 6)
N175 Y2.625 G98 (pt 7)
N180 X0.375
N185 G80 M09
N190 G91 G28 Z0 M19
N195 M01

(1/2 END MILL)
N200 T04 M06
N005 G54 G90 S800 M03 T05
N210 G00 X0.375 Y0.375 (pt 5)
N215 G43 H04 Z2.0
N220 M08
N225 G82 R0.1 Z-0.375 P500 F4.5 G98
N230 X4.625 G99 (pt 6)
N235 Y2.625 G98 (pt 7)
N240 X0.375 (pt 8)
N245 G80 M09
N250 G91 G28 Z0 M19
N255 M01

(15/64 DRILL)
N260 T05 M06
N265 G54 G90 S1150 M03 T06
N270 G00 X1.0 Y0.375 (pt 9)
N275 G43 H05 Z2.0
N280 M08
N285 G83 R0.1 Z-0.855 Q0.5 F3.4 G98
N290 X4.0 Y2.625 (10)
N295 G80 M09
N300 G91 G28 Z0 M19
N305 M01

(0.250 REAMER)
N310 T06 M06
N315 G54 G90 S800 M03 T07
N320 G00 X1.0 Y0.375 (pt 9)
N325 G43 H06 Z2.0
N330 M08
N335 G81 R0.1 Z-0.8 F4.5 G98
N340 X4.0 Y2.625 (pt 10)
N345 G80 M09
N350 G91 G28 Z0 M19
N355 M01

(5/16 DRILL)
N360 T07 M06
N365 G54 G90 S700 M03 T08
N370 G00 X2.5 Y1.0 (pt 12)
N375 G43 H01 Z0.1
N380 M08
N385 G81 R-0.15 Z-0.87 F5.0 G98
N390 Y2.0 (pt 11)
N395 G80 M09
N400 M01

(3/8-16 TAP)
N405 T08 M06
N410 G54 G90 S400 M03 T01
N415 G00 X2.5 Y1.0 (pt 12)
N420 G43 H08 Z0.1
N425 M08
N430 G84 R0 Z-1.0 F25.0 G98
N435 Y2.0 (pt 11)
N440 G80 M09
N445 G91 G28 Z0 M19
N450 G28 Y0
N455 M30

Lesson 17
Working With Subprograms

There are times when a series of CNC commands must be repeated – within one program – and sometimes among several programs. Whenever you find yourself writing a series of commands a second time – you should consider using a subprogram. The longer the series of commands and the more often they must be repeated, the more a subprogram can help.

You know that a CNC machining center will execute a program in sequential order. It will start with the first command in the program: read it – interpret it – and execute it. Then it will move on to the next command – read, interpret, execute. It will continue this process for the entire program.

You also know that there are times when a series of commands in your program must be repeated. We've shown two times so far. One time is when using cutter radius compensation to rough and finish mill using the same set of coordinates. You must program the finishing tool path coordinates twice, once for the rough milling cutter (using an offset value that is larger than the actual cutter size) and once for the finish milling cutter.

Another time when commands must be repeated is when machining holes that require multiple machining operations (like center drilling, drilling, and tapping a series of holes) – the more holes that must be machined, the more commands that must be repeated.

In Lesson Seventeen, you will learn about a way to change the order of program execution to some extent – and this will be especially helpful when commands must be repeated.

The difference between main- and sub- programs
A *main program* is the program that a setup person or operator will execute when they activate a cycle. *Every program shown to this point in the text has been a main program.* Main programs almost always end with M30 (or M02 with some machines).

A *subprogram* is a program that is invoked by a main program or by another subprogram. A subprogram ends with M99.

In a main program, when you reach a series of commands that you know will be repeated, you can invoke a subprogram that contains the repeated commands (you place the repeated commands in the subprogram instead of in the main program). Each time the commands must be repeated, you will invoke the same subprogram.

M98 is the word used to invoke a subprogram. A *P word* in the M98 command specifies the program number of the subprogram to be executed. When the machine is finished executing the subprogram, it will come back to the main program to the command after the calling M98.

Figures 6.11 and 6.12 show a simple application for a subprogram.

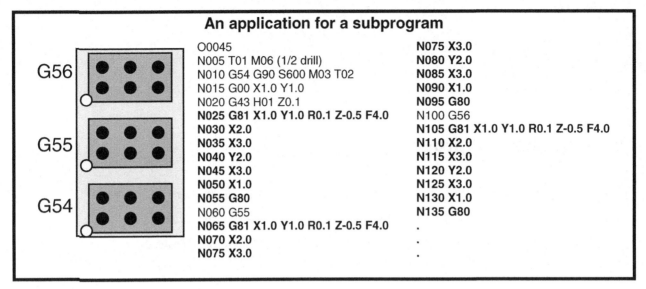

An application for a subprogram

O0045	N075 X3.0
N005 T01 M06 (1/2 drill)	N080 Y2.0
N010 G54 G90 S600 M03 T02	N085 X3.0
N015 G00 X1.0 Y1.0	N090 X1.0
N020 G43 H01 Z0.1	N095 G80
N025 G81 X1.0 Y1.0 R0.1 Z-0.5 F4.0	N100 G56
N030 X2.0	**N105 G81 X1.0 Y1.0 R0.1 Z-0.5 F4.0**
N035 X3.0	**N110 X2.0**
N040 Y2.0	**N115 X3.0**
N045 X3.0	**N120 Y2.0**
N050 X1.0	**N125 X3.0**
N055 G80	**N130 X1.0**
N060 G55	**N135 G80**
N065 G81 X1.0 Y1.0 R0.1 Z-0.5 F4.0	.
N070 X2.0	.
N075 X3.0	.

Figure 6.11 – Application for subprogram usage

In this example, three identical workpieces are being machined. The lower workpiece uses fixture offset number one, the middle workpiece uses fixture offset number two, and the upper workpiece uses fixture offset number three. In the program, after selecting the fixture offset (in lines **N010**, **N060**, and **N100**), the commands for the machining operation are specified (drilling six holes in this example). Notice that these are *identical* commands for each workpiece. So lines **N025** through **N055** are specified three times – once for each workpiece.

Now look at Figure 6.12.

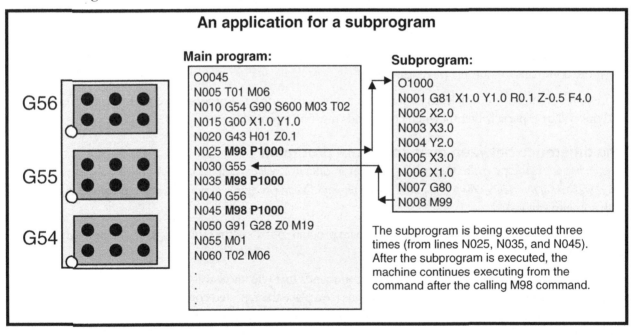

An application for a subprogram

Main program:

```
O0045
N005 T01 M06
N010 G54 G90 S600 M03 T02
N015 G00 X1.0 Y1.0
N020 G43 H01 Z0.1
N025 M98 P1000
N030 G55
N035 M98 P1000
N040 G56
N045 M98 P1000
N050 G91 G28 Z0 M19
N055 M01
N060 T02 M06
.
.
.
```

Subprogram:

```
O1000
N001 G81 X1.0 Y1.0 R0.1 Z-0.5 F4.0
N002 X2.0
N003 X3.0
N004 Y2.0
N005 X3.0
N006 X1.0
N007 G80
N008 M99
```

The subprogram is being executed three times (from lines N025, N035, and N045). After the subprogram is executed, the machine continues executing from the command after the calling M98 command.

Figure 6.12 – Now a subprogram is being used

This time we've placed all of the repeated commands in a *subprogram*. When these commands must be executed, an **M98** command is given with a P word specifying the program number for the subprogram. While this may not be the best application for a subprogram (there aren't many commands being repeated), it should nicely illustrate how subprograms can be used.

In line **N025**, the machine will execute the subprogram for the first time. The holes in the lower workpiece will be drilled. When the machine is finished executing the subprogram (notice the **M99** in line **N008** of the subprogram), it returns to the main program – to the command after the calling M98 (line N030). Line N030 selects fixture offset number two. Line N035 tells the machine to execute the subprogram again, causing the holes in the middle workpiece to be drilled. When finished, the machine will return (this time) to line N040. Fixture offset number three is selected, and in line N045, the subprogram is executed a third time – machining the holes in the upper workpiece.

A note about the P word: Look at the P word in lines N025, N035, and N045 in Figure 6.12. It is specified as P1000 (P one, zero, zero, zero) even though it is calling program number O1000. A common mistake is to *incorrectly* specify the P word as PO1000 (P oh, one, zero, zero, zero). This mistake will generate an alarm.

Subprograms provide two important benefits – one rather obvious one and the other not so obvious. The obvious benefit is that subprograms will minimize the number of commands that must be specified in order to machine a workpiece. So programming becomes simpler and programs become shorter.

A second, less obvious, benefit of using a subprogram has to do with program verification. If the subprogram is correct the first time it is executed, it will be correct *every time* it is executed. Consider the program in Figure 6.11. The commands to machine the six holes are specified *three different times.* Any one of these times you could make a mistake – either writing or typing the commands. When verifying the program, you must be very careful with them – during the machining of all three workpieces.

While you must still be careful when verifying the programs in Figure 6.12, if the holes are machined properly in the first workpiece, you can rest assured that they will also be machined properly in the second and third workpieces – *the same commands are used all three times.*

Words used with subprograms

Let's summarize the words used with subprograms.

> M98 – This word is used to call a subprogram from a main program – or another subprogram
>
> P word – This word specifies the subprogram to be called
>
> M99 – This word is used to end a subprogram – machine will return to the calling program to the command after the calling M98
>
> L word (part of the P word with Fanuc 0M and 3M controls) – if included in the M98 command, this word specifies the number of times the subprogram will be executed.

We've presented all but the *L word.* This word specifies how many times you want the subprogram executed. If left out of the M98 command (as in our examples so far), the subprogram will be executed once. With Fanuc 0M and 3M controls, the number of executions is part of the P word. Both of these commands will execute subprogram O1000 five times:

> N050 M98 P1000 L5 (for almost all Fanuc controls)
>
> N050 M98 P0051000 (for 0M and 3M Fanuc controls)

For Fanuc 0M and 3M controls, the first three digits of the P word specify the number of executions and the last four digits specify the program number. Note that with *all* Fanuc controls, the command

> N050 M98 P1000

will call subprogram O1000 and execute it once.

Know Your Machine From An Operator's Viewpoint

Key Concept Number Seven begins the setup and operation portion of this text. It parallels Key Concept Number One - but in Key Concept Number Seven, we'll look at a machining center from a setup person's or operator's perspective.

Key Concept Number Seven contains two lessons:
>   20: Tasks related to setup and running production
>   21: Buttons and switches on the control panels

We now begin discussions related to setup and operation. Throughout these discussions, we will be assuming that you have read and understand certain presentations from the programming-related lessons (in Lessons One through Nineteen). There are *many* setup- and operation-points made during the programming-related lessons. Without an understanding of these important points, much of what is presented from this point in the text will make little sense.

The need for hands-on experience

Hands-on experience is extremely important to fully mastering CNC machining center setup and operation. But unfortunately, no text can provide hands-on experience. If you are using this text as part of a course you are taking at a technical school, hopefully there are machines available for practicing what you learn in class. *We cannot overstate the importance of hands-on experience* – experience we cannot provide in this text.

That said, we can provide *all of the principles* needed to set up and run CNC machining centers. As a comparison, think of what it takes to learn to fly an airplane. A future pilot must, of course, spend a great deal of time in the cockpit at the flight controls in order to master skills needed to become a pilot. But *before* a person begins hands-on flight training, they *must* understand the basics of aerodynamics and flight. Without this understanding, controls in the cockpit will have no meaning. A flight student must first attend *ground school*. In ground school they will learn concepts that help understand the controls in the cockpit.

In similar fashion, a CNC setup person and/or operator *must* spend time at the CNC machine to fully master tasks related to setting and running production on a CNC machining center. But before a person can spend any meaningful time at the machine, they *must* understand concepts related to setting up and running the CNC machine. Without an understanding of the material presented from this point in the text, machine functions will have no meaning.

Think of Key Concepts seven through ten in this text as a kind of *ground school for learning to setup and run CNC machining centers.* While you will not gain the important hands-on experience you'll need to fully master running a CNC machining center from this text alone, you will learn the *concepts* required to begin practicing. An understanding of these concepts is mandatory *before* you begin working at the machine.

Procedures you must know about

Setting up and operating a CNC machining center requires that you master some basic procedures. While we show *step-by-step* procedures in Lesson Twenty-Three, we want to acquaint you with some of the most rudimentary machine-usage procedures and describe when they are required. We'll be referring to these procedures throughout Key Concepts Seven through Ten.

Machine power-up

You must, of course, be able to turn on your machining center/s. This procedure usually involves turning on a main power breaker (a switch that is usually located on the back of the machine), then pressing the control *power on* button, and finally pressing a button to turn on the machine's hydraulic system (if it has one).

Sending the machine to its zero return position

Most machining centers require that you perform this procedure when the machine is first powered up. You will also have to send the Z axis to its zero return position (the tool change position) for vertical machining centers prior to running a program.

This procedure varies from one machine tool builder to another. With some machines, it is as simple as selecting the zero return mode and pressing one button which causes each axis (first Z, then X, then Y) to move to its zero return position. With other (indeed most) machines, you must send each axis to it zero return position independently. After selecting the zero return mode, you select the axis you want to zero return first (usually Z for vertical machining centers). Next you press the plus button and hold it. The selected axis will rapid until it comes to within an inch or so of its zero return position. Then it will slow down, eventually reaching the zero return position. You can tell when this occurs, since an axis origin light will come for the axis as it reaches the zero return position. This must be repeated for the other axes.

Manually moving each axis

Manually moving an axis is often required. Some examples of when you must manually move an axis include measuring program zero assignment values, measuring tool length compensation values on the machine, and manually positioning a cutting tool into a position so that it can be inspected or replaced. Most machine tool builders provide you with two ways to manually move an axis (most machines allow you to manually move only one axis at a time).

One way to move an axis is to use the *jog* function of the machine. This allows rather crude control of motion, and will be used when there is no danger of an axis component colliding into an obstruction. To jog an axis, first you select the jog mode (also called the manual mode). Next you select the axis you want to move (X, Y, or Z). You then select the motion rate with a multi-position switch (commonly called jog feedrate). And finally, you activate the axis motion in the desired direction (plus or minus). With most machines there are two buttons used for this purpose, one labeled plus – the other labeled minus. As long as you hold the button in, the axis will move. Also, you can manipulate the jog feedrate switch to speed up or slow down the motion.

Another way to move an axis is to use the *handwheel*. This provides a much more precise way to move an axis. It is commonly required when there is danger of an axis component (like a cutting tool) colliding with an obstruction. With the handwheel, you can control how quickly the machine will move when you turn the handwheel. In the times-one (**X1**) position, each increment of the handwheel will be 0.0001 inch (or 0.001 mm). In the times-ten (**X10**) position, each increment will be 0.001 inch (or 0.010 mm). In the times one-hundred position (**X100**), each increment will be 0.010 inch (or 0.1 mm). To use the handwheel, first you select the handwheel mode, then you select the axis you want to move (X, Y, or Z). Finally, you select the rate of motion (**X1**, **X10**, or **X100**). When you turn the handwheel in the desired direction (plus or minus), the axis will move.

Manually starting the spindle

This procedure will be required when using a conventional edge-finder to measure program zero assignment values as shown in Lesson Five and if you will be performing any manual machining operations.

Machine tool builders vary when it comes to how much manual control they provide for the spindle. With some machines, you can simply select the manual mode, set a multi-position switch to select the desired spindle speed, select the desired direction (with a two-position switch), and press a button to activate the spindle. Another button can be pressed to turn off the spindle.

If your machine does not provide manual control for the spindle, you must use a function called *manual data input* (MDI) and enter and execute a program-like command to start the spindle (like **S500 M03**). Another command must be entered and executed to stop the spindle (**M05**). We'll discuss the use of manual data input in Lesson Twenty-Two.

Manually making tool changes

Machine tool builders vary when it comes to how much manual control they provide for the automatic tool changer. Frankly speaking, few will allow you to make tool changes in a completely manual fashion. Almost all will require that you use manual data input (MDI) and enter and execute a program-like command to change tools (like **T04 M06**). The machine must be at its tool change position before a tool change can be commanded (usually the Z axis zero return position). We'll discuss the use of manual data input in Lesson Twenty-Two.

Manipulating the display screen

The display screen will show you a great deal of important information. For setup people and operators, the four most important display screen mode are *position*, *program*, *offset*, and *program check*.

The position display screen

As you would expect, in position mode, the display screen mode shows information about machine position. Four position pages are involved. The *absolute* position display page shows the machine's current position relative to program zero. The *relative* position display page is used to take measurements and can be set or reset to specify a point of origin for the measurement. The *machine* position display page shows the machine's current position relative to the zero return position. And a fourth page shows a combination of the previous three – along with a special *distance-to-go* position display that lets you see how much further the machine will move in the current command.

More about the relative position display page
Again, *the relative position display page is the page you will use to take measurements*. In step three of the tool length measuring procedure shown above, you must reset (set to zero) the Z axis relative position display. This sets a point of reference for tool length measurements that follow. The relative position display is the *only* position page that can be set or reset in this manner.

More about distance-to-go
The multi-position display page includes a *distance-to-go* display. This is a very important function – one that is also displayed on the *program check* display screen. When you are verifying a program, it is often helpful to know just how much further the machine will move in the current command. Say a tool is making its first approach movement in the Z axis. You know the tool is *supposed* to stop 0.1 inch above the work surface. But you're worried. With *dry run* and *single block* turned on (two other program verification functions), you will have total control of the machine's motion rates (including rapid) with a multi-position switch (commonly the feedrate override switch). As the tool approaches you slow motion with feedrate override. When the tool is about 0.5 inch away from the work surface, you're getting worried, so you press the *feed hold* button to stop axis motion (feed hold is yet another program verification function). When you look at the *distance-to-go* display, the value shown in the Z axis should be under 0.5 inch. If it is not, the current command will cause the tool to crash into the work surface (and, of course, you've found a problem with the program or setup).

The program display screen

In edit or memory mode, this display screen shows you the active program. If you are running the program, this will allow you to see up-coming commands. If you need to modify the program, this display provides you with a kind of text editor. In manual data input (MDI) mode, this display screen shows the page used to enter MDI commands.

Lesson 18

Tasks Related To Setup And Running Production

We define setup time as the total time a machine is down between production runs. We define cycle time as the total time it takes to complete a production run divided by the number of good workpieces produced. When you think about it, there are only two general tasks that occur on CNC machining centers – machines are either in setup or they are running production.

It is important to understand the distinction between making a setup and running production. The tasks you perform during setup are *getting the machine ready* to run production. Only when the setup is completed and a workpiece has passed inspection is it possible to run production. The person making the setup is called the *setup person* – the person completing the production run is called the *CNC operator* (though in many companies, one person makes the setup *and* runs production).

Setup-related tasks

- Tear down previous setup an put everything away
- Gather the components needed to make the setup
- Make the workholding setup
- Assign program zero point/s
- Assemble cutting tools
- Measure and enter tool length compensation values
- Measure and enter cutter radius compensation values
- Load cutting tools into the machine's automatic tool changer
- Load the CNC program/s
- Verify the correctness of the program and setup
- Cautiously run the first workpiece – ensure that it passes inspection
- If necessary, optimize the program for better efficiency (new programs only)
- If changes to the program have been made, save corrected version of the program

Running production-related tasks

Done during every cycle:

- Load a workpiece
- Activate the cycle
- Monitor the cycle to ensure that cutting tools are machining properly (first few workpieces only)
- Remove the workpiece
- Clean/de-burr the workpiece
- Perform specified measurements (if required)
- Report measurement results to statistical process control (SPC) system

Not required in every cycle:

- Make offset adjustments to maintain size for critical dimensions
- Replace worn tools
- Remove chips from work area (if required)

In Lesson Twenty, we'll be going through this list of tasks and describing each one in detail.

There are certain setup-related tasks that must sometimes be repeated during a production run. If for example, a cutting tool gets dull and must be replaced, the same tasks required to initially assemble it, measure it, and enter offsets for it must be repeated.

Many of these tasks, of course, draw upon your basic machining practice skills. Tasks related to making workholding setups and assembling cutting tools, for example, require that you understand workholding- and cutting-tool-components. And these tasks are identical to those that must be performed on conventional (not CNC) machine tools. So, as with programming, setting up and operating a CNC machining center require a firm understanding of basic machining practices. And also as with programming, if you have experience working with conventional machine tools, you have a head start for mastering CNC machining center setup and operation.

As you look through the list of tasks shown above, notice how have been introduced during our discussions of programming. While we're going to expand our discussions from this point on, you have already been exposed to many setup- and operation-related concepts.

This lesson is truly at the heart of the setup and operation presentations in this text. It presents the most important concepts you must understand in order to be a successful setup person or operator.

A CNC job from start to finish

Figure 7.5 shows the job we're going to use to discuss every task related to setup and maintaining production. We're going to be describing this job as if we're actually running it. While this isn't as good as actually demonstrating a job being run, it's the next best thing.

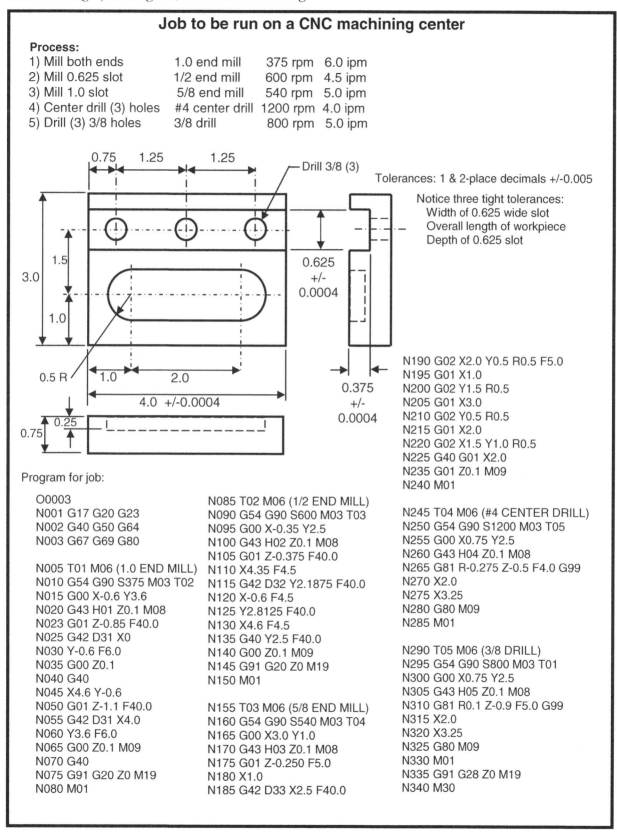

Job to be run on a CNC machining center

Process:

1) Mill both ends	1.0 end mill	375 rpm	6.0 ipm
2) Mill 0.625 slot	1/2 end mill	600 rpm	4.5 ipm
3) Mill 1.0 slot	5/8 end mill	540 rpm	5.0 ipm
4) Center drill (3) holes	#4 center drill	1200 rpm	4.0 ipm
5) Drill (3) 3/8 holes	3/8 drill	800 rpm	5.0 ipm

Tolerances: 1 & 2-place decimals +/-0.005

Notice three tight tolerances:
Width of 0.625 wide slot
Overall length of workpiece
Depth of 0.625 slot

Drill 3/8 (3)

0.75 1.25 1.25

3.0 1.5 1.0

0.5 R

1.0 2.0

4.0 +/-0.0004

0.25 0.75

0.625 +/- 0.0004

0.375 +/- 0.0004

Program for job:

```
O0003
N001 G17 G20 G23
N002 G40 G50 G64
N003 G67 G69 G80

N005 T01 M06 (1.0 END MILL)
N010 G54 G90 S375 M03 T02
N015 G00 X-0.6 Y3.6
N020 G43 H01 Z0.1 M08
N023 G01 Z-0.85 F40.0
N025 G42 D31 X0
N030 Y-0.6 F6.0
N035 G00 Z0.1
N040 G40
N045 X4.6 Y-0.6
N050 G01 Z-1.1 F40.0
N055 G42 D31 X4.0
N060 Y3.6 F6.0
N065 G00 Z0.1 M09
N070 G40
N075 G91 G20 Z0 M19
N080 M01
```

```
N085 T02 M06 (1/2 END MILL)
N090 G54 G90 S600 M03 T03
N095 G00 X-0.35 Y2.5
N100 G43 H02 Z0.1 M08
N105 G01 Z-0.375 F40.0
N110 X4.35 F4.5
N115 G42 D32 Y2.1875 F40.0
N120 X-0.6 F4.5
N125 Y2.8125 F40.0
N130 X4.6 F4.5
N135 G40 Y2.5 F40.0
N140 G00 Z0.1 M09
N145 G91 G20 Z0 M19
N150 M01

N155 T03 M06 (5/8 END MILL)
N160 G54 G90 S540 M03 T04
N165 G00 X3.0 Y1.0
N170 G43 H03 Z0.1 M08
N175 G01 Z-0.250 F5.0
N180 X1.0
N185 G42 D33 X2.5 F40.0
```

```
N190 G02 X2.0 Y0.5 R0.5 F5.0
N195 G01 X1.0
N200 G02 Y1.5 R0.5
N205 G01 X3.0
N210 G02 Y0.5 R0.5
N215 G01 X2.0
N220 G02 X1.5 Y1.0 R0.5
N225 G40 G01 X2.0
N235 G01 Z0.1 M09
N240 M01

N245 T04 M06 (#4 CENTER DRILL)
N250 G54 G90 S1200 M03 T05
N255 G00 X0.75 Y2.5
N260 G43 H04 Z0.1 M08
N265 G81 R-0.275 Z-0.5 F4.0 G99
N270 X2.0
N275 X3.25
N280 G80 M09
N285 M01

N290 T05 M06 (3/8 DRILL)
N295 G54 G90 S800 M03 T01
N300 G00 X0.75 Y2.5
N305 G43 H05 Z0.1 M08
N310 G81 R0.1 Z-0.9 F5.0 G99
N315 X2.0
N320 X3.25
N325 G80 M09
N330 M01
N335 G91 G28 Z0 M19
N340 M30
```

Figure 7.5 – A job that must be run on a CNC machining center

This workpiece is the practice exercise from Lesson Twelve with three holes added. Notice that there are three tight tolerances specified on the print. The process is shown, providing you with the machining order used in the program. All coordinates in the program have been specified as mean values for each tolerance.

Notice that the program is not documented nearly as well as most we have shown in this text. Most programmers don't provide much documentation in the program (remember – documentation takes up memory space in the machine) – so this is typical of what you'll see with the jobs you'll be running. Tool naming messages are included in the program at the beginning of each tool. This program uses the format for a vertical machining center shown in Lesson Fifteen.

Again, we'll use this job to describe the various tasks that must be completed in order to make a setup and complete a production run. Our descriptions will assume, however, that you have some basic machining practice experience. Be sure to question an experienced person in your company if any of these presentations require more basic machining practice skills than you currently possess.

Setup documentation

As described in Lesson Eight, a programmer must provide setup documentation. Most programmers will use a one-page setup sheet to describe everything that must be done to make the setup. Figure 7.6 shows the setup sheet for our example job. Most companies also supply the setup person with a copy of the print and a print-out of the CNC program.

Part No. A-5487		Program No.: O0003			Setup Sheet
Part Name: Bracket		Programmer: ML			
Machine: Mori MV-40		Date: 09/20/03			

Station	Tool	Offsets	Notes	Instructions:
1	1.0" end mill	L:1 R:31	min: 0.75", max: 1.2"	Assemble & measure cutting tools and load into specified stations. Enter offsets.
2	1/2" end mill	L:2 R:32	min: 0.375", max: 0.55"	
3	5/8" end mill	L:3 R:33	min: 0.5", max: 0.65"	Mount 3" vise in center of table and align. Workpiece must be centered in vise when loaded (raw material length is 4.2").
4	#4 center drill	L:4		
5	3/8 drill	L:5		
6				Measure program zero assignment values and enter them into fixture offset number one. Note that there is 0.1" stock to be removed from each end and that program zero in X is the left end of the *finished* workpiece. In Z, program zero is the top surface of the workpiece
7				
8				
9				
10				
11				
12				
13				**Workholding setup:**
14				
15				
16				
17				
18				
19				
20				

Notes about this job:

Since both ends of this workpiece will be machined, there is no end stop. The workpiece must be centered in the vise when loaded to ensure that the same amount of material will be removed on each side.

Workholding setup: 3" vise, Workpiece, Program zero

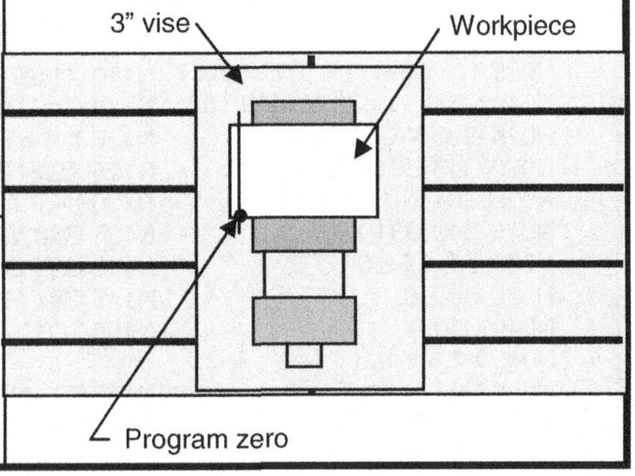

Figure 7.6 – Setup sheet for our example job

As is typical of setup documentation provided by most programmers, our setup sheet does not explain every detail of what must be done to get the job up and running. It assumes quite a bit of the setup person who will be making the setup. Many setup-related tasks are quite redundant from one setup to the next – and most programmers won't be very specific about the most redundant tasks. We'll point out common shortcomings of setup documentation as we continue with this discussion.

Tear down the previous setup and put everything away

We now begin describing the specific tasks you must complete in order to make the setup for the example job. We'll do so in the approximate order that setups are normally made.

Before you can begin making a new setup, you must remove components from the previous setup. Tasks include removing cutting tools that won't be used in the next job, removing the workholding device (assuming it's not used in the next job), and cleaning chips and debris from the machine table. The old CNC program should also be removed from the machine's memory (after saving a corrected version – if required). Once removed, tooling components should be put back where they belong – so you can easily find them in the future.

Gather the components needed to make the setup

The more organized you can be when making a setup, the easier the setup will be to make. Part of being organized is gathering the components you'll need to make the setup. Gathering everything at one time will keep you from having to walk to and from the tool crib every time you need something.

Companies vary with regard to how well they list the components needed for a given setup. Our setup sheet isn't very good in this regard. For example, we're assuming that the setup person can figure out which components are needed to assemble the cutting tools for the job. Indeed – we're assuming they can determine which *cutting tools* will be used. This kind of setup sheet is very common in workpiece- and tooling-producing companies – when the setup person is expected possess a high degree of skill.

Some companies do much better, providing a complete list of every needed component, including workholding components, cutting tool components, hardware (fasteners like tee-nuts and bolts), gauging tools and even hand tools. This makes the gathering of needed components much easier. With this kind of list, someone other than the (highly skilled) setup person will be able to gather components. This will free up the setup person to be doing other things while components are being gathered. This kind of component list is common in product-producing companies.

Make the workholding setup

Workholding devices vary from setup to setup. They commonly include fixtures, vises, and indexing heads. Our example job requires a table vise. Making a workholding setup for a CNC machining center requires the same skills needed to make workholding setups for conventional milling machines. If you've made workholding setups for manual mills, you already know what it takes to do so for CNC machining centers.

For our example job, the workholding setup requires mounting a common table vise on the machine table, like the one shown in Figure 7.7.

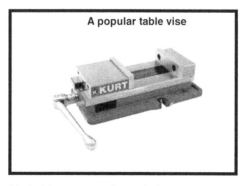

A popular table vise

Figure 7.7 – Table vise used to hold our example workpiece

We're using a *3" table vise*. This means the length of the jaws will be three inches. Our 4.2 inch workpiece will overhang these jaws by 0.6 inch on each end. Though the vise in Figure 7.7 doesn't show it, the jaws in most table vises have a small step machined to grip on a workpiece. Without this step, the workpiece must be placed on *parallels*.

While our example job doesn't need one, many vise setups require an end stop like the one shown in Figure 7.8. This end stop is held to the side of the vise with a magnet. Also, notice the machined steps in the jaws for workpiece gripping.

Figure 7.8 – Easily adjusted end stop for a table vise

After mounting the table vise on the table, the setup person will make sure it is *square with the table*. They'll mount a dial indicator in the spindle and run the stylus of the dial indicator along the fixed jaw (by using the handwheel in the X axis). If necessary, they will move the vise so that the jaws are perfectly parallel with the X axis. Ask an experienced person to show the squaring of a table vise if you've never seen it done.

While many vise setups require an end stop to locate one end of the workpiece, our workpiece requires the machining of *both ends*. For this reason, the workpiece will simply be centered in the vise. While this may be possible by estimating, the setup person and operator will probably use a six inch rule (scale) to be more accurate. For our 4.2 inch long piece of raw material, 0.6 inch of the workpiece must be protruding from each side of the three inch long vise jaws.

Remember that workholding setups vary from job to job. The setup documentation must be sufficient to describe how the workholding setup must be made – and directed at the lowest skill level of setup person in the company.

Assign the program zero point
In lessons four, five, and six, we describe program zero and how it is assigned. Remember that for qualified setups, program zero assignment may be done by the program, meaning the setup person may not have to assign program zero.

With our example job, there is only one program zero point. It is located at the lower-left corner of the finished workpiece in X and Y and at the top surface of the workpiece in Z. Our setup instructions specify that program zero assignment values must be placed in fixture offset number one.

Measure program zero assignment values
This vise setup is not qualified. That is, the setup instructions do not specify which table slots should be used in which to mount the vise. Indeed, this vise may not even have location keys that locate in table slots. (It surely does not if it must be squared with the table.) This means that program zero assignment values must be *measured* by the setup person every time this job is run.

The general procedures to measure program zero assignment values and enter them into fixture offsets is shown in lessons five and six. We'll review them here, showing specific techniques for our example job.

For the X and Y axis program zero assignment values, the setup person will use an edge-finder (like the one shown in Figure 7.9). Also, the work holding setup must be made and a workpiece must be loaded prior to performing this procedure.

Figure 7.9 – An edge-finder (also called a *wiggler*)

1) The edge-finder is placed in the spindle with the 0.2 diameter end down (pointed end in the tool holder).

2) The spindle will be started at about 500 rpm and the setup person will push the edge-finder out of alignment with its shank (so the 0.5" diameter is not concentric with the 0.5" shank).

3) Using the jog and handwheel functions of the machine, the setup person will bring the edge-finder's 0.2" diameter perfectly flush with the left side of the workpiece (the raw material edge in our case).

4) Without moving the X axis, the setup person will retract the edge-finder in the Z axis until it is well above the workpiece.

5) The setup person will reset (set to zero) the X axis relative position display register.

6) The edge-finder's 0.2" diameter currently is flush with the left edge of the raw material. The program zero point is now precisely 0.200 inch in the X plus direction from the spindle center (0.1" edge-finder radius plus the 0.1" raw material that is still on the workpiece). So, using the handwheel and by monitoring the X axis relative position display, the setup person will move the edge-finder in the X plus direction by 0.200 inch (actually the machine table will move to the left on most machines). At this point, the center of the spindle is perfectly aligned with the X axis program zero point.

7) The setup person will reset (set to zero) the X axis position display again.

8) The setup person will push the edge-finder out of alignment with its shank (the 0.5" diameter is not concentric with the 0.5" shank).

9) Using the jog and handwheel functions of the machine, the setup person will bring the edge-finder's 0.2" diameter perfectly flush with the lower side (Y minus side) of the workpiece.

10) Without moving the Y axis, the setup person will retract the edge-finder in the Z axis until it is well above the workpiece.

11) The setup person will reset (set to zero) the Y axis relative position display register.

12) The edge-finder's 0.2" diameter is currently is flush with the Y minus edge of the workpiece. The program zero point is now precisely 0.100 inch in the Y plus direction from the spindle center. So, using the handwheel and by monitoring the Y axis relative position display, the setup person will move the edge-finder in the Y plus direction by 0.100 inch (actually the machine table will move forward on most machines). At this point, the center of the spindle is perfectly aligned with the Y axis program zero point.

13) The setup person will reset (set to zero) the Y axis position display again.

14) The setup person will send the machine to its zero return position in the X and Y axes. At this point, the X and Y axis relative position display registers will be showing the program zero assignment values in X and Y.

For the Z axis program zero assignment value (distance from spindle nose to program zero in Z), there must be no tool in the spindle at the start of this procedure. We'll also place the three inch side of a 1-2-3 block on top of the workpiece.

1) Using the jog and handwheel functions, the setup person will bring the spindle nose into contact with the three inch side of the 1-2-3 block.

2) The setup person will reset (set to zero) the Z axis relative position display register.

3) The setup person will send the Z axis to its zero return position. At this point, the setup person will add three inches to the value currently shown on the Z axis relative position display (to account for the 1-2-3 block). This will be the Z axis program zero assignment value.

Enter program zero assignment values into fixture offset number one

Since the program zero assignment values have been measured during setup, they must be manually entered into the fixture offset (fixture offset number one in our case). To do so, the setup person will press the offset soft key until the fixture offset page appears. They'll scroll the cursor down to the X register for fixture offset number one, enter the X program zero assignment value, and press the input key. They'll do the same for the Y and Z registers. (Remember that these values are the distances *from* the zero return position *to* the program zero point– they're almost always *negative values*.)

Assemble the cutting tools needed for the job

In some companies, the CNC setup person is not responsible for assembling cutting tools. Some companies have a person in the tool crib (commonly called the *tool setter*) assemble cutting tools. Though this may be the case, all CNC setup people will have to assemble cutting tools from time to time – so it is important that you have the ability to perform this task.

Our setup documentation is a little sketchy when it comes to cutting tools. You haven't been told what components to use – so you're on your own to figure it out. While experienced setup people can easily do so, beginners may have some problems. Fortunately, most companies are a little more helpful when it comes to listing components needed to assemble cutting tools. They'll use some kind of identification system (usually of the company's own design) to specify cutting tool components. Or they may use the catalog numbers from their cutting tool manufacturers. The best kind of cutting tool documentation shows every component that makes up each assembled cutting tool – as is shown in Figure 7.10.

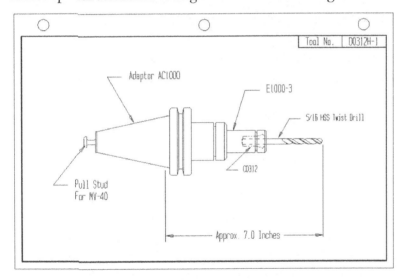

Figure 7.10 ~ Great cutting tool documentation

Another good system incorporates a special one or two letter abbreviation for each type of component along with four numbers to specify its size. While this system is not as easy to use as the documentation shown in Figure 7.10, it is still pretty easy to understand. Consider, for example, the following tool component identification numbers:

- DT0750 (a 3/4" twist drill)
- ME0375 (a 3/8" end mill)
- RM1000 (a 1.0" reamer)
- AC1000 (a 1.0" collet adapter)
- CT0500 (a 1/2" collet)
- AE0750 (a 3/4" end mill holder)

Fundamentals of CNC © CNC Concepts, Inc.

- AF30000 (a 3.0" face mill holder)
- EC0500 (a 1/2" collet extension)
- ET0250 (a 1/4" tap extension)

With a little ingenuity, you can come up with a naming convention for just about any cutting tool component. Instead of using one drawing per cutting tool (as is shown in Figure 7.10), a more concise tool list can be used. Using this kind of system, here is the tool list for our example setup.

Tool 1: ME1000, AE1000
Tool 2: ME0500, AE0500
Tool 3: ME0625, AE0625
Tool 4: DC0004, AC1000, EC0375, CT0375
Tool 5: DT0375, AC1000, EC0375, CT0375

While this is not nearly as explicit as the drawing in Figure 7.10, it adequately lists all cutting tool components needed for our example job.

A special bench-mounted assembly figure is used to assemble cutting tools. The tapered tool holder shank is placed into the fixture and is located from the keyways on the outside of its flange. This provides rigid support for the tool holder shank as cutting tool components are assembled and tightened. Figure 7.11 shows a bench-mounted assembly fixture.

Figure 7.11 – A bench mounted cutting tool assembly fixture

Most of the hand tools used to assemble cutting tools should be pretty familiar and include Allan wrenches, box-end wrenches, and spanner wrenches. And assembling most cutting tools is pretty straight forward. But if you are in doubt about how a particular cutting tool goes together, you must ask an experienced person in your company.

Measure tool length- and cutter-radius-compensation values

As with assembling cutting tools, this task may not be the responsibility of the setup person. But even if your company employs a tool setter to measure cutting tools, all setup people must still possess the ability to do so.

All cutting tools require a tool length compensation measurement. As is shown in Lesson Eleven, if you use our recommended method (offset is tool length), you can measure tool lengths either off line prior to setup or on the machine during setup. If you use the second method shown in Lesson Eleven (offset is the distance from tool tip to program zero), you must measure tool length compensation values on the machine during setup.

We show the procedures to measure tool length compensation values in Lesson Eleven, so we won't repeat them here. Our example job requires that you determine five tool length compensation values. If you do the measurements off line, write down each value as you measure it. If you measure on the machine, you can transfer the Z axis relative display screen value right into the tool length compensation offset (as is also shown in Lesson Eleven). This will eliminate the possibility for entry mistakes.

For milling cutters that perform contour milling operations (milling on the periphery of the cutter) and use cutter radius compensation, you must also measure the milling cutter's radius. To do so, simply use a micrometer to measure the milling cutter's diameter and divide the diameter by two to determine its radius. For Fanuc controls, it is the cutter's *radius* that is needed for cutter radius compensation offsets.

The static nature of cutting tool measurements

Remember that when a cutting tool begins machining, it will be under the influence of *tool pressure*. Tool pressure can influence the quality your initial measurements. Another tooling-related problem can also affect the quality of your initial cutting tool measurements. When an end mill is placed in an end mill holder, it may not be perfectly concentric with the spindle. If the milling cutter is not concentric with the spindle it will *run-out*, and will remove more material from the workpiece than it should. The smaller the tolerances you must hold on the workpiece, the more important it is that you consider using a technique called *trial machining* when you machine the first workpiece. Trial machining is presented in Lessons Ten, Eleven, and Twelve – and we'll discuss it in greater detail later in this lesson.

Enter tool length and cutter radius compensation values into offsets

If tool length compensation values are measured on the machine, you can automatically transfer them into offsets during the measuring procedure. There will be no need to enter them as is discussed here.

But if you measure tool length compensation offset values off line, they must be entered into offsets during setup. In similar fashion, cutter radius compensation values must be entered into offsets for milling cutters.

For our example job, the setup sheet explicitly specifies which offsets are related. Frankly speaking, most setup sheets will not. In most companies, there are unwritten rules related to how tool length- and cutter-radius-compensation offset numbers are determined. For tool length compensation, most programmers will make the tool length compensation offset number correspond to the tool *station number* (as we're doing).

For cutter radius compensation, offset numbering depends upon how the offset display screen displays offsets. With some machines, each offset will have two values – one for the tool's length and another for its radius. With these machines, the programmer will use the same offset number for both tool length- and cutter-radius-compensation. This makes it pretty easy to determine which offsets are related to a given cutting tool.

Unfortunately, the offset display screen for most machines has but *one* value per offset (as we're assuming in our example job). With this kind of offset display, the offset number corresponding to the tool station number is already being used for tool length compensation – so the programmer must choose another. Most programmers will add a constant number (like thirty) to the tool station number to determine the offset number used with cutter radius compensation (as we did).

If your machine has but one value per offset, you *must* know which offsets are used with cutter radius compensation when milling cutters are used in a job. If it is not explicitly specified in the setup documentation, you must ask the programmer.

For our example job, we'll enter tool length compensation values into offsets numbered one through five. Since we're using the recommended method for tool length compensation (offset is tool length), each will be a positive value – probably between about five and ten inches in length. And we'll enter cutter radius compensation values of 0.5 in offset thirty-one, 0.25 in offset thirty-two, and 0.3125 in offset thirty-three.

Load cutting tools into the machine's automatic tool changer magazine

If you measure cutting tools on the machine, you should perform this task *before* measuring and entering tool length compensation values. This will allow you to use the machine's automatic tool changing system to load each tool that must be measured. We haven't shown the procedure to manually use the automatic tool changer. It involves using the *manual data input* (MDI) mode – and will be shown in Lesson Twenty-Three.

Machine tool builders vary when it comes to how cutting tools must be loaded into the automatic tool changer magazine. Most provide a special and convenient *tool loading station*. They also allow the magazine to be easily rotated (with two push-buttons, one for clockwise and another for counter-clockwise) right from the magazine operating panel, so you don't have to walk back to the main control panel to rotate the magazine.

Even once you've rotated the magazine to the desired loading station, machine tool builders vary when it comes to how tools are actually put into the magazine. With smaller tools (CAT-40 and smaller), most machine tool builders use a *ball-detent* system that holds each tool by its pull stud. These tools are *snapped* into the magazine station's tool pot by simply pressing them in. A special tool may be required to pry the tool out of the tool pot. For larger tools, most machine tool builders use a more positive method of clamping tools in the magazine – and at the loading station, a lever is used to release the clamping mechanism so that tools can be removed and replaced.

Even tool numbering for each station of the automatic tool changer magazine will vary from one machine tool builder to another. Some builders use hard-and-fixed tool station numbers while others allow each station to be numbered through the control panel. You must reference your machine tool builder's operation manual to determine how your automatic tool changer functions in this regard.

For our example job, we place the one inch end mill in station one, the 1/2" end mill in station two, the 5/8" end mill in station three, the #4 center drill in station four, and the 3/8" drill in station five.

Load the CNC program

For our example job, program number O0003 must be loaded. Our setup documentation is assuming that the setup person knows where to find program number O0003. In most companies, each machine will have a special program storage folder (or directory) in the distributive numerical control (DNC) system – and setup people are expected to know the related folder names. Some companies will specify the storage location right in the setup documentation.

The actual task of loading a program depends upon the company's DNC system. Most require that the machine be made ready first, by placing the mode switch to *edit* and then pressing the *read* button. At this point, the machine waits for a program to be sent from the DNC system. Then the setup person goes to the DNC system and commands it to send the program. Some DNC systems do allow the entire process to be completed right from the CNC machine tool. You must talk to an experienced person in your company to learn the specific procedure.

The physical tasks related to setup are now completed

At this point, you're (finally) ready to run the program. However, there could still be problems. How many problems you'll have and how severe they are depend upon several things. Generally speaking, new programs written manually by new programmers tend to present the most problems. But even proven programs (programs that have been successfully run before) can still have problems. Additionally, the setup person can make mistakes that will cause even a perfectly written program to fail. For these reasons, *all programs must be cautiously verified*. While proven programs tend to be easier to verify than new programs, you should never skip the program verification steps.

Verify the correctness of a new or modified program

This task is only necessary for new programs – or for programs that have been modified since the last time they were run.

Verifying the correctness of the program is really the responsibility of the *CNC programmer*, and is usually done prior to making a setup. Indeed, it is usually done soon after the program is written. There are many *tool path verification systems* that allow a programmer to visually check the motions a CNC program will make even before the CNC program is loaded into the machine. Most are relatively inexpensive, so *there is really no excuse not to have one* – especially when you consider the expense of using a CNC machine to verify the correctness of a program.

Figure 7.12 shows a typical *solid-model-type* tool path verification screen for our example program run on a Windows computer. The system we used sells for about $200.00, which is extremely reasonable when you consider the amount of time and problems it can save during program verification on a CNC machining center.

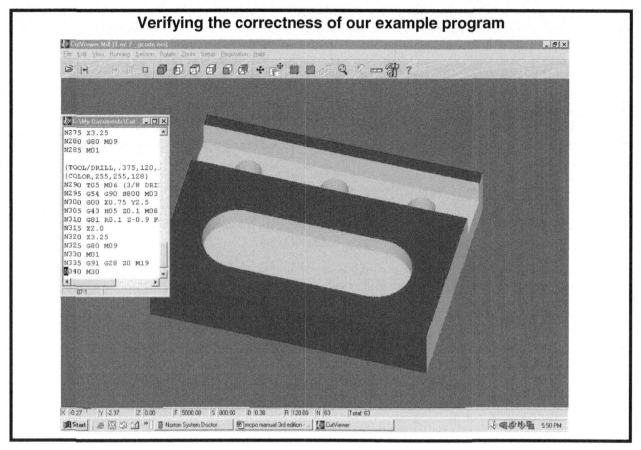

Verifying the correctness of our example program

Figure 7.12 – An example of tool path verification done on a Windows computer

Once again, keep in mind that CNC-using-companies vary when it comes to what they expect of their CNC people. If you are a setup person in a product producing company, you probably won't have access to the program verification system, so you won't see the program verification being done off line. You'll have to take the programmer's word that the program's tool path is correct.

Some CNC controls do include tool path verification capabilities that rival the best off line systems. If your machine has this capability, by all means, learn how to use it, especially if you can't see the program verification done off line. But if at all possible, this program verification step should be done *off line* (to save machine time). If mistakes are found while verifying the correctness of the program, the time required for *correcting* them can also be kept off line.

If you have neither of these tool path verification capabilities, you will not be able to see the movements that cutting tools in the program will make until you actually run the program on the machine. In this case, verifying the correctness of your programs will be extremely cumbersome, requiring that you cautiously check each movement the machine makes, using features like *dry run, single block,* and *feed hold.* When verifying programs in this manner, you may have to run the program several times just to begin to understand the program's movements. And you still won't be able to tell if all of the motions are correct.

Verify the correctness of the setup
Mistakes made during setup can cause even a perfect program to behave poorly. Consider a few of the possibilities: You could load a tool into the wrong station. You could measure or enter program zero assignment values incorrectly. You could measure or enter tool length compensation values incorrectly. You could forget to enter a cutter radius compensation offset value. You could load the wrong program.

And by the way, there are certain programming mistakes that will not show up during tool path verification. If, for example, the programmer forgets to start the spindle (or if they have it running in the wrong direction) few tool path verification systems will catch the mistake. And most tool path verification systems will ignore offset use. If the program invokes the wrong offset, most tool path verification systems won't show it.

For these reasons, *you should run the program at least once without a workpiece.* This is commonly called a *dry run.* In fact, one of the program verification functions you have to help with program verification is called *dry run.* When the dry run switch is on, you will have control of how quickly the machine moves (even during rapid movements). A multi-position switch (usually the feedrate override switch) is used for this purpose. You'll also have a *panic button* that will stop all axis motion if you get nervous (called the *feed hold* button).

When you do a dry run, again, there will be no workpiece in the setup and you'll have complete control of all motions. You'll be checking for very basic and serious mistakes – like collisions, spindle direction, tools loaded into the spindle, and general movements made by the program. If a mistake is found, you'll stop the cycle, diagnose and correct the problem, and do another dry run. You must not continue until you've seen the *entire program* run – and until you understand and agree with the movements made by the program.

A tip that will save a crash some day

Whenever you activate a cycle by pressing the cycle start button, always have a finger ready to press the *feed hold button.* If you don't have certain switches set properly, the machine will not behave as you expect, and you'll be ready to *immediately* stop the machine when something goes wrong. For example, say you forget to turn on the dry run switch. You have feedrate override turned down to its minimum position, so you're *expecting* the machine to crawl during its first motion. But in reality, it will move at rapid (assuming rapid traverse override is set to one-hundred percent). If you have a finger ready to press the feed hold button, you'll be ready to stop the rapid motion as soon as you see that something is wrong. You won't have to look for the feed hold button – which takes time – time that may allow the machine to crash.

Dry running our example program

We remove the workpiece from the vise, and turn on dry run. We confirm that the machine is at its tool change position. We turn the feedrate override switch down to its lowest setting (usually ten percent), we place the mode switch in the auto or memory position (to run a program). We press the program check display screen key (this is the page that shows the distance-to-go function), and confirm that program number O0003 is the active program.

With a finger ready to press the feed hold button, we press the cycle start button. The first thing that happens is that tool number one (the 1.0 end mill) is placed into the spindle. Then the spindle starts – sure enough, in the forward direction. From the program check page (which also shows absolute position of the cutting tool), we can see that the machine is moving, but barely. As we crank up the feedrate override switch, sure enough, we can see the X and Y axes moving. This first XY movement is pretty safe because the machine is still at its Z axis zero return position. So we crank the feedrate override switch up to its maximum setting and the X and Y axes move quickly into position.

When the end mill is over the vise, the Z axis begins to move. The end mill is still quite a distance from the vise, but we're getting a little nervous. So we slow the motion a little by turning down the feedrate override switch. That's better. Not so scary.

The end mill comes into position and makes the movements needed to mill the left end of the workpiece. While it is still at the work surface in Z, we press the feed hold button. Let's confirm that the tool looks like it is at the appropriate Z position. We look into the work area and sure enough, the tool's Z position looks good. So we press the cycle start button to reactivate the cycle.

The end mill finishes milling the left end, it retracts, and then moves into position to mill the right end. It drops back down in Z and mills the right end. Everything looks good.

The first tool is now finished and it begins retracting for the tool change. We increase the feedrate override switch to maximum. As the tool change occurs, we reduce the feedrate override switch again. The 1/2" end mill is placed into the spindle and the spindle starts, again, in the correct direction. It moves a little in X and Y to get to the left side of the slot and then begins moving in Z. As it gets closer, we slow the motion a little. It makes the first pass through the middle of the slot. Then it moves a little in the Y minus direction and makes the second pass. Then it moves in the Y plus direction and makes the third pass. Finally it starts retracting to the tool change position in Z.

The 5/8" end mill is placed into the spindle and the spindle starts. We cautiously monitor its movements as it mills the oval-shaped pocket. Everything looks fine.

We do the same for the center drill and 3/8" drill. No problems are found.

What if you do find a problem?
Our dry run went flawlessly. But here are a couple of common scenarios that signal a problem.

Say the very first tool, the 1.0" end mill, is making its first approach movement in the Z axis. It's seems to be going *too far*. It's well past what seems to be the work surface and it is still moving. So you press the feed hold button to stop the tool's motion and look at the distance-to-go display. Sure enough, it says the tool is still going to move another six inches or so in the Z negative direction. This will cause the tool to crash into the machine table. So you cancel the cycle (by pressing the *reset key*). You must find and correct the problem. What do you think it could be?

We saw the tool path display off line and it looked good. So the problem shouldn't be with the Z axis movements in the program. Since this is the *first* tool, the problem could be with your program zero assignment in the Z axis or with tool length compensation for this tool. (Note that if the first tool approaches properly and this problem occurs on any subsequent tool, the problem *must be* with tool length compensation.)

When we look at the tool length compensation offset for this tool (offset number one), say we find that there is a value of *zero* specified. We forgot to enter a tool length compensation value for this tool! The machine thinks we have a cutting tool that is zero inches long. It was bringing the spindle nose to the work surface – which is causing the tool tip to go way past its intended position.

After correcting the problem (by entering the tool length compensation value in offset one), we must re-run the tool. Since this is the very first tool, we'll rerun the entire program. Sure enough, this time, the tool runs properly.

During tool number three (the 5/8" end mill) an alarm sounds while the tool is in the pocket. We look at the program check page and find that the cursor is currently on line **N185**. This is the command that instates cutter radius compensation. The alarm says *over-cutting will occur during cutter radius compensation*. Now, what do you think could be wrong?

Frankly speaking, the over-cutting alarm can be generated for several reasons. But say we look at the cutter radius compensation offset value and find it to be 0.625. Oops. We entered the cutter's *diameter*, not its radius. This is what caused the alarm. So we correct the offset. Now we must rerun the program, starting with tool number two. (There is no need to rerun the first two tools – doing so will waste time.) To restart, we scan to line **N160** (the command after the tool change since tool number three is currently in the spindle), and restart the program. The further you are into a program when a problem is found, the more wasteful it will be to rerun the entire program once you find and correct a mistake.

Cautiously run the first workpiece

If the program is correct (you've seen it run on a tool path verification system or it is a program that has run before), if the process and cutting conditions are correct, and if you have made no mistakes during setup (the dry run looks good), you should be able to machine a good workpiece on your very first try.

Admittedly, there are a lot of *ifs* in the previous paragraph. And problems could still exist with *new* programs that you could not detect during the tool path verification and the dry run. If, for example, a small motion mistake is made in a tool path, you may not be able to spot it during the tool path verification. This, of course, will cause the workpiece to be machined incorrectly. This kind of mistake should not exist with proven programs – as long as the *current version* of the program is being used.

And there could also still be mistakes in your *setup* that went undetected during the dry run. If, for example, you make a small measurement mistakes during tool length compensation measurement or entry (say, a tool length compensation entry is off by 0.25 inch), you may not be able to detect it during a dry run.

The most dangerous time

You must still be very careful when running the first workpiece. During programming lessons, we recommend using a rapid approach distance of 0.1 inch (about 2.5 mm). While this is a safe and acceptable approach distance, it is difficult to tell during a dry run whether the tool tip truly stops 0.1 inch above the work surface (since there is no workpiece). If even a small mistake is made with tool length compensation, the result could be disastrous. For this reason, *you must be extremely careful with each tool's first Z axis approach movement.* This is the most dangerous movement of each tool. We recommend using a special procedure during the first few commands of each tool. It assumes that the format shown in Lesson Fifteen is used to write the program. This procedure allows you to take full control of each tool's first Z axis approach movement.

Once you confirm that the machine is at its tool change position and that the correct program is active, you can follow this approach procedure to approach with each tool. The first workpiece, of course, must be loaded at this time.

1) Turn on single block and dry run.

2) Set the feedrate override switch to its lowest setting.

3) Press the program check display screen key (this is the page that shows distance-to-go)

4) Repeatedly press cycle start until the first tool change occurs (because single block is turned on, each time you press cycle start only one command is executed).

5) Press cycle start (the spindle starts).

6) Press cycle start. The machine begins moving in the X and Y axes to the tool's first XY position. This motion is still pretty safe, since the Z axis is still at the zero return position. Crank up the feedrate override switch until the tool comes into position in X and Y.

7) Set the feedrate override switch to its lowest setting.

8) Press cycle start. Machine now begins moving in Z. *This is the dangerous movement.* Be very careful as the tool gets close to the workpiece. When the tool comes to within about 0.5 inch of the work surface, stop the motion by pressing the feed hold button. Look at the distance-to-go display. Confirm that the amount of motion left in the approach movement is less than the distance between the tool tip and the workpiece. As long as it is, press the cycle start button to reactivate the cycle. Let the tool come to its final approach position and stop.

9) Turn off dry run. (*Never* let cutting tools machine a workpiece under the influence of dry run. Dry run tends to slow down rapid motions, but it *speeds up* cutting motions.)

10) The tool has safely approached. What you do next will depend upon whether you're running a new or a proven program. For a new program you must carefully step through the cutting tools cutting motions. In this case, leave the single block switch on, and repeatedly press cycle start to step through the tool. For a proven program (are you *absolutely sure* that you're running the current version of the program?), turn off single block and press cycle start to let the tool machine the workpiece as it did the last time the program was run.

11) When the tool is finished, the machine will return to the tool change position. While the machine is moving to the tool change position, turn on dry run and, if you turned it off, single block. Repeat the procedure, starting at step four.

For each tool in the program, you must take control of the tool's first Z axis approach movement with single block and dry run. Once you have seen the tool approach properly, you need not use this procedure (after running the tool for the first time). But whenever you're worried about a tool's first approach motion, this procedure will keep the tool from crashing into the workpiece.

Making sure the first workpiece is a good one

While there is a lot to think about when you run the first workpiece, it must still be your goal to make the very first workpiece being machined a good one. If there are any tiny motion-mistakes in a new program that you did not detect when you verified the tool path, it may be impossible to machine a good workpiece on your first try. The program must be correct, of course, in order for the first workpiece to be machined properly (as a *proven* program is).

We do consider the running of your first workpiece to be a *setup-related task*. If the first workpiece doesn't pass inspection, we also consider any corrections made to the program as well the running of the *second* workpiece to be a setup-related task. Indeed, *the machine is still in setup until a workpiece passes inspection* and you can run acceptable workpieces.

Frankly speaking, if the program is correct, *there is no excuse for scrapping the first workpiece you machine*. If you use the techniques we recommend, each tool will be *forced* to machine properly – and when the program is finished, each workpiece surface being machined will be within it tolerance limits.

Look at step number four in the procedure shown above for cautiously approaching with each tool. This is the point at which you're going to *begin* each tool's first approach. The tool change has just occurred, and you know which tool is going to run next. At this point in the procedure, you must consider what the tool is going to do – and especially – consider the tolerance bands for dimensions that the cutting tool is going to machine. If tight tolerances must be held, you must use *trial machining* techniques.

Machining the first workpiece in our example job

Let's go through the entire example job and show how this is done.

Tool one: Tool number one is the 1.0" end mill. At step four of the approach procedure shown above, this tool is placed in the spindle. We must to and consider what this tool will be doing. This tool will be machining both ends of the workpiece. And notice the very tight tolerance for the workpiece overall length (4.000 +/- 0.0004). Even if we measure the diameter of the 1.0" end mill perfectly and enter the cutter radius compensation value accordingly (into offset thirty-one), this end mill may not machine the overall length within its tolerance limits. The end mill may be running out a little bit in its holder. If it is, it will machine too much stock. If this happens, we're going to scrap the first workpiece.

Since we're worried about whether our initial cutter radius compensation offset entry is accurate enough to machine the overall length of the workpiece within its tolerance limits, we must use *trial machining* techniques for this tool. This will force the end mill to leave additional stock on the two ends, ensuring that we don't machine the workpiece undersize on the first try. (Trial machining for cutter radius compensation is discussed in detail during Lesson Twelve.) Before running this tool, we'll increase offset number thirty-one by 0.01 (if its initial value is 0.500, we'll make it 0.510). This will cause the end mill to leave 0.01 inch more material on each end of the workpiece (0.02 overall).

Before continuing with the procedure, we'll turn on the *optional stop* switch. This will ensure that the machine will stop at the completion of the tool (though it will anyway if the single block switch is on).

We'll continue with the approach procedure, cautiously bringing the end mill to its approach position. Then we'll allow this tool to machine the workpiece (as is done in step ten). When the tool is finished and back at the tool change position, the optional stop function (or the single block function) will cause the machine to stop.

At this point, we'll measure the overall length of the workpiece. If our initial tool measurement is perfect, the overall length should be precisely 4.020 right now (remember that the end mill is leaving an additional 0.01 on *each* end). But when we measure, we find the overall length to be 4.0188. Sure enough, the end mill must be running out a little bit. And if we had tried to run the workpiece with the initial offset setting, the overall length would have come out to 3.9994 inches – and the workpiece would have been scrapped.

Assuming the target value for this dimension is 4.000 inches, we must reduce the current setting of the offset by 0.0094 (half of 0.0188, which is the amount the workpiece is currently oversize) and rerun the tool. This time we don't have to be so careful with the approach movement, since we've seen this tool successfully approach the workpiece. So we turn off the dry run and single block switches (leaving on the optional stop switch), set feedrate override to one-hundred percent and press cycle start. The end mill machines the overall length again. At the end of the tool, the optional stop function stops the machine. We measure the overall length again, and sure enough, it comes out to 3.9999, well with the tolerance limits.

Tool two: At step four of the approach procedure for the next tool, the 1/2" end mill is placed in the spindle. Again, we must stop and consider what this tool will do. This tool will machine the 0.625 wide slot. And notice the tight tolerances on both the slot width and its depth (+/-0.0004 in both cases). Again, we're worried that our initial offset settings are not accurate enough to machine the slot within its tolerance limits.

We increase the tool length offset value for this tool (offset number two) by 0.01 and increase the cutter radius compensation offset value for this tool (offset number thirty-two) by 0.01 and use the approach procedure to machine the workpiece for the first time. At the optional stop (or because of single block), the machine will stop when the tool is finished.

When we measure the pocket width, we find it to be 0.6058. The target value for the pocket width is 0.625, so we must decrease offset number thirty-two by 0.0096 (remember, the end mill is machining on both sides of the pocket). To calculate the adjustment amount in this case, we first subtract 0.6058 from 0.625. We then divide the result by two. When we measure the pocket depth, we find it to be 0.3653. The target value for pocket depth is 0.375, so we must decrease offset number two by 0.0097.

Rather than running the entire program again, which would waste time and could cause the previously run tool to scuff up surface finishes (or even machine more material), we'll re-run tool number two. Since tool number two is currently in the spindle, restart block in the program is N090 (the command after the tool change). And since we have seen this tool successfully approach, there is no need to use the approach procedure. We'll turn off the dry run and single block switches (leaving on the optional block switch), set feedrate override to one-hundred percent, and press the cycle start button. The 1/2" end mill machines the slot again. At the optional stop, the machine stops. We measure the slot – and this time the width is 0.6251 and the depth is 0.3749. We're ready to go on to the next tool.

Tool three: At step four of the procedure, the next tool, the 5/8" end mill, is placed in the spindle and we're ready to run it the first time. But before continuing with the approach procedure, we must consider what this tool will be doing. This tool will be machining the oval shaped pocket. Notice that there is nothing critical about this pocket. Tolerances for width and depth are +/-0.005 inch, and we're pretty sure that our initial tool length- and cutter-radius-compensation offset measurements are adequate to machine the pocket within tolerance limits.

Using the approach procedure, we allow the end mill to approach and machine the pocket. When it's finished, the optional stop (or single block) will cause the machine to stop. Even though we're pretty confident, we should check to see that this tool has machined properly – so we measure it. Sure enough, the pocket width is 1.0008 and its depth is 0.249. While are they well within tolerance limits, the width and depth dimensions are not at their target values. Most setup people will make offset adjustments to ensure that the pocket machined in the *next workpiece is perfect*. So we'll increase cutter radius compensation offset number thirty-three by 0.0004 (making the tool remove less material) and reduce tool length compensation offset number three by 0.001 (making the tool go deeper). There is no need to re-run this tool, so we proceed to the next tool using the approach procedure.

Tool four: At step four for the next tool, the #4 center drill is placed in the spindle. We stop to consider what this tool will be doing. According to the drawing, there is really nothing critical about the three holes. Tolerances are +/-0.005 inch. All we're going to care about is that the center drilled holes get machined in the correct position (which is controlled by the program). So we're not going to need trial machining techniques. We'll use the approach procedure and let the tool machine the three holes. When the tool is finished, the machine stops. At this point, we check to see that the center drill has gone deep enough for the up-coming drill and that the holes are in the correct locations. We now proceed to the next tool.

Tool five: At step four for the next tool, the 3/8" drill is placed in the spindle. We stop to consider what this tool will be doing. Again, there is nothing critical about the three holes. We use the approach procedure and allow the drill to machine the holes. When the machine stops (this time at the end of the program), we check to see that the holes are in the correct locations and that they have broken through the bottom of the workpiece.

At this point, we're finished running the first workpiece *and it is a good one* (it will pass inspection). We're (almost) ready to begin the production run.

What if the program isn't perfect?

Again, the program must be correct in order for you to be able to make the first workpiece a good one. Say that in our example (new) program, the depth of the 0.625 slot, which is supposed to be 0.375, is incorrectly specified as 0.395 deep (the programmer typed **Z-0.395** in the program instead of **Z-0.375**). It is unlikely that we'd catch such a small mistake during the tool path verification or during the dry run. It is likely that we'd scrap the first workpiece, even with trial machining techniques. But hopefully we would be able to find and correct the mistake while we're running the first workpiece – so the *second* workpiece will be correct.

If necessary, optimize the program for better efficiency

This task in only required with new programs and large lot sizes. If you have a large number of workpieces to machine, it may be necessary to optimize the motions in the program. As stated in Lessons Fourteen and Fifteen, the format we use to create programs are not as efficient as possible. As you watch your programs run, you will probably notice things that can be improved.

As you begin working with CNC machines, try not to be overly concerned about program efficiency. Concentrate on becoming familiar with running your machine and making good workpieces. As you gain confidence, by all means, apply the suggestions we offer in Lesson Fifteen to make your programs run faster.

If you make optimizing changes to the program at the machine right after verifying the program, you must verify that your changes are correct. You must dry run the tools that have been changed.

If changes have been made to the program, save the corrected version of the program

This is an important task that tends to be overlooked. While you are verifying the program – and whether it is a new program or a proven one, if you make any changes to the program, you *must* remember to save a corrected version of the program back to your distributive numerical control (DNC) system. If you forget to do so, you'll have to repeat the program verification steps the next time the job is run. Worse, you may remember running the job before and you may *think* that the program being loaded is a proven program. This, of course, will lead to serious problems.

The procedure for saving programs is similar to the procedure to load programs – and again – it will vary based upon the DNC system you use. With most systems, you first go to the DNC system and get it ready to receive the program. Then you go to the CNC machine and make the command to send the program to the DNC system. With some DNC systems, the entire process can be completed from the CNC machine.

Production run documentation

With the setup completed and the first workpiece passing inspection, you're ready to begin the production run. We're assuming at this point, that you have more than one workpiece to machine. As with just about every facet of manufacturing, companies vary with regard to how many workpieces they commonly run per job. This can even vary within one company.

Generally speaking, CNC machining centers are most often applied to small to medium sized lots – from one to about one-thousand workpieces. While there are companies that dedicate their CNC machining centers to running one workpiece – day in and day out – the vast majority of CNC using companies, run a variety of jobs of varying lot sizes.

The number of workpieces commonly run has a big impact on how you approach production runs. Indeed, it has a big impact on how companies utilize CNC people. If a company consistently runs small lots of say, under ten workpieces, it won't take much time to complete each production run. Companies in this situation tend to have one person setup the job and run the job out. On the other hand, if a company consistently runs larger lots of say, five-hundred workpieces, it will take some time to complete the production run. These companies tend to have one person make the setup and another (lesser skilled person) run out the job.

If one person is doing everything, including programming, setup, and operation, there won't be much of a need for production run documentation. This person will know what is intended since they planned the entire job. On the other hand, if tasks are divided (one person programs, another sets up, and yet another runs out the job), the need for good documentation is much greater.

As you have seen, most companies provide *setup* documentation in the form of a one-page setup sheet. However, many companies expect their setup people to relate what must be done to run out the job to the CNC operator. Admittedly, many of the tasks related to running production are quite basic – and very redundant if many workpieces must be produced. But adequate production-run documentation is essential if several people will be running out the job (like first, second, and third shift operator).

A note to programmers:

This text is providing you with information needed to master all three skills needed to use a CNC machining center – programming, setup, and running production. However, depending upon your company, you may be responsible only for programming. In this case, you must provide setup and production run documentation for other people to make setups and run production.

While you can pretty much rest assured that the setup person will have adequate basic machining practice skills to understand even minimal documentation, you can make no such assumption about CNC operators. Many companies hire people with limited (if any) basic machining practice skills to run CNC machines. Remember that you must direct your production run documentation to the *lowest skill level* of CNC operator in your company.

Many programmers do a terrible job in this regard. Even if your operators have pretty good basic machining practice skills, remember that they're not going to be nearly as familiar with the job as you are. Many programmers assume way too much of their CNC operators. While experienced operators may eventually able to figure out what is expected of them, good production run documentation will minimize the effort required to do so.

Figure 7.13 shows the production run documentation for our example job in the form of a one-page form. This form must answer any questions a CNC operator will have about the job. We'll be referring to this form as we describe the tasks related to completing a production run.

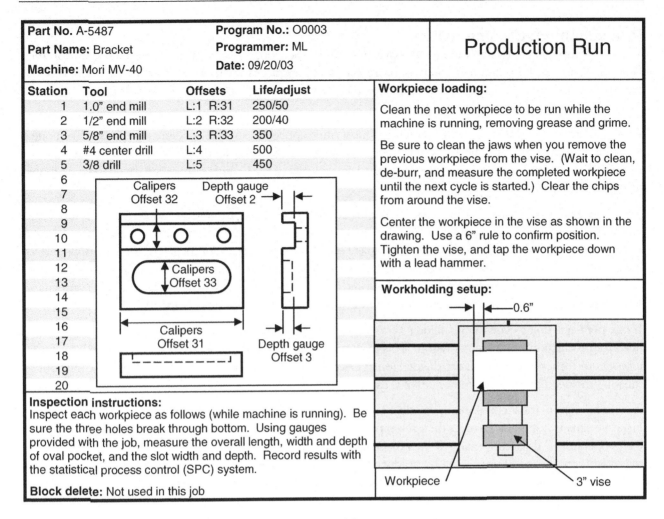

Part No. A-5487 | **Program No.:** O0003 | **Production Run**
Part Name: Bracket | **Programmer:** ML |
Machine: Mori MV-40 | **Date:** 09/20/03 |

Station	Tool	Offsets	Life/adjust
1	1.0" end mill	L:1 R:31	250/50
2	1/2" end mill	L:2 R:32	200/40
3	5/8" end mill	L:3 R:33	350
4	#4 center drill	L:4	500
5	3/8 drill	L:5	450
6			
7			
8			
9			
10			
11			
12			
13			
14			
15			
16			
17			
18			
19			
20			

Calipers Offset 32   Depth gauge Offset 2

Calipers Offset 33

Calipers Offset 31

Depth gauge Offset 3

Workpiece loading:

Clean the next workpiece to be run while the machine is running, removing grease and grime.

Be sure to clean the jaws when you remove the previous workpiece from the vise. (Wait to clean, de-burr, and measure the completed workpiece until the next cycle is started.) Clear the chips from around the vise.

Center the workpiece in the vise as shown in the drawing. Use a 6" rule to confirm position. Tighten the vise, and tap the workpiece down with a lead hammer.

Workholding setup:

0.6"

Workpiece   3" vise

Inspection instructions:
Inspect each workpiece as follows (while machine is running). Be sure the three holes break through bottom. Using gauges provided with the job, measure the overall length, width and depth of oval pocket, and the slot width and depth. Record results with the statistical process control (SPC) system.

Block delete: Not used in this job

Figure 7.13 ~ production-run documentation for our example job

Remove the previous workpiece

The task of workpiece loading begins with the removal of the most recently completed workpiece from the work area. This requires the operator to understand the workholding device being used to secure the workpiece – and should be described in the production run documentation.

For our example job, the operator must remove the workpiece from the table vise, which requires the use of a vise handle. Workpiece removal usually requires that the location surfaces of the workholding device be cleaned. In most cases, the work area near the workholding device should also be cleaned to remove chips and debris.

Note that some operators get into the (bad) habit of beginning to work with the workpiece just removed before loading the next workpiece. While this workpiece must be cleaned, de-burred and measured, the operator must set it aside until the machine is in cycle for the next workpiece.

Load the next workpiece

While loading workpieces is usually pretty simple, the setup documentation should include workpiece loading instructions. Notice that our example production run documentation form includes explicit loading instructions requiring the use of a 6" rule. Like workpiece removal, workpiece loading requires an understanding of the workholding device being used.

Once the workpiece is loaded, the doors (if there are any) to the work area must be closed.

Activate the cycle

If several workpieces have just been run, this task simply involves pressing the *cycle start* button. The machine will begin running the next workpiece in the normal manner.

But if the operator has been away from the machine for any length of time (at lunch or on break), or if they are just beginning a shift, they must confirm that the machine is truly ready to run a workpiece. This involves checking (and setting) the current condition of several switches on the control panel. While this may not be a complete list, here are some of the switches that must be set.

- Dry run: off
- Single block: off
- Optional stop: off
- Machine lock (if available): off
- Block delete: As requested by programmer (probably off)
- Rapid traverse override: 100%
- Feedrate override: 100%
- Spindle override (if available): 100%
- Mode switch: Memory or Auto
- Display screen: Program mode (and correct active program is shown and cursor is at the beginning of the program)
- Machine position: At the tool change position

Again, with the machine ready to run a program, activating the cycle simply requires pressing the cycle start button. Whenever you press the cycle start button, *always* have a finger ready to press the feed hold button. If the machine behaves in an unexpected manner, you'll be ready to stop it.

Monitor the cycle

This may not be necessary for proven programs that you have seen run many times before. But if you're new to a program, it is a good idea to get familiar with the cycle – especially if you did not program the job or make the setup. This will help you understand the tools that are being used and the general machining order.

For new jobs, you'll also want to confirm that cutting tools are machining properly for several workpieces. Certain cutting tool related problems, especially with speeds and feeds, may not present themselves when the first workpiece is being machined during setup. You'll need to stay alert while running the first few workpieces.

Clean and de-burr the workpiece

Once you have loaded the next workpiece and activated the cycle, you can begin working on the workpiece that has just been removed. It will probably be covered with coolant and debris. And it will probably have some razor sharp edges, so be careful handling it.

Most companies expect their CNC operators to clean and de-burr the workpieces they produce. Cleaning usually involve wiping the workpiece with a rag or shop towel. If you use any kind of air-blowing system to blow off the workpiece, be *extremely careful* – chips can fly anywhere. While you must wear eye-protection (all shops require it), nothing is protecting your ears, nose, mouth, etc.

While most companies strive to remove sharp edges in the machining cycle, there will almost always be some sharp edges on workpiece you remove from the machine. Again, be careful not to cut yourself as you handle newly removed workpieces.

A variety of hand tools is available to de-burr sharp edges, like files and hole de-burring tools. If any are unfamiliar to you, ask an experienced person to demonstrate them.

For our example workpiece, just about all machined surfaces will have sharp edges since we didn't use any de-burring tools in the program. These surfaces include all surfaces on each end of the workpiece, the top surfaces of the slot and pocket, and the top and bottom surfaces of the three holes.

Perform specified measurements

Most companies expect their CNC operators to inspect the workpieces they machine. They also expect the results of these measurements to be recorded in some fashion. But companies vary when it comes to specific methods used to measure and record.

Many companies require one-hundred percent inspection on critical dimensions. And they use some kind of statistical process control (SPC) system to record the measurements. An SPC system usually incorporates a computer screen and keyboard at the measuring station for data entry.

Taking measurements, of course, requires basic machining practice skills. The operator must know how to use the various gauging tools needed to take the measurements. Production run documentation must specify the gauging tools to be used (as our example documentation does). If you are unfamiliar with the required gauges, you must ask to be shown how to use them. And you must practice with them to ensure that you can take accurate measurements.

For our example job, notice that production run documentation specifies that the overall length of the workpiece must be measured. So must the width and depth of the 0.625 slot must be measured on every workpiece. The operator is also asked to check the width and depth of the 1.0" pocket. And they're asked to visually check that the holes break through. They must record measured dimensions with the company's SPC system.

Make offset adjustments to maintain size for critical dimensions (sizing)

The tasks shown so far must be performed in every cycle. The tasks shown from this point are only performed if and when they are required.

The tighter the tolerances a cutting tool must hold, the more likely it will be that it will not machine surfaces within the tolerance band for its entire life (when it must be replaced). With small lots, this probably won't present a problem. Every cutting tool will machine properly until the production run is completed.

But with larger lots, the wear a cutting tool experiences may cause the surfaces it machines to change by a small amount. And with tight tolerances, this may place the workpiece in jeopardy. In this case, the CNC operator must make an offset adjustment to keep the cutting tool machining properly.

With our example job, there are three very tight tolerances to hold: the overall workpiece length, the 0.625 slot width, and the 0.625 slot depth. Say we have a large lot of two-thousand workpieces to produce. After the setup is made and the first workpiece passes inspection, each of these critical dimensions will be being machined at its target value (its mean value in our example).

As the operator continues to run workpieces, they notice that the overall workpiece length is *growing*. It started out at 3.9999 (slightly smaller than its target value), but after twenty workpieces, they find it to be 4.0001. After forty workpieces, it is 4.0002. And after fifty workpieces, it is 4.0003. This growth in overall length, of course, is being caused by wear of the 1.0" end mill. If this trend continues, the overall length of a future workpiece will be out of tolerance (scrap).

An adjustment must be made to the 1.0" end mill's cutter radius compensation value (offset number thirty-one). If the overall length of the workpiece is currently 4.0003, the current offset value *should* be reduced by 0.00015 to bring the dimension to its target. But as you know – in the inch mode – you are only allowed to make offset adjustments in 0.0001 increments. We'll reduce offset number thirty-one by 0.0002. This will cause the end mill to machine 0.0002 more stock from each end of the next workpiece, making the overall length of the next workpiece 3.9999 – back to a value slightly smaller than its target value.

The same situation will exist with the width (and possibly the depth) of the 0.625 wide slot. As the operator continues to run workpieces, they must be alert for sizing adjustments required because of tool wear.

If you expect the operator to make sizing adjustments during a production run, they must be told *which* offsets are related to each tool. In our production run documentation, we list the offsets related to each tool in the tool list. But we further clarify with a nice drawing which shows the offsets related to each workpiece surface. This allows the operator to make sizing adjustments without having to know which tool machines each surface.

Unfortunately, most production run documentation doesn't provide this level of clarity. Most operators are on their own to figure out which tool machines each critical surface and which offsets are related to the tool.

Since the 1.0" wide pocket and the holes have no tight tolerances, the tools that machine them will last for their entire lives without requiring sizing adjustments (with drills, there is nothing that can be done with offsets to change the diameter they machine).

If possible, production run documentation should help the CNC operator know *when* sizing adjustments are needed. This, of course, requires experience running the job. During the first production run for a new program, the operator will learn if sizing adjustments are necessary. If they are, this information should be added to the production run documentation. Notice in the tool list for this job, we include the approximate number of workpieces that can be machined before a sizing adjustment is necessary (in the Live/adjust column). This lets the operator know if sizing adjustments will be required before a tool must be replaced.

Replace worn tools

As with sizing adjustments, this task will only be required with larger lot sizes – when cutting tools will wear out before the job is completed.

When a cutting tool must be replaced, the same tasks required during the initial setup must be repeated. These tasks include assembly, measurement of tool length compensation values, measurement of cutter radius compensation values for milling cutters, offset entry, and placement into the machine's automatic tool changer magazine. All of these tasks are presented during our discussion of setup and will not be repeated here.

Keep in mind that if a cutting tool requires trial machining during the initial setup, it will require trial machining during replacement. Tools one and two in our example job (the 1.0" end mill and the 1/2" end mill) both require trial machining during setup. If they are replaced during the production run, they'll require trial machining again to ensure that the surfaces they machine come out to size.

All of this means, of course, that if a CNC operator is responsible for replacing tools during a production run, they must possess many of the same skills possessed by the setup person. For this reason, some CNC using companies do not expect their CNC operators to replace worn tools. Instead, they have the setup person do so.

If possible, production run documentation should specify the expected life for each tool. Again, this requires some experience with the job (or at least, experience with the related cutting tools and the material being machined). In our example documentation, we specify an expected tool life for each tool – so the operator will know if the job can be completed before any tools wear out.

Clean the machine

Most companies expect their CNC operators to keep their machines clean. Every so often (commonly at the end of each shift), the operator will remove all chips from the work area and clean the machine table. Chips machined during the shift will be dumped (from the chip disposal drum).

Preventive maintenance

Some CNC using companies expect their CNC operators to perform basic preventive maintenance – like maintaining coolant levels, way lube levels, hydraulic oil levels, and filters. If this is required, instructions must be provided that describe the required procedures and their frequency.

Key points for Lesson Twenty:

- There are only two general tasks that occur on CNC machining centers – machines are either in setup or they are running production.
- You must understand the tasks needed to make setups – we've provided a complete explanation of each task in the approximate order that setups are made.
- You must know how to tell when trial machining is required and know how to perform trial machining.
- You must know how to perform each task that is required to complete a production run – we've provided a complete explanation of each task.

Practice running the first workpiece

Instructions: Study the following drawing, paying particular attention to the dimensions and tolerances. Note that this drawing is not fully dimensioned. Only the dimensions and tolerances that are related to this exercise are provided. Next, read the description of the process. Finally, answer the questions that follow.

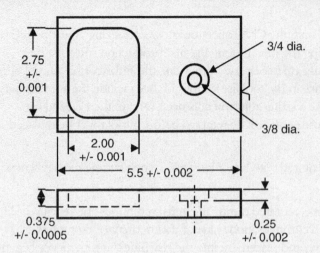

Process:

Description	Tool	Station
Mill right side	1.0 end mill	1
Mill 2 x 2.75 pocket	1.25 end mill	2
Drill 3/8 hole	3/8 drill	3
Counter-bore 3/4 hole	3/4 end mill	4

After measuring cutting tools, you enter the offsets as follows:

#	Value:
1	6.4435 (length of tool 1)
2	5.5443 (length of tool 2)
3	7.3342 (length of tool 3)
4	6.9448 (length of tool 4)
31	0.5000 (radius of tool 1)
32	0.6250 (radius of tool 2)

Questions:

1) You're ready to run tool number one (the 1.0 end mill that mills the right side) for the first time. You notice that there is a pretty close tolerance (+/- 0.002 inch) on the 5.5 inch dimension that this milling cutter machines. You're worried that this tool may machine too much stock on its first try, so you decide to trial machine. You increase the value of offset number 31 by 0.01 inch (making it 0.510) and allow this milling cutter to machine the right end. When the tool is finished, you stop the machine (with the optional stop switch) and measure the 5.5 dimension. You find that it comes out to 5.509. What must you do to offset 31 to make this milling cutter machine properly when you re-run the tool? Be specific.

2) After re-running the 1.0 end mill, the 5.5 dimension comes out right on size (to 5.5). You're ready to run tool two, the 1.25 end mill that machines the pocket. Again you notice the tight tolerance (+/- 0.001) and decide to trial machine. You increase the value of offset number 32 by 0.010 (making it 0.635). You also notice the even tighter tolerance on the pocket depth (+/- 0.0005) and decide to trial machine this surface as well. So you increase the value of offset number two by 0.010 (making it 5.5543). After this tool machines you stop the machine. You measure the pocket size in XY and it comes out to 1.982 along X and 2.732 along Y. You measure the pocked depth and find it to be 0.366. What must you do to offset 32? What must you do to offset 2? Be specific.

3) After re-running the 1.25 end mill, you're ready to run tool number three (the 3/8 drill). Does this tool require trial machining? (yes or no) Why?

4) You move on to tool number four (the 3/4 end mill that plunges the 3/4 counter-bore). This tool machines in a plunging fashion, just like a drill. Again, the depth of this pocket (0.25) has a tight tolerance (+/- 0.002), so you decide to trial machine. You increase the value of offset number four by 0.01 (making it 6.9548) and run the tool. After machining, you stop the machine. When you measure the depth of the counter-bored hole, you find it to be 0.238. What must you do to offset number four?

Answers: 1) Reduce offset number thirty-one by 0.009. 2) Reduce offset thirty-two by 0.009 and reduce offset number two by 0.009. 3) No, this tool does not require trial machining because none of the dimensions it machines have tight tolerances. 4) Reduce offset number four by 0.012.

Lesson 18

Buttons And Switches On The Operation Panels

While there are many buttons and switches on a CNC machining center, you must try to learn the reason why each one exists. If you don't, you may be overlooking a helpful – if not necessary – machine function.

You now know the tasks required to setup and run a CNC machining center. During our discussions of these tasks, we have mentioned several specific buttons and switches. So you also know the function of some of the most important buttons and switches on the machine. But there are still some buttons and switches that you have not yet been exposed to. In Lesson Twenty-One, we're going to introduce *all* of the buttons and switches that are found on a typical CNC machining center.

The two most important operation panels

Most machining centers have at least two distinct operation panels. We'll be calling them the *control panel* (designed by the control manufacturer) and the *machine panel* (designed by the machine tool builder).

For machining centers that have a Fanuc control, the control panel will be remarkably similar from one machining center to the next. Indeed there are not too many variations, even among different Fanuc control models.

But since the machine panel is designed by the machine tool builder, machining centers that have been manufactured by different machine tool builders will have substantially different machine panels. To compound this problem, machine tool builders can't seem to agree on the functions needed by CNC setup people and operators.

While we can be pretty specific about the function of buttons and switches found on the control panel, we will be a little vague about machine panel buttons and switches. Also, there may be buttons and switches on your machining center's machine panel that we don't cover in this text. If you find one, be sure to reference the machine tool builder's operation manual to determine the function for the button or switch.

And by the way, *a proficient setup person or operator knows the function of all buttons and switches* on their machine/s. While some may be seldom or never used, you must not consider yourself fully capable of running a machine until you know the function of all buttons and switches. Again, if we don't cover a given button or switch in this text, you must read the machine tool builder's operation manual. Don't stop until you know why every button and switch is on your machining center.

The control panel buttons and switches

Figure 7.14 shows the control panel for one popular control model. Again, this operation panel is made by the control manufacturer (Fanuc in our case). For the most part, it is used in conjunction with the display screen (on the left side of this control panel). Notice the power buttons on the left side. The power-on button is usually the *second* switch used to power up the machine (the first is usually a main breaker switch that is placed behind the machine). The power-off button is used, of course, as part of the procedure to turn off the machine.

As you know, the display screen is used to display all kinds of important information. We've already introduced the four most important display screen modes (position, program, offset, and program check). In this lesson, we'll discuss them in much greater detail.

On the right side of the control panel is the keyboard. Notice that this particular keyboard does not resemble the keyboard used with personal computers (some control panel keyboards do). With this keyboard, only the functions needed for CNC machining center usage are provided – along with some special functions that you will not find on a computer keyboard.

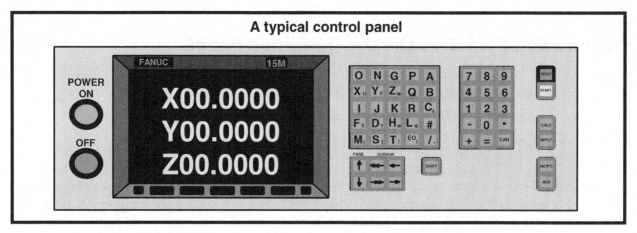

Figure 7.14 – The control panel for a popular control model

Display screen control keys (soft keys)

With most current CNC controls, these keys are located under the display screen (see the seven buttons under the display in Figure 7.15). They are called *soft keys*, and are used to manipulate what is shown on the display screen. They are a little difficult to describe because their functions change from one display page to another (which is why they're called *soft* keys). In most cases, when you press one, the functions of all of them change. Their current functions will always be shown at the bottom of the display screen. Look at Figure 7.15 for an example.

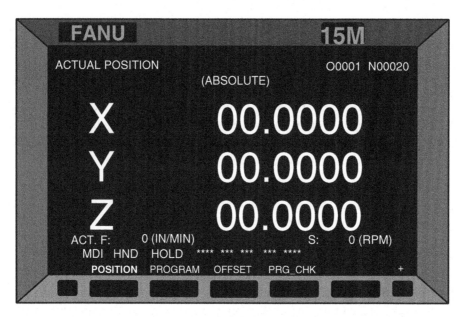

Figure 7.15 – The display screen of a CNC control – notice the soft keys at the bottom of the screen

At the present time, the position mode of the display screen is selected (notice that the word POSITION is bold above one of the soft keys). Right now, other display screen choices are PROGRAM, OFFSET, and PRG_CHK (program check).

Notice the two smaller keys on the right and left end of the soft keys. The key on the left will always bring back your most basic display screen choices (as are shown now). The key on the right will (with the plus sign above it) will allow you to see more of what is in the current display screen mode.

While mastering the soft keys on the display screen takes practice, they are not at all difficult to master. Actually, the display screen modes are quite intuitive. Let's look at some of the most important display screen pages.

Position display pages

When you press the POSITION soft key, you will be shown the current position display page. With most controls, there are four position display pages, *absolute* (as is shown in Figure 7.15), *relative*, *machine*, and *all* (a combination of all pages along with the distance-to go display). To toggle from one position display page to another, simply continue to press the soft key under POSITION. Figure 7.16 shows the various position display pages (except the absolute position display page, which is shown in Figure 7.15).

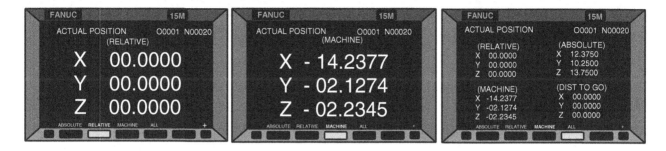

Figure 7.16 – Position display screen pages

If you've read the entire text to this point, you already know the meaning of each position display screen page. The *absolute* position display page shows the current position relative to program zero. The *relative* position display page is used to take measurements on the machine (like tool length compensation measurements), allowing you to set a point of reference for the measurement. The *machine* position page shows the current position relative to the zero return position. And the *all* position page additionally shows the distance-to-go display, which is extremely important when you are verifying programs.

Program display pages

When you press the soft key under PROGRAM, you'll be shown the current program display screen page. And again, to toggle through them, simply keep pressing the program soft key. What you'll see on each program display screen page depends upon which mode the machine is in. In the *edit* or *memory* mode, you'll see the currently active program. In the manual data input (MDI) mode, you'll see a special page that allows you to enter and execute MDI commands. Figure 7.17 shows the pages available in the program mode

Figure 7.17 – Program display screen pages

The left-most illustration shows what you'll see in the edit or memory mode. The middle illustration shows what you'll see in the MDI mode. And the right-most illustration shows the directory page, which allows you to see the programs that are currently in the machine.

Offset display pages

When you press the soft key under OFFSET, you'll see the current offset display screen page. And keep pressing this button to toggle through the various offset display screen pages. Figure 7.18 shows them.

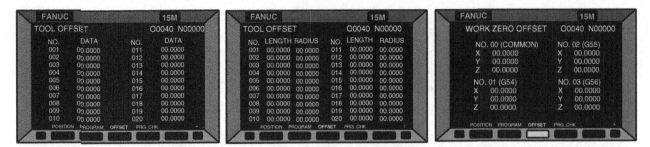

Figure 7.18 – Offset display screen pages

The left-most illustration shows the tool offset page for a machine that has one value per offset. The middle illustration shows the tool offset page for a machine that has two values per offset. And the right-most illustration shows the fixture offset page.

And again, if you have read the entire text to this point, you know the functions of these display screen pages. The tool offset page is used to enter and modify tool length and cutter radius compensation offsets. The fixture offset page is used to enter and modify program zero assignment values.

Program check display pages

When you press the soft key under PRG_CHK, you will see the program check display screen page, as Figure 7.19 shows.

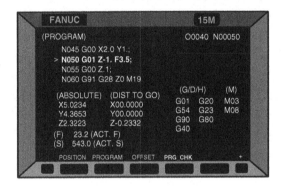

Figure 7.19 – Program check display screen page

As you know, this display screen page is used when you are verifying CNC programs. It shows a portion of the program, including the currently active command. It shows the most important currently instated G and M words, the machine's current position relative to program zero, and the very important distance-to-go display. Again, if you have read the entire text to this point, you know the importance of all of these functions.

Other display screen modes

There are some other display screen modes, but they are not of immediate importance to setup people and operators. From the main display screen page (that shows POSITION, PROGRAM, OFFSET, and PRG_CHK at the bottom, if you press the left-most soft key, you'll see display screen modes labeled SETTING, SERVICE, and ALARM.

The SETTING display screen pages are used to set the default modes for the machine. You can, for example, choose whether the machine will power up in the inch or metric mode from this page. The SERVICE display screen pages will allow you to view and change control parameters. And while the ALARM display screen pages will be important to setup people and operators, if the machine does go into alarm state, this display screen page will be *automatically* shown (there is no need to press the soft key under ALARM when an alarm is sounded).

The keyboard

Now let's turn our attention to the keyboard (shown to the right of the display screen in Figure 7.14).

Letter Keys

This keypad allows character entry. Some control panel keyboards allow only those alpha keys needed for CNC programming (N, G, X, Y, Z, etc.), while others provide a full character set (A through Z). With the keyboard shown in Figure 7.14, you must use a special *shift key* in order to gain access to lesser used characters (like parentheses and alphabet letters that are not used with CNC programming).

The slash key (/)

This key allows the entry of the slash code into programs. The slash code is the block delete character. Block delete is described in Lesson Eighteen.

Number keys

These keys allow numeric data entry, as is required when entering and modifying offsets and programs. Normally located close to the letter keypad, most CNC controls have these keys positioned in much the same way as they are on the keypad of an electronic calculator.

Decimal point key

This key allows numeric entry with a decimal point. Setting offsets and entering CNC programs are examples of when it will be needed.

The input key

This key is pressed to enter data. Examples of when this key is pressed include entering tool offsets and fixture offsets. This key is *not* used for entering words and commands into a program. Instead the *insert* key (a soft key that is shown when the display screen is in the program mode) is used fro inserting words and commands.

Cursor control keys

The display screen often shows a *cursor* that designates the current entry position. Examples include when working on the active program or when entering offset data.

How the cursor is moved will vary from one control model to another. With the control panel shown in Figure 7.14, the up and down arrows will move the cursor from page to page. The double arrows left and right will move the cursor from line to line. And the single arrows left and right will move the cursor left and right in the current line.

Program Editing Keys

Programs can be modified right at the machine using these keys. With most controls, they are soft keys that only appear when the display screen is in the program mode (and when the mode switch is set to edit). So you won't find them on the keyboard in Figure 7.14. See Lesson Twenty-Three for step-by-step procedures for entering and modifying programs.

Insert key

Not to be confused with *input*, this program editing key allows new words and commands to be entered into a program. Most controls will insert the entered word or command *after* the cursor's position.

Alter key

This program editing key allows words in a program to be altered. After positioning the cursor to the *incorrect* word in the CNC program, enter the correct word, and press the alter key.

Delete key

This program editing key allows program words and commands to be deleted. Most CNC controls allow you to delete a word, a command, a series of commands, or an entire program.

Reset key

This very important key has three functions. First, in the *edit* mode, the reset key will send the cursor to the beginning of the program.

Second, in the *memory* or *auto* mode while executing programs, the reset key will clear the *look-ahead buffer* and stop execution of the program. This cancels the cycle's execution and is required when there is something wrong in the program and you wish to stop executing the program. This is common when verifying a new program. *Be careful with the reset key.* Again, the reset key will clear the look-ahead buffer. If the program is reactivated (by pressing cycle start) immediately after the reset key is pressed, the control will skip the commands that were in the look-ahead buffer. This, of course, can lead to serious problems.

Third, when the control is in alarm state, the reset key will clear the alarm as long as the problem causing the alarm has been corrected.

The machine panel

As stated, the machine panel is designed and built by the machine tool builder. And machine tool builders vary dramatically with regard to the functions they provide on the machine panel. They vary most with regard how much manual control they provide for machine functions (like spindle activation and automatic tool changer control). Figure 7.20 shows a typical machine panel for a vertical machining center. As you look at this operator's panel, notice how many of its functions have been introduced in previous lessons.

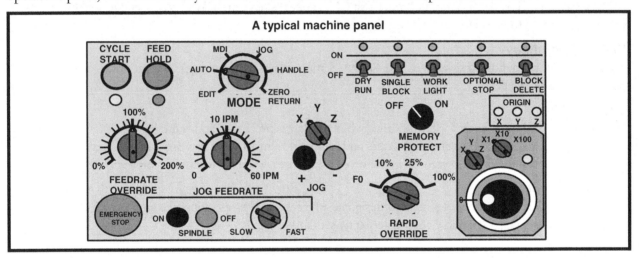

Figure 7.20 – A typical machine panel

Mode switch

The mode switch is the heart of a CNC machining center. It is so important that we devote an entire Key Concept to discussing its use. It must be *the very first switch you set when performing any function* on the machine. The mode switch must be positioned properly in order for the desired function to be performed. If it is not, the control will not respond to your action. Again, we discuss the mode switch in great detail during Lesson Twenty-Two.

Cycle start button

This button has two functions. First, it is used to activate the active program when the machine is in the memory or auto mode. Second, the cycle start button can be used to activate manual data input (MDI) commands. We recommend that you to get in the habit of having a finger ready to press the feed hold button (which is usually in close proximity to the cycle start button) whenever you press the cycle start button.

Feed hold button

While a program (or MDI command) is being executed, this button allows you to halt axis motion. The cycle start button can be used to reactivate the cycle. Note that all other functions of the machine (coolant, spindle, etc.) will continue to operate even when the machine is in feed hold state.

You should think of feed hold as your *first panic button.* You should *always* have a finger on this button whenever you press the cycle start button. You should also keep a finger ready to press the feed hold button for the entire time that you are verifying a program.

Some people feel the *emergency stop button* should be considered the panic button. However, the emergency stop button will actually turn the power off to the machine tool, which sometimes causes more problems than it solves. For example, when the emergency stop button is pressed and the machine power is turned off, the axes of the machine will drift until a mechanical break can lock them in position. Axes bearing a great deal of weight (like the Z axis of most vertical machining centers) are most prone to drift. While the amount of drift is usually quite small (less than .010 inch in most cases), if a cutting tool is actually machining a workpiece when the emergency stop button is pressed, the drift could cause damage to the tool and workpiece.

Feedrate override switch

This multi-position switch usually has two functions. Under normal circumstances, this switch allows you to change the programmed feedrate during cutting motions (like G01, G02, and G03). Notice we said *feedrate during cutting motions*. Under normal conditions, this switch allows no control of *rapid motion*. The feedrate override switch usually allows the programmed feedrate to be changed in 10% increments and will usually range from 0% through 200%. This means the operator can actually stop cutting motions (at the 0% setting) at one end of the feedrate override switch and double the programmed feedrate at the other.

If the program is written correctly, the entire cycle can run at 100% feedrate, meaning once a program is verified, you should be able to set this switch at 100% and leave it alone. But when you are verifying a program, running the very first workpiece, this switch can be very helpful to confirm that the programmed feedrate for each tool is correct.

For example, say a cutting tool has been cautiously brought into its first cutting position under the influence of single block and dry run (as is described in Lesson Twenty). Now the cutting tool is ready to make its first machining motion. If you are in doubt as to whether the cutting conditions are correct (feeds and speeds), you can turn the feedrate override switch down to its lowest value. With single block still on but with dry run turned off, you can press the cycle start button to allow the tool to begin its cutting motion. As the tool begins to machine the workpiece, you can slowly crank up the feedrate override switch – shooting for the 100% setting. If the 100% position cannot be achieved – or if you end up with a position greater than 100%, the programmed feedrate must be changed. Again, a correct program will let you run the entire cycle at the 100% setting.

Most machine tool builders use the feedrate override switch for a *second* purpose. When the dry run switch is turned on, the feedrate override switch is used to take control of the motion rate for all movements (including rapid motions). This function is necessary during program verification, when you're running the program for the first time without a workpiece to check for setup-related mistakes. Some machine tool builders use the *jog feedrate switch* for this purpose.

Rapid traverse override switch

As the name implies, this function is used to slow the machine's rapid rate. As you know, the rapid rate on current model machining centers is very fast (some machines rapid at well over 1,500 inches per minute), and this switch brings some welcomed control over the very fast rapid rate.

With most machines, the rapid traverse override switch is a multi-position switch (as is the case with the one on our example machine panel). But we have seen some that are simple on/off switches. When this switch is on, rapid rate is slowed to about twenty-five percent of its normal rate. When it is off, rapid motions will at 100%.

Notice that our rapid override switch has four settings, F0, 10%, 25%, and 100%. At the F0 setting, the machine will move at about 1% of its rapid rate when rapid movements are commanded.

Generally speaking, the rapid traverse override switch is used for program verification. It's nice to leave it at 10% or 25% when you're verifying programs. You can rest assured that the machine will never achieve its true rapid rate. This will give you more time to react should something go wrong.

Emergency stop button

This button will turn power off to the machine tool. Usually, power remains on to the control. See the description of the feed hold button for more information about how emergency stop is used.

Conditional switches

There are several on/off switches on the machine panel that control how the machine behaves during automatic and manual operation. They could be toggle switches (as our machine panel shows), locking push-buttons, or lighted buttons (a light within the button comes on when the function is on).

Frankly speaking, these functions are more related to the control panel than they are the machine panel, so they remain amazingly similar from one machining center to another, regardless of the machine tool builder. These switches are very important. If one improperly set, the machine will not perform as expected. You must get in the habit of checking each of these switches before a CNC program is executed.

Dry run on/off switch

As described in Lesson Twenty, this conditional switch is used during program verification. When this switch is on, it gives you full control of the motion rate for all movements the machine makes (including cutting movements and rapid movements).

When the dry run switch is turned on, another switch (usually feedrate override) will act as a rheostat – and allow you control of all motion rates. Dry run will slow rapid motions, but it will tend to speed up cutting motions. This means you should *never allow a cutting tool to machine a workpiece when dry run is turned on* (you'll have no idea as to what the actual feedrate is).

Single block on/off switch

As described in Lesson Twenty, this conditional switch will force the machine to stop after it executes each command. To reactivate the cycle (execute the next command), you must push the cycle start button.

This switch is helpful during program verification. With a new program, you must cautiously check each motion the machine makes, especially movements that have a cutting tool approach the workpiece. With single block turned on, you can rest assured that the machine will stop at the end of each motion, giving you a chance to check the motion just made and to check the program to see what is going to happen next. It is commonly used with the distance-to-go display, which shows how much further the machine is going to move in the current command.

Block delete on/off switch (also called optional block skip)

The applications for block delete are described in Lesson Eighteen. The block delete switch works in conjunction with slash codes (/) in the program. If the control sees a slash code in the program, it will look to the position of the block delete switch. If the switch is on, the control will ignore the words to the right of the slash code. If the block delete switch is off, the control will execute the words to the right of the slash code. This function is used to give the operator a choice between one of two possibilities. (See Lesson Eighteen to learn more about block delete applications.)

Optional stop on/off switch

This conditional switch works in conjunction with an M01 word in the program. When the control sees an M01, it looks to the position of the optional stop switch. If the switch is on, the control will stop the execution of the program and turn off certain machine functions (like spindle and coolant). You must press the cycle start button to reactivate the program. If the optional stop switch is off, the control will ignore the M01 and continue executing the program.

In the programming format shown in Lesson Fifteen, we recommend placing an M01 in the program at the end of each tool. This gives you the ability to stop the machine at the end of each tool by simply turning on the optional stop switch. Stopping at the end of each tool is important during program verification (as described in Lesson Twenty) to let you check what one tool has done before going on to the next.

Buttons and switches for manual functions

The machine panel for all CNC machining centers will include some buttons and switches related to manual control of the machine's functions. These buttons and switches vary dramatically from one builder to the next. It seems no two builders can agree on what manual functions a setup person or operator needs.

Axis jogging controls

On our example machine panel (Figure 7.20), notice the multi-position switch labeled jog feedrate, the two push-buttons (plus and minus) and the axis selector (X, Y, and Z). These functions are active when the machine is in a manual mode (jog or zero return on the mode switch). To jog an axis, after selection a manual mode, you will first select the axis to be jogged (with the X, Y, Z switch). Then you will select the motion rate (with the jog feedrate switch). Finally, you will press the desired direction button (plus or minus). As you press the button, the selected axis will begin moving in the selected direction at the selected feedrate. This, of course, requires that you know which axis you want to move and which way is plus and which way is minus (as is described in Lesson One and at the beginning of the discussions for Key Concept Number Seven).

Jogging an axis is often required. Applications include measuring program zero assignment values and tool length compensation values on the machine, sending the machine to its zero return position, and moving an axis into a convenient position so the cutting tool in the spindle can be inspected or replaced.

Positioning is not very precise with jog functions. For this reason, most machine tool builders also provide a *handwheel* on the control panel.

Handwheel controls

On the example machine panel, notice the handwheel in the lower-right corner of the panel. The handwheel is active when the mode switch is set to handwheel (with many machines, a light next to the handwheel will come on when the handwheel is active). Also notice the axis selector switch (X, Y, Z) and the rate selector switch (X1, X10, and X100).

To use the handwheel, after you select the handwheel mode, you first place the axis selector to the desired position (X, Y, or Z). Then you select the motion rate. In the X1 (times one) position, each increment of the handwheel is 0.0001 inch (or .001 mm), and motion is very slow – barely detectable except with the position displays. In the X10 (times ten) position, each increment is .001 inch (or 0.010 mm), and motion is faster – easily detectable. In the X100 (times one hundred) position, each increment is 0.010 inch (or 0.100 mm), and quite fast.

Using the handwheel is often required after using the jog function to manually move an axis. Again, the jog function does not allow precise positioning control. So you will commonly use jog to bring an axis close to its intended position, and then use the handwheel for the rest of the motion. Applications include measuring program zero assignment values and tool length compensation values on the machine.

Spindle control

Manual spindle control is not often required. About the only time we've needed to manually activate the spindle is when using an edge-finder to measure program zero assignment values.

Some machining centers do allow manual control of the spindle. With our example machine panel, for example, you have a speed selector rheostat and buttons to start and stop the spindle. But notice that you cannot select the spindle direction (with this machine panel). Some machine panels don't provide *any* manual spindle control.

For any machine function that you must activate manually – but for which there are no manual controls – you must use the manual data input mode (described in Lesson Twenty-Two).

Automatic tool changer control

Notice that there are no manual controls for the automatic tool changer on our example machine panel. Very few machine tool builders provide manual controls that allow you to make tool changes. Yet you will probably have to make manual tool changes on a regular basis. If, for example, you measure tool length compensation values on the machine during setup, it is usually best to load the tools into the automatic tool changer magazine and then use the automatic tool changer system to load each tool into the spindle for measuring.

For any machine function that you must activate manually – but for which there are no manual controls – you must use the manual data input mode (described in Lesson Twenty-Two).

Indicator lights and meters

Most CNC machining centers have a series of lights and meters that allow the operator to monitor the condition of important machine functions.

Spindle rpm and horsepower meters

Most machining centers have two meters that show you key information about the spindle. The first meter is the rpm meter, which shows you how fast the spindle is currently rotating. Some machines show spindle speed on the display screen. The second spindle-related meter is a *load meter*. If the machine tool has this meter, you can monitor the amount of stress on the spindle drive system. Usually this meter shows a percentage-of-load, ranging from 0% through 150%. This means you can easily tell to what extent a machining operation is taxing the spindle of the machine.

Axis drive-motor horsepower meter

This meter, if equipped, allows the operator to see how much horsepower is being drawn by any of the machine axes drive motors. Usually there is only one meter and you must select which axis (X, Y, or Z) you wish to monitor by a three-position switch. Like the spindle horsepower meter, this meter allows you to see how much stress is on a drive motor during a machining operation.

Cycle indicator lights

Most CNC machines have two indicator lights to show whether the machine is in cycle. One is above (or close to) the cycle start button and will stay on as long as the machine is in cycle. The other is above (or close to) the feed hold button and comes on when the machine is in feed hold state (when you press the feed hold button).

Zero return position indicator lights

All machining centers have indicator lights that come on when an axis is at its zero return position. These lights are commonly labeled *axis origin lights*. If a program is planned to start from the zero return position, these lights can be helpful to an operator. They will tell you whether it is safe to activate the cycle.

Optional stop indicator light

Some machining centers have an indicator light that is close to the optional stop switch. If the machine has been halted by an optional stop in the program (M01), this indicator light comes on to tell you why the machine has halted operation.

Other buttons and switches on the machine panel

Remember that the machine panel on your particular machine may have more buttons and switches than we have described – especially if the machine is equipped with special accessories (like a pallet changer or automatic loading system). When you come across a button or switch that you don't recognize, be sure to reference the machine tool builder's operation manual to find out what it does.

Other operation panels on your machining center

Many machine tool builders include additional operation panels. Because the automatic tool changer is mounted quite a distance from the main operation panels, many machine tool builders place a special automatic tool changer control panel right next to the automatic tool changer. Buttons and switches on this panel will commonly allow magazine rotation, tool station numbering, and if required, clamp & unclamp of the tool in the loading station.

If your machine has a pallet changer, it will likely have a special operation panel mounted near the pallet changer to allow manual pallet changing, and to allow you to press a button (commonly called the *standby button*) to inform the machine that you have finished loading the workpiece and that it is safe to make a pallet change when it is commanded.

Key Concept

8

Know The Three Basic Modes Of Operation

When it comes right down to it, every button and switch on the machine can be divided into one of three categories. It is related to manual mode, manual data input mode (MDI), or program operation mode.

Key Concept Number Eight is a short, one-lesson Key Concept:

    22: The three modes of operation

We mentioned the mode switch in Lesson Twenty-One. You know it is the most important switch on a CNC machining center – and it is the first switch you should set when you perform any operation on the machine.

Lesson 19
The Three Modes Of Operation

The most common operation mistake is having the mode switch in the wrong position. Fortunately, this mistake will not cause serious problems. The machine will not respond to your action.

You know the mode switch of a machining center is a multi-position switch. The actual positions of a typical mode switch include edit, memory (or auto), jog, zero return, and handwheel. Though most mode switches have at least five actual positions, there are really only three basic modes of operation. Mastering these three modes will be the focus of Lesson Twenty-Two.

The manual mode
In the manual mode, a CNC machining center behaves much like a conventional milling machine. The manual mode positions of the mode switch include *jog* (sometimes called *manual*) mode, *handwheel* mode, and *zero return* mode.

You know from previous lessons that you will often need to perform manual functions. Examples include measuring tool length compensation offset values, measuring the program zero assignment values, and when replacing dull tools.

With the manual mode, you will press a button, turn a handwheel, or activate a switch that will cause an *immediate response* from the machine. An axis will move, the spindle will start, the coolant will come on, or some other machine function will respond to your action.

As you know from Lesson Twenty-One, the *jog* mode switch position will allow you to manually move a selected axis. After selecting the jog mode, you select the axis to move (X, Y, or Z), select the motion rate (with the jog feedrate switch) and press a button corresponding to the direction you want the axis to move (plus or minus). The machine will respond by moving the selected axis at a selected motion rate in the selected direction.

In similar fashion, the *handwheel* mode switch position will allow you to move an axis with the handwheel. Again, after placing the mode switch to the handwheel position, you select the axis to move (X, Y, or Z), select the rate of motion (with **X1, X10,** or **X100**), and turn the handwheel. The machine will respond by moving the selected axis.

In the *zero return* mode switch position, you can manually send the machine to its zero return position. Machining centers vary when it comes to the actual procedure to send an axis to its zero return position. With most machines, after placing the mode switch to the zero return position, you select the axis to be zero returned (X, Y, or Z) and then press the plus button. Hold it until the axis reaches its zero return position (and the axis origin light for the axis comes on).

In any of the manual mode switch positions, certain other buttons and switches may be active. Some machining centers allow manual control of the spindle and coolant. Some (but not many) allow manual control of the automatic tool changer. The related buttons and switches will be aptly named and placed on the machine panel.

But as you know from Lesson Twenty-One, three are usually machine functions that you need to control manually, that the machine tool builder has not provided buttons or switches to control. The automatic tool changer is a classic example. Few machining centers allow you to manually activate the automatic tool changer.

For functions that you need to activate manually but for which you have no manual buttons and switches, you must use the manual data input (MDI) mode switch position.

The manual data input mode

This mode includes two positions on the mode switch, the *edit* position, and the *manual data input* position (MDI). With both positions, the operator will be using the display screen and keyboard to enter data.

Though these two mode switch positions have substantial differences, we consider them together for two reasons. First, both mode switch positions provide manual capabilities that can be done in a more automatic way. With the edit mode switch position, an operator can enter CNC programs into the controls memory. This can also be accomplished by loading the program a distributive numerical control (DNC) system. With the MDI mode switch position, CNC commands are entered and executed just like CNC program commands.

Second, both mode switch positions involve entering information through the control panel keyboard. With the edit position of the mode switch, a program is entered or modified with the keyboard. With the MDI position, CNC commands are entered and executed with the keyboard. Let's discuss each MDI-related mode switch position in detail

The manual data input (MDI) mode switch position

Manual data input (MDI) mode switch position is used for two reasons. First, it is used to perform manual functions that cannot be done by any other means. Again, machine tool builders vary with regard to what can be done in a completely manual manner. For those machine functions of which you have no manual control, you must use the MDI mode.

Second, MDI mode can be used to perform certain manual function of which you do have manual control, but can do faster or easier in the MDI mode. A manual zero return on most machines, for example, takes much more time and effort to complete than it does in the MDI mode.

Commanding an MDI zero return

The MDI mode requires that you know the CNC command to activate the manual function you want to perform. This is the same command used in a CNC program to activate the manual function. If, for example, you want to use the MDI mode to command a zero return in all three axes, you must know the program command needed to do so, which happens to be

G91 G28 X0 Y0 Z0;

When this command is entered and executed in the MDI mode, the machine will rapid to the zero return position in all three axes (simultaneously).

Notice the semi-colon (;) at the end of the **G28** command example. This is the character most (Fanuc) controls use to represent an end-of-block. There is a key on the keyboard labeled **EOB**, which stands for end-of block. When you enter an MDI command for most controls, it must end with the end-of-block character. With these control models, when you are finished entering an MDI command, you must remember to press the EOB key prior to inserting the command into the MDI buffer.

If you want only the Z axis to be sent to the zero return position (which is the tool change position on most vertical machining centers), the command will be

G91 G28 Z0;

The complete procedure to give an MDI command

The step-by-step procedure to use the MDI mode with current model Fanuc controls is as follows:

1) Place the mode switch to MDI

2) Press the soft key under PROGRAM (the MDI display screen page is shown)

3) Using the keyboard, enter the MDI command you wish to execute (for the zero return, this is G91G28X0Y0Z0). There is no need for spaces between each word.

4) Press the EOB key (this places a semi-colon at the end of the command).

5) Press the soft key under *Insert* (as soon as you started typing the MDI command, the soft keys changed to include the insert key). When you press insert, the command moves up into the active MDI buffer.

6) Press the *start button* on the control panel keyboard or the *cycle start button* on the machine panel. Because the cycle start button is so close to the feed hold button, and since we recommend that you always have a finger ready to press feed hold when you execute an MDI command, you should get in the habit of using the *cycle start button* to activate MDI commands. As soon as you press this button, the machine will perform the commanded action. In this case, the machine will rapid to the zero return position in all three axes.

Commanding an MDI tool change

We've mentioned several times to this point that most machining centers provide no manual control of the automatic tool changer. If you want to make a manual tool change, you must use the MDI mode to do so. Just as in a CNC program, however, the machine must be at its tool change position. And again, you must know the CNC words related to your automatic tool changer. For most machining centers, a *T word* places the tool in the ready station of the magazine and an **M06** word actually makes the tool change. As long as the machine is at the tool change position, the command

T04 M06;

will place tool number four into the spindle. This command can be entered in step number three of the procedure just shown.

Commanding spindle activation with MDI

While some machining centers provide adequate manual control of the spindle, others do not. If yours does not, you can use the MDI mode to activate the spindle. As you know from the programming lessons, an *S word* is used to specify the desired speed in rpm. An **M03** will start the spindle in the forward direction (needed for right-hand tools) and an **M04** will start the spindle in the reverse direction (needed for left-hand tools). When you want to stop the spindle, an **M05** must be commanded. The command

S700 M03;

will start the spindle at 700 rpm in the forward direction. The command

M05;

will stop the spindle. Again, enter these commands in step three of the procedure shown above.

Other times when MDI is used

Again, any time you need to perform a manual function, you can use MDI to do so. Any machine function can be activated in MDI mode – as can any G code. Truly, if a command works in a program, it will work in the MDI mode. Some examples include

- Coolant (M08 and M09)
- Pallet changer (M60 on most machines)
- Door open and door close (if the machine has automatic doors)
- Chip conveyor on and off (if the chip conveyor is programmable)
- Switching between inch and metric modes (G20 and G21)

Can you make motion commands with MDI?

The MDI mode can even be used to machine a workpiece. Since just about all CNC commands that run in a program will work in the MDI mode (including G00, G01, G02, G03, G90, and G91), you can make machining commands in the same way they are commanded in a CNC program. You can even enter several commands at a time – just remember to enter the EOB key at the end of each command.

However, you must be extremely careful when using MDI to make motion commands. While some operators get very good at making motion commands with MDI, your command/s will be executed just as you enter them. If you make a mistake while entering a CNC command in the MDI mode, the results could be disastrous. You will have no chance to verify your MDI commands as you can with a CNC program. And MDI commands cannot be saved. As soon as you press the cycle start button to execute MDI commands, they will be lost.

The edit mode switch position

With the *edit* mode switch position, you can enter new programs (which can be time-consuming) or modify programs that are currently in the machine's memory. Aptly named, editing functions include *insert*, *alter*, and *delete*. *Insert* allows you to enter new words and commands into a program. *Alter* allows you to modify words in the program. And *delete* allows you to delete a word, a command, a series of commands, or an entire program.

Some machines have a feature called *memory protect* that can be used to keep the operator from making changes in a program. A special *key* (like the key to your home) is used to turn on and off this function. If the memory protect function is turned on and the key is removed from the machine, you will not be able to modify programs.

In order to modify programs, you must understand programming words and commands. If you have read the programming lessons in this text, you should have a good understanding of programming. Without this understanding, an operator cannot make safe and correct changes to a CNC program (which is the reason many companies use the memory protect function – they have CNC operators that are unfamiliar with programming).

The step-by-step procedures to use program editing functions is shown are Lesson Twenty-Three. But we want to give a few examples.

All program editing procedures begin with:

1) Place the mode switch to edit.

2) Press the soft key under program (the active program is shown). Be sure the active program is the one you want to edit.

3) Turn off the memory protect function (if the machine has this function). This requires a key.

You will notice a cursor somewhere on the page (a highlighted or back-lit word in the program). Before performing any editing function you must to move the cursor to the desired location. One way to do so is with the cursor control keys (described during the control panel discussions in Lesson Twenty-One). Cursor control keys consist of a series of arrow keys that allow you to move the cursor – one word, one command, or one page at a time.

Another way to move the cursor is to use the *forward-search* and *backward search* functions. You can type the CNC word to which you want the cursor positioned and then press the soft key under forward-search or backward-search (depending upon the cursor's current position). The machine will bring the cursor to the next occurrence of the word you typed. Again, procedures for editing programs, including cursor movement, are shown in Lesson Twenty-Three.

Once the cursor is at the desired position, you will be able to insert a new word or command after the current cursor position, modify the word on which the cursor is placed, or delete words or commands starting from the cursor's current position.

Say, for example, the cursor is currently at the beginning of the program (on the O word). You want to change the first feedrate word in the program to **F4.5**. If you can see the F word on the current page of the display screen, the easiest way to position the cursor may be to use the arrow keys. Or you could type **F** and press the forward search soft key. The control will scan to the next occurrence of an F word – which is the F word you want to change. With the cursor on the F word to be changed, type **F4.5** and press the *alter* soft key. The feedrate word will be changed to **F4.5**.

To make a program in memory the active program (to call up a program)
We have been referring to the *active program*. CNC controls can hold several programs, but only one of them will be the active program. A program must be the active program in order for you to modify it or run it. To make a program the active program, first follow the three steps above. Then:

> 4) Type the letter address O (letter O *not* number zero) and the program number for the program in memory that you want to make active.

> 5) Press the soft key under forward search. As long as the program you've typed is in the machine's memory, it will be shown on the display screen. It is now the active program. If the program number you've typed is not in the machine's memory, an alarm will be sounded.

To enter a new program
Entering programs using the display screen and keyboard is time-consuming. Hopefully your company has a distributive numerical control (DNC) system to eliminate this time consuming task.

First, follow the three steps given above. Then:

> 4) Press the letter address O key (letter O *not* the number zero) and type program number to be entered.

> 5) Press the soft key under insert.

> 6) Press the EOB key and press the soft key under insert.

> 7) Type the first command of your program followed by the EOB key and then press the soft key under insert.

> 8) Enter the rest of the commands in the program, ending each by pressing the EOB key and then the soft key under insert.

What if I make a mistake when typing?
There will always be a kind of back-space key. With most Fanuc controls, it is the *cancel* key. When you press it while entering a command, the cursor on the command entry line will back up one space, deleting the last character you typed.

Remember that a CNC control makes a very expensive typewriter. Many machines cannot be running a workpiece at the same time a program is entered (unless the machine has a feature called *background edit*). Even with those controls that do have background edit, it is somewhat cumbersome to enter programs while the machine is running production. Also, the control panel for most machines is not mounted in a comfortable position (most are mounted in a vertical attitude). Your arm will soon get tired when typing in this position.

The program operation mode

The third mode of operation involves actually *running programs*. With current CNC controls, there is only one program operation mode. It will be called *memory*, *auto*, or *automatic* mode on the mode switch. The machine must be in this mode in order to run the active program in memory. Note that older machines have another program operation mode switch position called *tape mode*. This mode switch position only applies to machines that have a *tape reader*. No CNC machines made today do.

When in the memory or auto mode, the *cycle start button* is used to activate the program and *feed hold button* can be used to stop axis motion at any time during the cycle. (Again, keep a finger ready to press the feed hold button whenever you press the cycle start button.)

Several conditional switches (discussed in Lesson Twenty-Two) determine how the machine will behave in the program operation mode. The *dry run* switch gives the operator control of the motion rate. *Single block* forces the control to execute only one command at a time. *Optional stop* (when on) will cause the control to stop the program's execution when an M01 word is executed in the program. *Block delete* (when on) will cause the control to skip words to the right of the slash code (/). (Again, see Lesson Twenty-Two for more information about these conditional switches.)

To run the active program from the beginning

This procedure will only make sense if you understand the presentations made in Lessons Twenty and Twenty-Four. You must thoroughly understand these presentations in order to safely run a CNC program. If you don't, it is likely that you will incorrectly set one of the switches listed in this procedure – and the results will be disastrous.

1) Place the mode switch to the program operation mode (memory or auto)

2) Press the soft key under program. The active program will be displayed. Be sure the active program is the one you want to run and that the cursor is on the program number.

3) Check the position of all conditional switches (dry run, single block, optional stop, etc.).

4) Place the feedrate override switch to 100% (assuming the program is proven).

5) Place the rapid override switch to the desired position (100% if running production).

6) Be sure the machine is in an appropriate position for the program to begin (the tool change position).

7) With a finger ready to press the feed hold button, press the cycle start button. The program will run, behaving based upon how the conditional switches are positioned.

Key points for Lesson Twenty-Two:

▪ Though there are more than three positions on the mode switch, there are really only three basic modes of operation – manual, manual data input, and program operation. All buttons and switches can be placed into one of these three categories.

▪ Manual mode is used to perform an immediate action.

▪ Manual data input mode is used to use the display screen and keyboard to enter and modify commands.

▪ Program operation mode is used to run programs.

Understand The Importance Of Procedures

Running a CNC machining center requires little more than following a series of step-by-step procedures. The trick lies in knowing when a given procedure is required. From the material presented to this point, you should now know when each procedure is needed.

Key Concept Number Nine is another short, one-lesson Key Concept:

    20: The key operation procedures

From what has been presented to this point, you should be pretty comfortable with what you must do to make setups and run production – at least in theory. But when you step up to a CNC machining center for the first time, you'll probably be quite intimidated. Things you thought you knew won't seem so clear.

This will happen because you are still lacking hands-on experience – experience that we cannot provide in this text. Though we can't provide hands-on training, we can provide you with a description of the *procedures* that you are going to need in order to run your machining center.

Procedures will help you get familiar with your machining center. For example, if you need to power-up the machine at the beginning of your shift and make it ready to run production, what will you do?

Without a machine start-up procedure – or at least someone to demonstrate how this procedure is performed – you'll be lost. And trying to operate a CNC machining center without being sure of what you are doing can be very dangerous.

Even with a person available to help, there can be problems. This person may not be available every time you need them. And they will soon tire of repeating procedures that they feel you should have memorized by now. While you will *eventually* memorize often-used procedures, it may take you longer than it took the person helping you. And this can be frustrating – for you and for them.

When you are shown a procedure for the first time, *write it down*. This will keep you from having to keep asking someone for help the next time you need to perform the procedure. In Lesson Twenty-Two, we do provide a quick reference sheet for four of Fanuc's most popular control models. Though you may find them very helpful, they're just intended to get you started. You will surely come across machine functions for which we have not provided a procedure. You'll have to develop one – for yourself – and for others that come after you.

Lesson 20
The Key Operation Procedures

Step-by-step procedures can keep you from having to memorize every function that you must perform on your CNC machining center. You will soon memorize procedures for task that you perform on a regular basis – but written procedures will help you perform lesser used tasks.

We divide the procedures needed for CNC machining center usage into five categories:

- Manual procedures
- Setup procedures
- Manual data input (MDI) procedures
- Program editing procedures
- Program operation procedures

On the pages that follow, we're providing four *quick reference sheets* to help you with the most commonly used procedures. These quick reference sheets will help with some of Fanuc's most popular control models. Here is a list of procedures shown in each:

Manual procedures:
>   To power up the machine
>   To do a manual zero return
>   To manually start the spindle
>   To manually jog the axes
>   To use the handwheel
>   To manually load and remove cutting tools in the spindle

Setup procedures:
>   To load tools into the automatic tool changer magazine
>   To set or reset the relative position display
>   To enter and modify tool offsets
>   To enter and modify fixture offsets

MDI procedures:
>   To use MDI to change tools
>   To use MDI to start the spindle
>   To use MDI to do a zero return

Program editing procedures:
>   To enter a program through the keyboard
>   To load a program from a DNC system
>   To save a program to a DNC system
>   To see a directory of programs
>   To delete a program
>   To call up a program (make it the active program)
>   To search within a program
>   To alter a word in a program
>   To delete a word in a program
>   To insert a word in a program

Fundamentals of CNC © CNC Concepts, Inc.

Program operation procedures:

    To run a verified program

In addition we have shown some other important procedures in this text, including:

    To measure program zero assignment values – Lesson Six
    To measure tool length compensation values – Lesson Eleven
    To verify programs – Lessons twenty and twenty-four
    To re-run tools – Lesson Twenty

In Lesson Twenty-Four, we show two more important procedures:

    To cancel the cycle
    To re-run a tool

The goal with any step-by-step procedure is to keep you from having to memorize. But you must, of course, understand *when* and *why* to perform each procedure. If you have read the entire text, you should be pretty comfortable with the when and why. For example, say you're beginning a new setup and you've just placed the workholding device on the machine table. Now you must measure program zero assignment values. According to the general procedure in Lesson Six, you must place an edge-finder in the spindle and start the spindle at 500 rpm or so.

As you're standing in front of the machine – scratching your head – you remember that we've provided procedures to manually load tools into the spindle and to manually start the spindle (or use MDI to do so). With these procedures, even a newcomer can perform the needed tasks. But again, you must know when and why to perform them.

If you come across a task for which we have not supplied a step-by-step procedure, get someone to help you. As they demonstrate what it takes to perform the tasks, write down – in step-by-step fashion – what they do. You'll have a procedure for the *next time* you must perform the task – and you won't have to ask for help. Figure 9.1 shows a form you can use to create your own procedures.

Procedure name: _____	Machine: _____
Description: _____	
_____	
Procedure:	
1) _____	9) _____
2) _____	10) _____
3) _____	11) _____
4) _____	12) _____
5) _____	13) _____
6) _____	14) _____
7) _____	15) _____
8) _____	16) _____
Notes: _____	
_____	

Figure 9.1 – Procedure form

The two eight pages provide you with quick reference sheets for a popular Fanuc control model. But if you have other Fanuc controls, or if you have machining centers with controls that are not made by Fanuc, *make your own quick reference sheet*. It is not at all hard to do – make step-by-step procedures for the same list of tasks we have shown in our quick reference sheets.

Fanuc 16M and 18M Quick Reference For Key Procedures

To power-up the machine:

Required before doing anything else on the machine.

1. Turn on main breaker (Usually located at rear of machine).
2. Press control power on button.
3. Press machine ready or hydraulic on button. (not necessary on some machines.) Be sure that emergency stop button is *not locked in*.
4. Follow sequence to do a manual zero return. (All Fanuc controls require that you do a manual zero return as part of the power-up procedure.)

To manually jog the axes:

Required when taking measurements on the machine, when performing manual machining operations, and in general, whenever you must move an axis.

1. Place the mode switch to a manual mode (jog, manual, zero return, etc.).
2. Place axis selector switch to desired axis (X, Y, or Z).
3. Place the jog feedrate switch to the desired motion rate.
4. Using the plus or minus buttons or joystick, move the axis in the desired direction and amount.

Jogging the machine does not provide precise positioning. Use the handwheel when more precise positioning is needed.

To load tools into the ATC magazine

Required during setup and when tools must be replaced during a production run.

1. Place the mode switch to a manual mode (jog, manual, zero return, etc.).
2. Go to the tool changer and rotate it to the desired position.
3. Load the tool into the desired position by unclamping a lever or simply snapping it into position.
4. Continue rotating magazine and loading tools for all tools to load.

Machines vary when it comes to how the ATC magazine is loaded. If you have problems, consult your machine tool builder's operation manual.

To enter and modify fixture offsets

Required during setup to enter program zero assignment values.

1. The mode switch can be in any position.
2. Press the key labeled offset/setting.
3. Press the soft key under fixture offset (if needed).
4. Using the cursor control up and down arrow keys, position the cursor to the fixture offset register to enter.
5. Type the value of the offset and press input.
6. Repeat steps four and five for all axes.

To do a manual zero return:

Required right after power up. The zero return position in at least one axis is also the tool change position for most machines. The machine must be at this position prior to running a program.

1. Place mode switch to zero return.
2. Using the axis select switch select the X axis.
3. Using the minus direction push button or joystick, move the machine about two to three inches in the minus direction.
4. Using the plus direction push button or joystick, hold plus until zero return origin indicator light comes on.
5. Repeat steps 2-4 for all other axes.

To use the handwheel

Required when precise axis motion is needed - as when measuring program zero assignment values and tool length compensation values.

1. Place mode switch to handwheel.
2. Place axis select switch to desired position (X, Y, Z, or B/C).
3. Place handwheel rate switch to desired position (X1, X10, or X100).
4. Using handwheel rotate plus or minus to cause desired motion.

Use the relative position display when taking measurements.

To set or reset the relative position display

Required when taking measurements on the machine.

To reset (set to zero)

1. Mode switch can be in any position.
2. Press the key labeled position until the relative display screen is shown.
3. Type the letter address of the axis you wish to reset (X, Y, or Z). That letter will start flashing on the axis display.
4. Press the soft key under origin.

To set to a specific value

1. Mode switch can be in any position.
2. Press the key labeled position until the relative display screen is shown.
3. Type the letter address for the axis you wish to set and type its value.
4. Press the soft key under preset.

To manually start the spindle:

Required whenever you use a conventional edge-finder to measure program zero assignment values or when you will be performing manual milling operations.

1. Place the mode switch to a manual mode (jog, manual, zero return, etc.).
2. Select the desired direction (forward or reverse) with the appropriate switch.
3. If available, select the desired speed in rpm with the appropriate switch.
4. Press the spindle on button.
5. To stop spindle, press the spindle stop button.

Not all machines have manual controls for the spindle. This procedure will only work for those that do. You can also use the MDI procedure to start the spindle. Some machines allow you to manually start and stop the spindle, but not select speed or direction.

To manually load and remove tools in the spindle

Required when you want to inspect the cutting tool.

1. Place the mode switch to a manual mode (jog, manual, zero return, etc.).
2) If the machine has an air-blow switch, be sure it is turned off.
3. If there is already a tool in the spindle, hold the tool with one hand and press the unclamp button with the other. The tool will drop out of the spindle – so be ready.
4. To load a tool into the spindle, place the tool into the spindle with one hand and press the clamp button with the other.

With most machining centers, a tool holder will only fit into the spindle in one orientation.

To enter and modify tool offsets

Required during setup to enter initial offset values and when trial machining. Required during a production run when sizing is needed.

1. The mode switch can be in any position.
2. Press the key labeled offset/setting.
3. Press the soft key under wear (if needed).
4. Using the cursor control up and down arrow keys, position the cursor to the offset number to enter.
5. Type the value of the offset and press input.

You can overwrite the current offset setting by pressing the soft key under input – or you can modify its current value by your entry by pressing the soft key under input +

Fanuc 16M and 18M Quick Reference For Key Procedures

To use MDI to change tools
Required whenever a manual tool change is needed – as when measuring tool length compensation values on the machine.

1. Place the mode switch to MDI.
2. Press the key labeled program until MDI appears at the top of the display screen.
3. Type T and the number of the tool you wish to load into the spindle.
4. Press the EOB key and press the soft key under insert.
5. Press the start key or cycle start button (Tool rotates into waiting position.)
6. Type M06, the EOB key, and press the soft key under insert.
7. Press the start key or cycle start button (Tool change takes place).

This procedure assumes the machine is at its tool change position.

To enter a program using the keyboard
A program must reside in the machine's memory before it can be run. This is one way to load a program – but it is time consuming.

1. Place the mode switch to edit.
2. Press the key labeled program until the word program appears at the top of the display screen.
3. Type the letter O and the program number of the program to be entered.
4. Press the soft key under insert.
5. Press the EOB key and press the soft key under insert.
6. Typing one command at a time and ending each command with the EOB key, enter the balance of the program.

To see a directory of programs
Required when you want to know which programs are in the machine.

1. Place the mode switch to edit.
2. Press the key labeled program until the word program appears at the top of the display screen.
3. Press the soft key under program again.

To search within a program
Required when you must edit a program.
1. Place mode switch to edit.
2. Press the extreme left soft key until program appears at the bottom of the screen – then press the soft key under program.
3. Press the reset key to return the program to the beginning (not always required).
4. Type the word you wish to search to.
5. Press soft key under fwd search. The control will find the first occurrence of the word you typed.

To use MDI to activate the spindle
Required whenever a manual tool change is needed – as when measuring tool length compensation values on the machine.

1. Place the mode switch to MDI.
2. Press the key labeled program until MDI appears at the top of the display screen.
(example: S500 = 500 rpm)
3. Press the EOB key and press the soft key under insert.
4. Type M03 for forward or M04 for reverse, press the EOB key, and press the soft key under insert.
5. Press start key or cycle start button. (spindle starts)
6. To stop spindle, type M05, the EOB key, and press the soft key under insert. Then press start key or cycle start button.

To load a program from a DNC system
Required during setup to load the program/s needed for the job.
1. Place the mode switch to edit.
2. Press the key labeled program until the word program appears at the top of the display screen.
3. Be sure the DNC system is connected the machine.
4. Type O (letter, not zero) and the program number to be loaded.
5. Press the soft key under read.
6. Go to the DNC system and send the program.

To delete a program
Required when you're finished using a program.
1. Place mode switch to edit.
2. Press the key labeled program until the word program appears at the top of the display screen.
3. Type letter O and the program number to be deleted.
4. Press the soft key under delete.
You won't be asked for confirmation, so be careful with this procedure.

To alter a word in a program
1. Place mode switch to edit.
2. Press the key labeled program until the word program appears at the top of the display screen.
3. Search to the word to be altered.
4. Type the new word.
5. Press soft key under alter.

To insert a word in a program
1. Place mode switch to edit.
2. Press the key labeled program until the word program appears at the top of the display screen.
3. Search to the word just prior to the word you want to insert.
4. Type the word you wish to insert.
5. Press soft key under insert.

To use MDI to do a zero return
The zero return position in at least one axis is usually the tool change position. The machine must be at this position when the program is activated. This procedure is faster than the manual procedure.
1. Place the mode switch to MDI.
2. Press the key labeled program until MDI appears at the top of the display screen.
3. Check that the machine can go straight home with no interference.
4. Type G91 G28 X0 Y0 Z0, press the EOB key and then press the soft key under insert.
5. Press start key or cycle start button.

You can omit axes that you don't want to zero return.

To save a program to a DNC system
Required after program verification if changes have been made to the program.
1. Get the DNC system ready to receive a program.
2. Place mode switch to edit.
3. Press the key labeled program until the word program appears at the top of the display screen.
4. Type the letter O and the program number of the program to be sent.
5. Press the soft key under punch.

To call up a program (make it the active program)
Required when you want to work on another program in the control.
1. Place mode switch to edit.
2. Press the key labeled program until the word program appears at the top of the display screen.
3. Type the letter O and the program to be searched to.
4. Press the soft key under fwd search.

To delete a word in a program
Required when you must edit a program.
1. Place mode switch to edit.
2. Press the key labeled program until the word program appears at the top of the display screen.
3. Search to the word to be deleted.
4. Press the soft key under delete.

To run a verified program
Required during a production run.

1. Load part into setup.
2. Send the machine to its starting point (usually the tool change position).
3. Place the mode switch to edit.
4. Press the key labeled program until the word program appears at the top of the display screen.
5. Press the reset key (check the active program number).
6. Check position of all conditional switches (dry run, etc.).
7. Place the mode switch to memory or auto.
8. Place the feedrate override switch to 100 per cent.
9. Press cycle start button.

Other important procedures:
To measure program zero assignment values: Lesson Five
To measure tool length compensation values: Lesson Eleven
To verify programs: lessons twenty and twenty-four
To re-run tools: Lesson Twenty

Key points for Lesson Twenty-Three:

- Running a CNC machining center is little more than following a series of procedures. The trick is knowing when and why to perform a given procedure.

- Writing down procedures will keep you from having to ask someone for help every time you need to perform the procedure.

- Write down a procedure for every task you must perform on your machining center.

You Must Know How To Safely Verify Programs

You cannot begin a production run until the CNC program is verified and a workpiece passes inspection. Safely verifying programs will be the focus of Key Concept Number Ten.

Key Concept Number Ten is a short, one-lesson Key Concept:

21: Program verification

You know that program verification is part of setup. Indeed, we provide some pretty good explanations in Lesson Twenty about how programs are verified. But since program verification is the most dangerous part of running a CNC machine, we want to spend more time discussing it. Lesson Twenty-Four will show you how to safely verify programs, even if serious mistakes have been made while making the setup and writing the program.

Safety priorities

When verifying new programs, remember CNC machines will follow programmed instructions *precisely as they are given*, even if there are mistakes in the program. In Lesson Eight, we present some of the most common programming mistakes. With the exception of basic syntax (program formatting) mistakes, the machine will rarely alert you when a mistake has been made. While verifying any new program, you must be ready for just about anything. If you make a mistake in the program which tells the machine to rapid a tool into the workpiece, the machine will follow your commands and do so, causing what is commonly referred to as a *crash*.

And even with a proven program, remember that you can make serious mistakes during setup. Mistakes with tool loading, tool length measurement, program zero assignment value measurement, and offset entry can be every bit as serious as programming mistakes. So you must also be very careful when running programs that you have run before.

We cannot overstress the need for using safe procedures and staying alert when working with CNC equipment. While we are not trying to scare you, we do want to instill in you a very high level of respect for your very powerful and potentially dangerous machine tool.

There are three levels of priority that you must adhere to when you work with any machine tool, including CNC machining centers.

Operator safety

The first priority must be your safety. You must use every opportunity to ensure your safety and the safety of the people around you. The verification procedures we provide stress operator safety as the number one priority. As time goes on and you start gaining experience, your tendency will be to *short cut* these procedures in order to save some time. We urge you to avoid this tendency. When you begin relaxing your guard, you open the door to very dangerous situations.

Compare this to a person that enjoys snow skiing. The first few times a person goes skiing, they tend to be very careful, and rarely does a new skier out for the first time get seriously injured. It is only after a skier gains confidence that they become bold and careless. More experienced skiers sustain serious injuries than do beginners.

In similar fashion, most entry level setup people and operators tend to be very careful when running a CNC machine. It is only after they gain some experience that some people become bolder and less careful. This tendency is inspired by the need to work faster. Again, don't compromise safety in order to go faster. As you gain experience, you'll naturally gain the ability to perform more efficiently. You don't have to shortcut safety-related procedures to do so.

Machine tool safety

The second safety priority is the CNC machining center itself. Every operator must do their best to ensure that no damage to the machine can occur. Obviously, CNC machine time is very expensive. When a CNC machine goes down for any reason, the actual cost of repairing the machine is usually very small compared to the lost production time.

There is no excuse for machine downtime caused by operation mistakes. If the verification procedures we give are followed, you can ensure that the machine will not be placed in dangerous situations. While no method is completely failsafe, our recommendations will truly minimize the potential for machine damage.

Workpiece safety

The third safety priority is making all of your workpieces to size. The effort that companies put forth to ensure zero scrap varies from one company to another. One company may be machining extremely expensive material, like titanium or stainless steel. The raw material for large workpieces can be expensive –especially for workpieces that require machining operations *prior to* the CNC machining center operation. For this company, *anything* they can do to minimize the potential for even one scrap workpiece will be done.

In another company, the raw material cost may be very low. Consider machining a small 5/8 inch long, 1/2 inch diameter piece of steel bar. The total cost of raw material may be less than 0.50 cents. In this company, the setup person's time may be more valuable than the time it takes to run one (the first) workpiece. This company may be less concerned with attaining zero scrap. Some companies in this situation even supply the setup person with extra pieces (commonly called *practice parts*). They don't expect every workpiece machined to be a good one.

While things can happen during a production run that will cause scrap workpieces, the most critical time is during the machining of the very *first* workpiece. Knowing this, and by knowing which workpiece attributes are the most critical, a CNC setup person can minimize the potential for scrap workpieces by using *trial machining techniques* (as described in Lessons Eleven, Twelve, and Twenty). If you consider what each tool in the program will be doing when you come to it – and if you use trial machining techniques when appropriate – each tool will machine the workpiece properly. When you're finished with the last tool in the program, you'll have a good workpiece that will pass inspection.

There are people in our industry that feel that trial machining is wasteful. People that feel this way tend to come from companies that machine very *inexpensive* raw material. They do not care about scrapping the first few workpieces as long as they *eventually* learn enough about offset settings to get a workpiece to come out to size.

But regardless of whether your company believes in trial machining or not, *all* setup people should have the ability to machine the very first workpiece correctly for the times when it is necessary to do so. Even if workpiece material cost is very low, the day will come when you have five pieces of raw material and you must machine five good workpieces.

Lesson 21

Program Verification

The most dangerous time for a CNC setup person is when verifying programs. You must stay alert and be ready for just about anything. You must master the program verification procedures – they must truly become second nature.

In Lesson Twenty, we show all of the tasks related to making setups and running production. We do so by using an example job to stress how each task is done – including program verification. In Twenty-Four, we're going to do so again – but we'll only discuss tasks related to program verification. The setup and program we show in Lesson Twenty is perfect – there are no mistakes. By comparison, the setup and program we use *in this lesson* contain many mistakes. Most of the mistakes are typical of mistakes a beginner is likely to make.

While you may be able to spot some of the mistakes in our example job as soon as we show them, remember that when *you* make mistakes, you won't know it (if you did, you wouldn't make the mistake). Our objective is to show you how to catch even very serious mistakes as you verify CNC programs – and of course – to catch them before a crash can occur.

Two more procedures

In Lessons Twenty and Twenty-Three, we show several procedures that are related to verifying programs, including how to perform a dry run, how to cautiously run the first workpiece using a special approach procedure for each tool, how to trial machine, and how to run a verified program. These procedures are extremely important, and you must master them. But they don't show you how to *handle problems* when mistakes are found. Here are two procedures that are needed when you *do find mistakes*.

Canceling the CNC cycle

You know that feed hold is your panic button. As you're running a program, you can press it any time you are worried. It causes all axis motion to stop. If you find that nothing is wrong, you can press the cycle start button to continue. *But what if something is wrong* – something that is so serious that you cannot allow the program to continue?

Say for example, during the running of the first workpiece, you are allowing a tool to approach the workpiece for the first time. You are using the approach procedure shown in Lesson Twenty, so you have dry run and single block turned on, and you're controlling the tool's motion rate with feedrate override. As the tool gets within an inch or so of the work surface, you stop the cycle. You're worried that the tool will not stop at a position 0.1 inch above the work surface. So you look at the distance to go page. It says the Z axis is still going to move another 1.3 inches. As you look at the tool's position relative to the work surface, you can easily tell that the tool cannot move another 1.3 inches without contacting the workpiece (you've just saved a crash).

You cannot allow the program to continue. You must, of course, find and correct the problem. But before you go any further, you must *cancel the cycle*. Here is the procedure to do so:

> 1) Press the reset key on the control panel. For most machines, this will stop the spindle and coolant (if they're on).
>
> 2) Place the mode switch to edit (if the spindle and coolant don't stop when the reset button is pressed, they will now).
>
> 3) Select the program display screen page. The cursor is somewhere in the middle of the program.

4) Press reset again (this sends the cursor back to the program number at the beginning of the program).

5) Send the machine to its tool change position (usually the zero return position). This can be done manually or by using an MDI command.

6) Find and correct the problem that caused you to have to cancel the cycle.

7) When the problem is corrected, follow the procedure to re-run the tool.

To re-run a tool

This procedure assumes that you can identify the appropriate *restart command* for the tool you want to re-run. It also assumes the programmer has used the format shown in Lesson Fifteen. As we show in Lesson Fifteen, the restart command depends upon whether the tool you want to re-run is currently in the spindle. If the tool *is* in the spindle, the restart command will be the command *after* the tool change command that places the tool in the spindle. If the tool *is not* in the spindle, the restart command is the tool change command that places the tool in the spindle. This series of commands should help to clarify which command is the restart block:

N135 G40 Y2.5 F40.0

N140 G00 Z0.1 M09

N145 G91 G20 Z0 M19

N150 M01

N155 T03 M06 (5/8 END MILL) <---- Restart command if tool three **is not** in the spindle

N160 G54 G90 S540 M03 T04 <----- Restart command if tool three **is in** the spindle

N165 G00 X3.0 Y1.0

N170 G43 H03 Z0.1 M08

N175 G01 Z-0.250 F5.0

This procedure also varies based upon whether you have seen the tool run before. If you *have* seen the tool run properly (as is the case when you re-run a tool after trial machining), you need not be careful with its first approach movement and the balance of its cutting motions.

But if you *have not* seen the tool successfully complete its operation (as is the case in the scenario above), you must still be very careful with its first approach movement and all of its motions.

Here is the procedure to re-run a tool. The machine must be at its tool change position when this procedure is used (if you canceled the cycle with the procedure just shown, it will be):

1) Consider the condition of single block (if you've seen the tool run before, turn it off, if not, turn it on).

2) Consider the condition of dry run (if you've seen the tool run before, turn it off, if not, turn it on).

3) Consider the condition of feedrate override (if you seen the tool run before, set it to 100%. If not, set it at its lowest setting.

4) Consider the condition of optional stop (if you want the machine to stop when the tool is finished, as is normally the case when verifying programs, turn it on. If not, turn it off).

5) Place the mode switch to edit.

6) Press the program display screen key. The display will show the active program.

7) Press the reset key. This places the cursor on the first word in the program (the program number).

8) Scan to the restart command. To do so, type the sequence number for the restart command and press the forward search key.

9) Place the mode switch to memory or auto.

10) Press the cycle start button to activate the cycle. If you've seen the tool run before (single block is off), the tool will perform its machining operation. As long as you have the optional stop switch turned on, the machine will stop when the tool is finished so you can check what the tool has done.

11) If you have never seen this tool run before (single block is on), you must press the cycle start button repeatedly until the machine begins to move. You'll use the feedrate override switch to control motion rate.

12) Once the tool has successfully approached the workpiece, turn off dry run. (*Never* let cutting tools machine a workpiece under the influence of dry run. Dry run tends to slow down rapid motions, but it *speeds up* cutting motions.)

13) The tool has safely approached. What you do next will depend upon whether you're running a new or a proven program. For a new program you must carefully step through the cutting tool's cutting motions. In this case, leave the single block switch on, and repeatedly press cycle start to step through the tool. For a proven program (are you *absolutely sure* that you're running the current version of the program?), turn off single block and press cycle start to let the tool machine the workpiece as it did the last time the program was run.

14) When the tool is finished, the machine will return to its tool change position.

CNC Turning Centers

This begins the discussion of CNC turning centers. Note that only the differences from machining centers are discussed, meaning this section is dramatically abreviated. Additionally, several redundant lessons have been removed.

Know Your Machine From A Programmer's Viewpoint

It is from two distinctly different perspectives that you must come to know your CNC turning center/s. Here in Key Concept One, we'll look at the machine from a programmer's viewpoint. Much later, during Key Concept Seven, we'll look at the machine from a setup person's or operator's viewpoint.

Key Concept Number One is the longest of the Key Concepts. It contains eight lessons:

      1: Machine configurations
      2: Understanding turning center speeds and feeds
      3: General flow of programming
      4: Visualizing program execution
      5: Understanding program zero
      6: Determining program zero assignment values
      7: How to assign program zero
      8: Introduction to programming words

A CNC programmer need not be nearly as intimate with a CNC turning center as a setup person or operator – but they must, of course, understand enough about the machine to create programs – to instruct setup people and operators – and to provide the related setup and production run documentation.

First and foremost, a CNC programmer must understand what the CNC turning center is designed to do. That is, they must understand the machining operations a turning center can perform. They must be able to develop a workable process (sequence of machining operations), select appropriate cutting tools for each machining operation, determine cutting conditions for each cutting tool, and design a workholding setup. All of these skills, of course, are related to basic machining practice – which as we state in the Preface – are beyond the scope of this text. For the most part, we'll be assuming you possess these important skills.

Though this is the case, we do include some important information about machining operations that can be performed on CNC turning centers throughout this text. For example, we discuss how to develop tool paths for machining operations in Lessons Nine and Ten. We provide a description of rough and finish turning and boring in Lesson Eighteen. Threading is discussed in Lesson Twenty. And in general, we provide suggestions about how machining operations can be programmed when it is appropriate. This information should be adequate to help you understand enough about machining operations to begin working with CNC turning centers.

If you've had experience with conventional (non-CNC) machine tools…

A CNC turning center can be compared to an engine lathe (or any kind of conventional lathe). Many of the same operations performed on a conventional lathe are performed on a CNC turning center. If you have experience with manually operated lathes, you already have a good foundation on which to build your knowledge of CNC turning centers.

This is why machinists make the best CNC programmers. With a good understanding of basic machining practice, you can easily learn to program CNC equipment. You already know *what* you want the CNC machine to do. It is a relatively simple matter of learning *how to tell the CNC machine* to do it.

If you have experience with machining operations like rough and finish turning, rough and finish facing, drilling, rough and finish boring, necking and threading – and if you understand the processing of machined workpieces – believe it or not, you are well on your way to understanding how to program a CNC turning center. Your previous experience has prepared you for learning to program a CNC turning center.

We can also compare the importance of knowing basic machining practice in order to write CNC programs to how important it is for a speaker to be well versed with the topic they will be presenting. If not well versed with their topic, the speaker will not make much sense during the presentation. In the same way, a CNC turning center programmer who is not well versed in basic machining practice will not be able to prepare a program that makes any sense to experienced machinists.

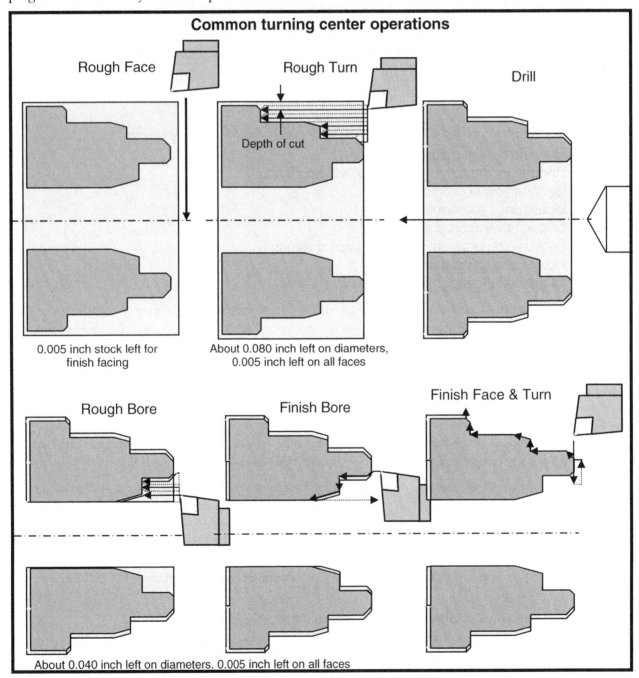

Common turning center operations

Rough Face
— 0.005 inch stock left for finish facing

Rough Turn
Depth of cut
— About 0.080 inch left on diameters, 0.005 inch left on all faces

Drill

Rough Bore
Finish Bore
Finish Face & Turn
— About 0.040 inch left on diameters. 0.005 inch left on all faces

Lesson 1:

Machine Configurations

As a programmer, you must understand what makes up a CNC turning center. You must be able to identify its basic components – you must understand the moving components of the machine (called axes) – and you must know the various functions of your machine that are programmable.

Most beginners tend to be a little intimidated when they see a turning center in operation for the first time. Admittedly, there will be a number of new functions to learn. The first point to make is that you must not let the machine intimidate you. As you go along in this text, you will find that a turning center is very logical and is almost easy to understand with proper instruction.

You can think of any CNC machine as being little more than the standard type of equipment it is replacing with very sophisticated and automatic motion control added. Instead of activating things manually by hand-wheels and manual labor, you will be preparing a *program* that tells the machine what to do. Virtually anything that needs to be done on a true CNC turning center can be activated through a program – meaning anything you need the machine to do can be commanded in a program.

Types of CNC turning centers

There are several types of CNC turning centers. While at first glance there may appear to be substantial differences among the various types, all turning centers share several commonalities. We'll begin by describing the most popular type of CNC turning center – the *universal style slant bed turning center*. Because it is so popular, this is the machine type we will use for all examples in this text. We will then introduce several other types of turning centers, comparing them to the universal style slant bed turning center.

Universal style slant bed turning center

This style of turning center is called a *universal style* turning center because it can perform all three forms of turning applications – chucking work, shaft work, and bar work. This explains why it is the most popular type of turning center – it provides the most flexibility to CNC turning center users.

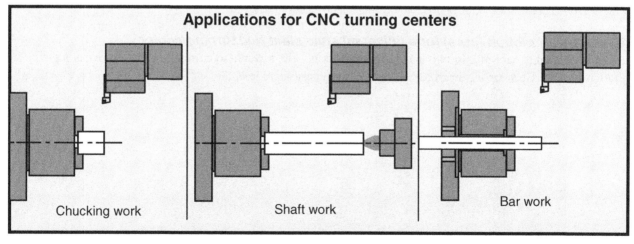

Applications for CNC turning centers
Chucking work   Shaft work   Bar work

When raw material comes to the machine in the form of short slugs (like round bars cut to length), the application is called chucking (or chucker) work. The raw material is secured solely by the workholding device (commonly a three-jaw chuck).

With longer slugs (longer than about three to four times the raw material diameter), the workholding device by itself will not be sufficient to secure the workpiece for machining. For these applications, some form of work support device/s must be used (commonly a tailstock and/or steady-rest). This application is called shaft work.

With bar work, the raw material comes to the turning center in the form of a long bar (from four to fifteen feet long [1.2-5 meters], depending upon the type of bar feeder being used). Bar work requires a special bar support and feeding device (called a bar feeder). The bar is fed through the headstock and spindle into the working area. A workpiece is machined and cut off from the bar. The bar is then fed again for another workpiece to be machined.

Figure 1.1 shows a universal style slant bed turning center. The headstock houses a spindle to which the workholding device is mounted. Our illustration shows a three-jaw chuck, but other types of workholding devices can be used (collet chuck, expanding mandrel, etc.). To the right of the workholding device is the tailstock, which is used to support the right end of long workpieces – again, for shaft work. The turret of the turning center is used to hold cutting tools and it can be quickly rotated from one tool station to another. Current turning centers have turrets that hold from six to twelve cutting tools.

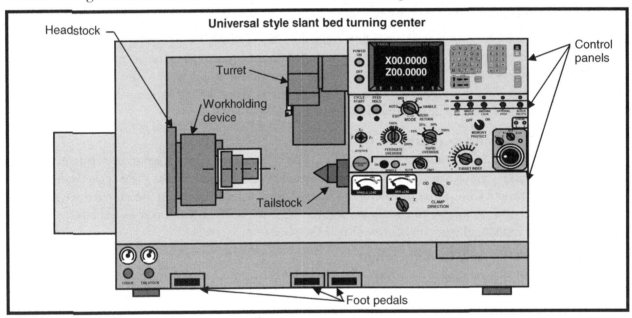

Figure 1.1 – A universal style slant bed turning center with its door removed

Directions of motion (axes) for a universal style slant bed turning center
All turning centers have at least two linear *axes* of motion. The turret (and cutting tool) will move along with these two axes. By *linear*, we mean the axis moves along a straight line.

Fundamentals of CNC © CNC Concepts, Inc.

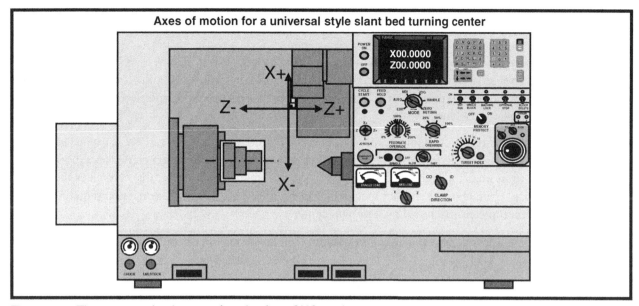

Figure 1.2 – The two most basic axes of motion for a CNC turning center

The *diameter-controlling* axis (up/down motion of the turret as shown in Figure 1.2) is the X *axis*. The *length-controlling* axis (right/left motion of the turret as shown in Figure 1.2) is the Z *axis*. Figure 1.2 shows these directions of motion along with the polarity (+/-) for each.

These two most basic directions of motion will remain exactly the same for almost all types of turning centers (only a handful of turning center manufacturers stray from what we show in Figure 1.2.) *The X axis will always be the diameter-controlling axis* – and X minus is always the direction that causes the cutting tool to move to a smaller diameter (toward the spindle centerline). *The Z axis will always be the length controlling axis* – and Z minus will always be the direction that causes the cutting tool to move toward the workholding device.

X is specified in diameter

Though we may be a little ahead of ourselves, the X axis is designated in *diameter* for almost all turning centers. That is, if a diameter of 3.0 inches must be machined, the designation for the X axis will be **X3.0**. There are some (especially older) turning centers that require the X axis to be specified with radial values. For these machines, the word **X1.5** will cause the tool to be positioned to a 3.0 inch diameter. Note that it is much easier to work with a turning center when the X axis if it is designated in diameter – which is why most current model turning centers do so.

What's in a linear axis?

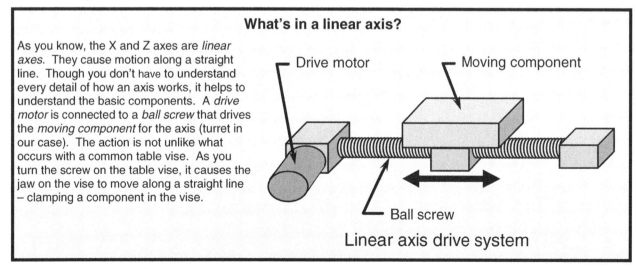

As you know, the X and Z axes are *linear axes*. They cause motion along a straight line. Though you don't have to understand every detail of how an axis works, it helps to understand the basic components. A *drive motor* is connected to a *ball screw* that drives the *moving component* for the axis (turret in our case). The action is not unlike what occurs with a common table vise. As you turn the screw on the table vise, it causes the jaw on the vise to move along a straight line – clamping a component in the vise.

Live tooling for a universal style slant bed turning center

We have just described the most basic form of a universal style slant bed turning center. Again, this machine has two axes (X and Z) – and it can perform all three kinds of turning work (chucking work, shaft work, and bar work). The majority of universal style slant bed turning centers that are in use today are of this configuration.

There is, however, a special accessory called *live tooling* that can be equipped on all types of CNC turning centers (including the universal style slant bed turning center). This accessory, which is becoming quite popular, makes it possible for a turning center to perform machining operations that are more commonly associated with CNC *machining centers* (or milling machines).

These operations include drilling, tapping, reaming, and milling (among others). In essence, turning centers equipped with live tooling can perform both turning center operations and machining center operations – giving it the ability to more completely machine a workpiece. For many applications, this eliminates the need to perform secondary operations on another machine tool. Figure 1.3 shows a workpiece that requires live tooling if it is to be completely machined on a turning center.

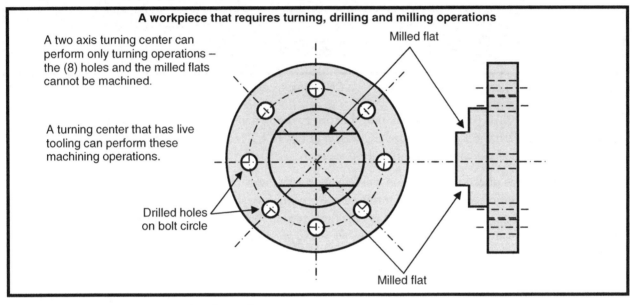

Figure 1.3 – A workpiece that requires live tooling

Turning centers that are equipped with live tooling have two additional features. First, as the name implies, they have a special device mounted within the turret that makes it possible to rotate cutting tools (again, like drills and end mills). Second, they have a special rotary axis (called the C axis) built into the spindle drive system. Figure 1.4 shows a universal style slant bed turning center that has live tooling. These turning centers are sometimes referred to as mill/turn machines.

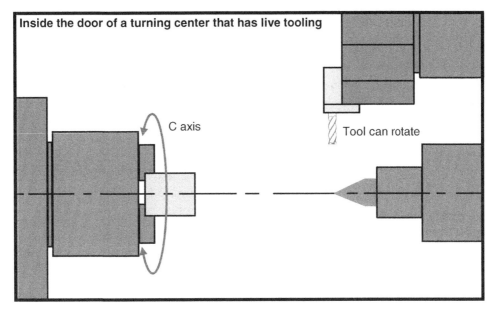

Figure 1.4 – Live tooling on a universal style slant bed turning center

The cutting tool depicted in Figure 1.4 points along the X axis. Another style of tool holder is usually available that allows the cutting tool to point along the Z axis (toward the chuck).

When the programmer selects the *live tooling* mode, the rotary axis within the spindle is engaged and the spindle within the turret will be used to rotate cutting tools. When the programmer selects the *normal turning mode*, these devices are disengaged.

The bulk of this text describes the programming of the normal turning mode (not using live tooling). We describe live tooling in detail in the Appendix after Lesson Twenty-Three.

Programmable functions of turning centers

A true CNC turning center will allow you to control just about any of its functions from within a program. There should be very little operator intervention during a CNC cycle. Here we list some common functions that can be programmed on all *true* turning centers. While we do show the related CNC words used to command these functions, our intention here is not to teach programming commands (yet). It is to simply make you aware of the kinds of things a programmer can control through a program.

Spindle

The spindle of all turning centers can be programmed in at least three ways, activation (start/stop), direction (forward/reverse), and speed (in either surface feet/meters per minute or revolutions per minute). Many turning centers additionally provide multiple power ranges (like the transmission of an automobile).

Spindle speed

You can precisely control how fast the spindle of a turning center rotates. An *S word* is used for this purpose. There are two ways to specify spindle speed. When the spindle is in rpm mode, an S word of **S500** specifies a speed of 500 revolutions per minute (rpm). When the spindle is in *constant surface speed mode*, an S word of **S500** specifies a speed of 500 surface feet per minute (sfm), assuming you are working in the inch measurement system. (If you work in the Metric measurement system, **S500** will specify 500 meters per minute when in constant surface speed mode.)

We will describe the two spindle speed modes – as well as how to determine how and when to use them – in Lesson Two.

Spindle activation and direction

You can also control which direction the spindle rotates – *forward or reverse*. The forward direction is used for right hand tooling (when machining occurs toward the workholding device). It will appear as counter-clockwise when viewed from in front of the machine. The reverse direction is used for left hand tooling and will appear as clockwise when viewed from in front of the spindle.

Three *M codes* control spindle activation. M03 turns the spindle on in the forward direction (used with right-hand tools). M04 turns the spindle on in a reverse direction (for left-hand tools). M05 turns the spindle off.

What is an M code?

M codes control many of the programmable functions of a turning center. In many cases, you can think of them as being like programmable on/off switches. "M" stands for "miscellaneous" or "machine" function.

M codes are created by the machine tool builder, and will often vary from one turning center to another. Here we show some common M codes, but you must look in your machine tool builder's programming manual to find the complete list of M codes for a given turning center.

These M codes don't vary:
M00: Program stop
M01: Optional stop
M03: Spindle on (forward)
M04: Spindle on (reverse)
M05: Spindle off
M08: Flood coolant on
M09: Coolant off
M30: End of program

These M codes vary:
M____: Low spindle range
M____: High spindle range
M____: Tailstock quill forward
M____: Tailstock quill reverse
M____: Automatic door open
M____: Automatic door close
M____: Chip conveyor on
M____: Chip conveyor off
M____: _____
M____: _____
M____: _____
M____: _____
M____: _____

Spindle range

Many, especially larger turning centers, have two or more spindle ranges. Spindle ranges are like the gears in an automobile transmission. Generally speaking, lower ranges are used for power – higher ranges are used for speed. With most turning centers, spindle range selection is done with M codes. While the specific M code numbers for spindle range selection will vary from one machine tool builder to another, many turning center use M41 to select the low range and M42 to select the high range. We'll use these two M codes (M41: low and M42: high) to specify spindle range selection throughout this text.

Turning centers vary when it comes to what will actually happen when the spindle range is changed. Some, especially older machines use a mechanical gearbox that must be engaged for the low range. These machines commonly require that the spindle be stopped during the range change. While the spindle stoppage, range change, and spindle restart will occur automatically, these machines can take from three to ten seconds or more to change ranges.

Some, especially newer machines have spindle motors with multiple windings. Two or more sets of windings within the motor itself control range selection. With these machines, range changing is almost instantaneous – and the spindle does not have to stop when the spindle range is changed.

It is important to know the power characteristics for the turning center/s you will be working with in order to make the correct spindle range selection for a given machining operation. Every turning center manufacturer will provide a power curve chart like the one shown in Figure 1.13 to document a machine's spindle power characteristics. You will normally find this power curve chart in the machine tool builder's operation or programming manual.

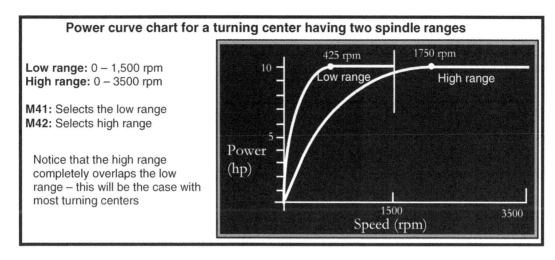

Figure 1.13 – An example spindle power curve chart

With this example spindle power curve chart, notice that the low range runs from zero to fifteen hundred rpm and achieves full power at 425 rpm. The high range runs from zero to thirty-five hundred rpm, completely overlapping the low range. Full power is achieved in the high range at 1,750 rpm.

Again, knowing your machine's spindle power characteristics is important for selecting the appropriate spindle range. We give a general rule-of-thumb for spindle range selection: *perform roughing operations in the low range and finishing operations in the high range.*

While this is a good rule of thumb, there are times when it isn't correct. Consider, for example, machining a steel workpiece that is less than 1.0 inch in diameter. Say the cutting tool manufacturer recommends a speed of 600 sfm for the rough turning operation. The formula to calculate rpm is as follows:

Rpm = 3.82 times speed in sfm divided by the diameter to be machined

In this case, the *slowest* speed needed for the roughing operation will be 2,292 rpm (3.82 times 600 sfm divided by 1.0). In this case, since full power is achieved at 1,750 rpm in the high range, and since the low range will peak out at 1,500 rpm – slowing the machining operation – the high range should be selected for this operation.

Feedrate

You know that all turning centers have at least two linear axes, X and Z. You also know that the cutting tool (for most turning center) moves along with these two axes. It is the motion of the cutting tool while it is in contact with the workpiece that causes machining to occur. It is important that the motion *rate* (how quickly the tool moves) be appropriate to the machining operation being performed. In CNC turning center terms, this motion rate is called *feedrate.*

An F word is used to specify feedrate. And like spindle speed, feedrate can be specified in two ways. It can be specified in *per minute* fashion or in *per revolution* fashion. As the names imply, when feedrate is specified in per minute fashion, it specifies how far the cutting tool will move during one minute. When feedrate is specified in per revolution fashion, it specifies how far the cutting tool will move during one spindle revolution.

Also as with spindle speed (at least in constant surface speed mode), feedrate specification is related to the measurement system you use. In the inch mode, feedrate is specified in either inches per minute (ipm) or inches per revolution (ipr). In metric mode, feedrate is specified in either millimeters per minute (mmpm) or millimeters per revolution (mmpr). Feedrate selection is discussed in much greater detail in Lesson Two. For now, we'll simply introduce the related words and give a few examples.

- F word – Feedrate specification
- G20 – Inch mode
- G21 – Metric mode
- G98 – Feed per minute mode

- G99 – Feed per revolution mode

Here are a few examples of feedrate specification:

N010 G20 G98 F4.0 (4.0 inches per minute)

N020 G20 G99 F0.015 (0.015 inches per revolution)

N030 G21 G98 F100.0 (100.0 millimeters per minute)

N040 G21 G99 F0.5 (0.5 millimeters per revolution)

What is a G code?

G codes are called *preparatory functions*. They prepare the machine for what is coming up – in the current command and possibly in up-coming commands. In many cases, they set *modes,* meaning once a G code is *instated* it will remain in effect until the mode is changed or cancelled.

Here we list a few common G codes, but don't worry if they don't make much sense yet. Upcoming discussions will clarify.

Common G codes:

G00: Rapid motion	G42: Tnr comp. right
G01: Straight line motion	G50: Spindle limiter
G02: Circular motion (CW)	G70: Finishing cycle
G03: Circular motion (CCW)	G71: Rough turning cycle
G04: Dwell	G72: Rough facing cycle
G20: Inch mode selection	G76: Threading cycle
G21: Metric mode selection	G96: CSS mode
G28: Zero return command	G97: RPM mode
G40: Cancel tool nose radius comp.	G98: Feed per minute
G41: Tool nose radius comp. left	G99: Feed per revolution

Look in your control manufacturer's manual for a full list of G codes.

Turret indexing (tool changing)

With the exception of gang style turning centers, all turning center introduced in this lesson have a turret into which cutting tools are placed (twin spindle turning centers have two turrets). Specific turret design will vary from one machine tool builder to another. All turning centers (even gang style turning centers) must provide a way to specify cutting tool selection. Figure 1.14 shows the turret of a typical turning center.

Turret of a turning typical center

Figure 1.14 – A typical turning center turret

The turret shown in Figure 1.14 is currently holding turning tools (external machining tools) as well as boring tools (internal machining tools). Notice that the internal tools protrude quite a distance from the turret face (they stick out). This can cause interference problems with a large workpiece and/or the workholding device. You must always be concerned with the potential for interference problems as you choose the turret stations into which you place cutting tools.

With many turrets, for example, you cannot place a long drill or boring bar in a station that is adjacent to a turning tool that machines a small diameter (like a facing tool) without interference problems. As the turning tool faces a workpiece to center, the adjacent boring bar will be driven into the chuck or workpiece.

Turret station and offset selection

A T word specifies which cutting tool will be used. For turning centers that have a turret, the T word will actually cause the turret to index to the specified turret station. But there's a little more to the T word than turret index.

For most machines, the T word is a four-digit word. The first two digits specify the turret station and *geometry offset* to be used with the tool (geometry offsets assign *program zero* – and will be discussed in Lessons Seven and Twelve). The second two digits of the T word specify the *wear offset* to be used with the tool (wear offsets allow the operator to make minor adjustments – and will be discussed during Lesson Thirteen).

The command

N020 T0404

will cause these three things to occur:

- the turret to index to station number four (first two digits)
- geometry offset number four will be selected (first two digits)
- wear offset number four will be selected (second two digits)

Almost all current model turning centers have *bi-directional turrets*. That is, the turret can rotate in either direction. When a T word is given, most machines will cause the turret to automatically rotate in a direction that that provides the shortest rotational distance to the specified tool

With gang style turning centers, of course, there is no turret to index. Only two things will happen with the previous command: geometry and wear offset number four will be selected.

Again, offset use is the topic of future lessons (again, Lessons Seven, Twelve, and Thirteen). For now, just remember that most programmers will make the wear offset number the same number as the turret station number and geometry offset number.

Coolant

All turning centers allow programmable control of *flood coolant*. Coolant is commonly used to cool the workpiece during machining and to lubricate the machining operation. Two M codes are used to control coolant. Almost all turning center manufacturers use M08 to turn on flood coolant and M09 to turn it off.

Other possible programmable functions

The programmable functions introduced to this point are available for all current model turning centers. Those we list from this point are related only to certain machine types – or they are optional functions that are not supplied with all turning centers.

Tailstock

Turning centers that can perform shaft work (like universal style slant bed turning centers) are equipped with a tailstock. The tailstock is used to support the right end of a long workpiece during machining (the end opposite the workholding device). Most machinists would agree that when the length of the workpiece exceeds about three to four times its diameter, a tailstock should be used to provide support during machining.

Though most current model turning centers have *programmable* tailstocks, machine tool builders vary with how they cause the tailstock body and quill to move. Many use a series of M codes. M16 and M17 may be used to cause quill movement forward and reverse while M28 and M29 may be used to cause tailstock body movement forward and reverse.

Some machine tool builders actually cause tailstock motion by engaging the turret to the tailstock and pulling it along with the Z axis. If your machine has a programmable tailstock, you must reference your machine tool builder's programming manual to determine how it is programmed.

What else might be programmable?

While this text will acquaint you with the most common programmable functions of turning centers, you must be prepared for more. Other programmable devices that may be equipped on your turning center include chip conveyer, automatic door open and close, automatic loading devices, and a variety of other application-based accessories. If you have any of these functions, you must reference your machine tool builder's programming manual to learn how these special features are programmed.

> Check with an experienced person in your company or school to find out what other programmable features are available on the turning center you will be working with.

Key points for Lesson One:

- There are several types of CNC turning centers.
- With all types of CNC turning centers, X is the diameter controlling axis and X minus is the toward spindle center, Z is the length controlling axis and Z minus is toward the workholding device.
- X axis positions are specified in diameter.
- Some turning centers – those that have live tooling – can additionally perform machining-center-like machining operations).
- You must understand the functions of your turning center that can be programmed.
- Spindle can be controlled in at least three ways (activation, direction, and speed). Additionally, many turning centers have more than one spindle range.
- Spindle speed can be specified in rpm or in surface feet/meters per minute.
- Feedrate specifies the motion rate for machining operations.
- Feedrate can be specified in per revolution fashion or per minute fashion.
- Coolant can be activated to allow cooling and lubricating of the machining operation.
- Most turning centers have a turret in which cutting tools are placed.
- Machines in the United States allow the use of inch or metric mode.
- You must determine what else is programmable on your turning center/s.

Lesson 2
Understanding Turning Center Speeds and Feeds

Speed and feed selection is one of the most important basic-machining-practice-skills a programmer must possess. Poor selection of spindle speed and feedrate can result in poor surface finish, scrapped parts, and dangerous situations. Even if speeds and feeds do allow acceptable workpieces to be machined – if they're not efficient – productivity will suffer.

You now know the basic configurations available for CNC turning centers. You know the main components, the directions of motion (axes), and the polarity for each axis. And you know that all turning centers have certain programmable functions – machine features that you can control from within a program. Two of the programmable functions introduced in Lesson One are spindle speed and feedrate. In Lesson Two, we're going to elaborate on these two important programmable functions.

As you know, spindle speed is the rotation rate of the machine's spindle (and workpiece). Feedrate is the motion rate of the cutting tool as it machines a workpiece. These two *cutting conditions* are extremely important to machining good workpieces. You will select a spindle speed and feedrate for every machining operation that must be performed.

Cutting tool manufacturers supply technical data including the recommendation of a cutting speed and feedrate for the cutting tools they supply. Recommendations are based on three important factors:

> 1) The machining operation to be performed
> 2) The material to be machined
> 3) The material of the cutting tool's cutting edge

The machining operation to be performed
This criterion determines the style of cutting tool that must be used to perform the machining operation. As the programmer, *you* will be the person making this decision. It requires that you draw upon your basic machining practice experience.

As stated in the Preface and Lesson One, turning centers can perform a wide variety of machining operations, and specific cutting tools are used to perform each operation. The specified spindle speed and feedrate must be appropriate to the cutting tool selection. Some cutting tools, like rough turning tools, perform very powerful machining operations and can remove a great deal of material from the workpiece per pass – while others, like small boring bars, perform lighter machining operations and can only remove a small amount of workpiece material per pass.

The material to be machined
This criterion determines (among other things) how quickly material can be removed from the workpiece. Soft materials, like aluminum can be machined faster than hard materials, like tool steel. So generally speaking, you'll use faster spindle speeds and feedrates for softer materials.

The material of the cutting tool's cutting edge
The material used for some cutting tools comprises the entire tool. A standard twist drill or end mill, for example, is made entirely of high speed steel (or cobalt, or some other material).

With other tools, like many turning tools, boring bars, grooving tools, and threading tools, the shank of the tool is made from one material (like steel), while the very cutting edge of the tool is made from another material (like carbide or ceramic). This lowers the cost of the cutting tool. With these kinds of cutting tools, only the cutting edge of the cutting tool is made from the (expensive) cutting tool material. The cutting edge component of the tool is called an insert. Figure 1.15 shows this kind of cutting tool.

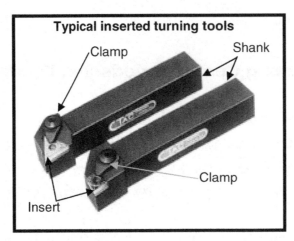

Figure 1.15 – Examples of cutting tools that use inserts

The cutting tool material also determines how quickly machining can be done. Generally speaking, the harder the cutting tool material, the faster machining can be done.

Based upon knowing these three criteria, spindle speed and feedrate selection is usually quite simple. You simply reference the cutting tool manufacturer's recommendations. These recommendations – along with other cutting condition recommendations like depth of cut – are usually provided in the tooling manufacturer's technical manuals – or they may be right in the sales catalog for the cutting tools.

- Speed will be recommended in **surface feet per minute** (sfm) if the data is given in inch, or **meters per minute** if the data is given in metric (most cutting conditions provided in the United States are given in inch). This is the amount of material (in feet or meters) that will pass by the tool's cutting edge in one minute.

- Feedrate will be recommended in **per revolution** fashion (either inches or millimeters per revolution – ipr or mmpr).

Say for example, you must finish turn 1018 cold-drawn steel with a finish turning tool that has a carbide insert. You look in the tooling manufacturer's technical handbook and find they recommend a speed of 700 surface feet per minute and a feedrate of 0.005 inches per revolution.

With conventional turning equipment like engine lathes and turret lathes, there is only *one way* to specify spindle speed (in rpm) and *one way* to select feedrate (in per revolution fashion). Since the speed for conventional lathes must be specified in rpm, a machinist must to convert the 700 sfm to the appropriate speed in rpm based on the diameter of the workpiece that is being machined. Here is the formula to do so:

Rpm = sfm times 3.82 divided by the diameter to be machined

If, for example, the workpiece diameter to be machined is 4.0 inches, the necessary speed for 1018 cold-drawn steel (if it is to be run at 700 sfm) will be 668 rpm (700 times 3.82 divided by 4.0).

As you can imagine, calculating spindle speed in rpm can be pretty cumbersome, especially if several diameters must be machined by the same tool. And consider a facing operation. As soon as a facing tool begins moving, the diameter it is machining will change.

The two ways to select spindle speed

CNC turning centers allow you to specify spindle speed in *two* ways. You can do so in rpm, or you can specify spindle speed directly in surface feet per minute (or meters per minute if you are working in the metric mode). In our previous turning example, this means you can specify the spindle speed to be used as 700 sfm – eliminating the rpm calculation.

This second method of selecting spindle speed for CNC turning centers is called *constant surface speed* (css). As the name implies, constant surface speed will cause the machine to constantly update (change) the spindle speed in rpm to maintain the specified speed in surface feet per minute (or meters per minute if you work in the metric mode). As a cutting tool moves in the X axis (changing diameter), spindle speed in rpm will also change. As the cutting tool moves to a smaller diameter, speed in rpm will increase. As it moves to a larger diameter, spindle speed in rpm will decrease.

Two preparatory functions (G codes) specify which of the spindle speed modes (rpm or css) you want to use for speed selection.

> G96 – select constant surface speed mode
> G97 – select rpm mode

An S word specifies the actual speed. Three M codes are used for spindle activation.

> M03 - Spindle on forward (right hand tools)
> M04 - Spindle on reverse (left hand tools)
> M05 - Spindle off

Here are two examples (working in the inch mode):

> N050 G96 S500 M03 (Turn spindle on forward at 500 sfm)
> N050 G97 S500 M04 (Turn spindle on reverse at 500 rpm)

As stated, constant surface speed mode (**G96**) will cause the machine to constantly and automatically update the spindle speed in rpm based on the current diameter the cutting tool is machining.

If for example, you have selected a speed of 500 sfm and the tool is currently machining a 3.5 in diameter, the machine will run the spindle at 546 rpm (3.82 times 500 divided by 3.5). There will be no need for you to perform the spindle rpm calculation – the machine does this for you.

If facing a workpiece (machining direction is X minus), the machine will constantly increase the spindle rpm as the facing operation occurs. See Figure 1.16 for a graphic illustration of how the diameter being machined changes during a facing pass– and the impact this has on spindle speed in rpm.

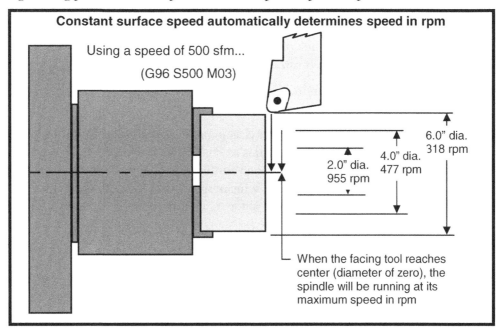

Figure 1.16 – Drawing illustrates a facing tool that is machining to the center of a workpiece

When to use constant surface speed mode

You should use constant surface speed mode (**G96**) whenever a single point cutting tool will be machining more than one diameter on the workpiece. Examples include rough and finish facing, rough and finish turning, rough and finish boring, necking (grooving), and cutting off (parting).

For these kinds of operations, constant surface speed provides three important benefits:

1) Constant surface speed simplifies programming. As you have seen, it eliminates the need for rpm calculations.

2) Since the appropriate rpm will be used as machined diameters change, the witness marks (finish) on the workpiece will be consistent from one surface to another. This is also related to the fact that feedrate will be specified in inches per revolution (ipr) or millimeters per revolution (mmpr). As the spindle speed changes, so does the feedrate per minute (feedrate selection is presented later in this lesson).

3) Since spindle speed in rpm will be correct during *all* machining operations, tool life will be extended to its maximum.

When to use rpm mode

There are three times when constant surface speed cannot be used – so spindle speed in rpm must be calculated and specified in the rpm mode (G97).

First, you must specify spindle speed in rpm for any cutting tool that machines right on the spindle's centerline. We call these tools *center-cutting tools*. Examples include drills, taps, and reamers. These tools machine a hole right in the center of the workpiece.

Again, these tools are sent right to the spindle's centerline (a diameter of zero). If you specify speed in the constant surface speed mode (G96) for a center cutting tool – even as just *one* surface foot per minute – the spindle will run at its maximum speed in rpm when a cutting tool is sent to a diameter of zero. (3.82 times one divided by zero is infinity).

When you machine with a center cutting tool, you must calculate and specify spindle speed in rpm. If for example, you must drill a 0.75 diameter hole and the drill manufacturer recommends a speed of 80 sfm based upon the material you are machining, the required speed will be 407 rpm (3.82 times 80 divided by 0.75).

Second, rpm mode must be used when chasing threads. The machine must perfectly synchronize the spindle speed with the feedrate motion during the multiple thread-passes required for machining a thread. With most machines, this cannot be done in the constant surface speed mode.

And third, if your machine has live tooling (introduced in Lesson One), spindle speed must always be specified in rpm when live tool are being used.

We'll mention one more time that some programmers elect to use the rpm mode even though constant surface speed *could* be used. When a turning tool or boring bar is machining but one diameter (or even several diameters that are close together), there is not much of an advantage to using constant surface speed – other than eliminating the need for the rpm calculation. So some programmers will calculate an rpm based upon the largest diameter being machined and program the operation in the rpm mode. See Figure 1.17 for an example.

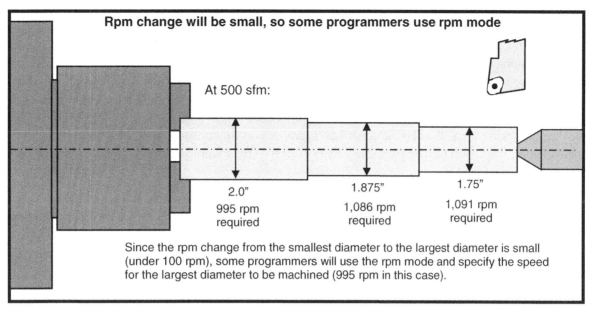

Figure 1.17 – With small diameter changes, rpm changes will also be small

How fast will the spindle be running when constant surface speed is used?

Say you're going to be rough turning a mild steel workpiece with a cutting tool that has a carbide insert. The cutting tool manufacturer recommends that you run the spindle at 500 sfm. In the program at the beginning of the tool, you give this command:

N050 G96 S500 M03

This command will start the spindle at 500 sfm (assuming the inch mode is being used) in the forward direction. But when the spindle starts, how fast will it be running in rpm?

Based upon the information just provided, you cannot answer this question. Prior to answering, you must know the diameter at which the cutting tool is currently positioned. And even then, you must perform the rpm calculation (3.82 times sfm divided by diameter).

For people that have experience running a conventional lathe, this can be a little unnerving. Machinists are accustomed to specifying spindle speed directly in rpm – so when the spindle starts on a conventional lathe, they will know precisely how fast the spindle will run in rpm. If you want to know the precise rpm at which the spindle will start with CNC turning centers (when specifying speed in the constant surface speed mode), you must perform the rpm calculation – and you must know the diameter position of the cutting tool in order to perform this calculation.

How fast can the spindle rotate?

As stated in Lesson One, you must reference the machine tool builder's documentation (commonly the programming manual) in order to determine your machine's spindle characteristics. Say, for example, you find that your turning center has two spindle ranges. The low range runs from 0 – 1,500 rpm. The high range runs from 0 – 5,000 rpm. This means, of course, that when the spindle is in the low range, it cannot run any faster than 1,500 rpm. When it is in the high range, it cannot run faster than 5,000 rpm.

When you specify spindle speed in the constant surface speed mode, the current spindle range (low or high) will determine the maximum spindle speed. Using the rpm calculation, the machine will attempt to run the spindle at the appropriate rpm. If the rpm calculation renders a speed in rpm that is higher than the maximum speed allowed in the current spindle range, the spindle will simply *peak out* at the maximum rpm of the spindle range and run at this speed. Consider these scenarios for the machine just described.

Scenario number one: Say you are rough turning a workpiece from an 8.0" diameter down to a 1.0" diameter. Based upon the cutting tool being used and workpiece material being machines, the cutting tool manufacturer recommends a speed of 800 sfm. Since this is a powerful machining operation, you select the low spindle range.

When rough turning begins, the spindle will be running at about 334 rpm (3.82 times 800 divided by 8.0). As the rough turning tool makes roughing passes, the spindle speed in rpm will increase. When the rough turning tool reaches 2.0" in diameter, the spindle will attempt to run at 1,528 rpm (3.82 times 800 divided by 2.0). Since the maximum rpm in the low spindle range for this machine is 1,500 rpm, the machine will not be able to achieve the appropriate rpm. It will peak out at 1,500 rpm. If you continue to rough turn the workpiece in the low spindle range, this rough turning pass – as well as the rest of the roughing passes – will be performed at 1,500 rpm.

Scenario number two: Whenever you face a workpiece to center (a common machining operation), the spindle will run up to the maximum speed of the current spindle range. For our example machine, this means it will run up to 1,500 rpm if the low range is selected or 5,000 rpm if the high range is selected.

For small, perfectly round workpieces, this will be acceptable. The workpiece will run true in the spindle all the way up to the machine's maximum speed – even in the high range. But you must exercise extreme caution when workpieces are larger – and especially when they are not perfectly round.

Castings, for example, are notorious for being out-of-round. An out-of-round workpiece will wobble in the workholding device when the spindle rotates. The faster the spindle runs, the more machine vibration this wobbling will cause. Of course, wobbling is caused by the fact that the workpiece is not truly concentric with the spindle – and it will place stress on the workholding device used to secure the workpiece.

If this stress is excessive, the workpiece will actually be released by the workholding device. This makes for a very dangerous situation. A workpiece that is rotating at a very high rate will be bouncing around inside the machine – and could actually come right through the door of the machine.

When you must machine large workpieces that are not truly round, you must be very careful not to allow the spindle to reach a speed in rpm that causes the machine to vibrate. A test can be made during the machine's setup to determine this maximum spindle speed. Once you know how fast the spindle/workpiece can safely rotate, you can specify a spindle limiting command in the program that will keep the spindle from exceeding this speed.

The maximum spindle speed test: The setup person will load a workpiece and start the spindle at a very slow rpm. They will continue to increase the spindle speed in small increments until they *start* to feel vibration. The speed at which the machine begins to vibrate will be reduced by about twenty percent – and will be the maximum speed for this workpiece while it is in its rough state.

How to specify a maximum speed for the constant surface speed mode
Here is a way to limit the maximum speed in rpm that the spindle can achieve. In essence, you will be superceding the maximum rpm of each spindle range. The spindle limiter is specified with a **G50** command. The command

>   N055 G50 S2000 (Limit spindle speed to 2,000 rpm)

specifies that the spindle will not be allowed to exceed 2,000 rpm, even if the constant surface speed mode is being used and the machine has calculated a speed that is greater than 2,000 rpm. If this command is specified at the beginning of the program for our example machine (maximum speed in the high range is 5,000 rpm), the spindle will not be allowed to run faster than 2,000 rpm, even if the high range is selected. If, after specifying the spindle limiter shown above in line **N055**, you program a facing tool to face to center (zero diameter) in the high spindle range, the spindle will peak out when it reaches 2,000 rpm.

The two ways to specify feedrate
As with spindle speed, there are two ways to specify feedrate for CNC turning centers. As mentioned in Lesson One, feedrate can be specified in per revolution fashion (inches or millimeters per revolution) or in per

minute fashion (inches or millimeters per minute). Again, two G codes are used to specify which feedrate mode will be used.

> G98 – feed per minute
> G99 – feed per revolution

When you power up on a CNC turning center, the feed per revolution mode (G99) is automatically selected. (By the way, CNC words that are automatically selected at power-up are called *initialized* words.) This means that if you do not include a feedrate mode specifying G code in a program, the machine will assume the feed per revolution mode. You'll notice that many of the example programs provided in this text make this assumption (they don't include a G99).

Also as mentioned in Lesson One, an F word actually specifies feedrate. Here are two examples (assuming you are working in the inch mode).

> N060 G98 F30.0 (Feedrate of 30.0 ipm)
> N060 G99 F0.012 (Feedrate of 0.012 ipr)

When to use the feed per revolution mode

We recommend using the feed per revolution feedrate mode (G99) for almost all machining operations you perform on CNC turning centers. In the per revolution feedrate mode, feedrate specifies how far the cutting tool will move during one spindle revolution.

Per revolution feedrate mode is especially helpful with cutting tools that use the constant surface speed mode (like rough and finish turning tools, rough and finish facing tools, rough and finish boring tools, and grooving tools). As the spindle changes speed in rpm based upon the current diameter position of the cutting tool, the feedrate in inches or millimeters per minute will also change. This will cause witness marks (finish) to be consistent for all surfaces being machined.

Even if you're using the rpm mode (possibly for a drill that machines a hole in the center of the workpiece), it is always easier to specify feedrate in per revolution fashion. Again, most cutting tool manufacturers recommend feedrates for their cutting tools in per revolution fashion.

When to use the feed per minute feedrate mode

Frankly speaking, about *the only time we recommend using the feed per minute mode is when you must cause a controlled motion with the spindle stopped.* If the spindle is stopped, of course, the axes will not move regardless of how large a feedrate is specified in the per revolution mode.

If your turning center has a bar feeder, for example, and if the bar feed operation requires the spindle to be stopped prior to feeding a bar, the feed per minute mode must be used for the bar advance motion (bar feeder programming is shown in the Appendix which is after Lesson Twenty-Three).

If your turning center is equipped with live tooling, operations performed with live tools require that the machine be in the live tooling mode (not the normal turning mode). In live tooling mode, the machine's main spindle is off (at least from a turning operation standpoint), meaning that feedrate for live tools must be programmed in per minute fashion.

Calculating feedrate per minute – To calculate feedrate in per minute fashion, multiply the desired feedrate per revolution times the previously calculated spindle speed in rpm. For example, say you must use the live tooling mode and drill a 0.5 diameter hole. For the material you must machine, the drill manufacturer recommends a speed of 80 sfm and a feedrate of 0.008 ipr. First, calculate the speed in rpm – 3.82 times 80 divided by 0.5 – or 611 rpm. Now multiply 611 times 0.008 – which renders a per-minute feedrate of 4.88 ipm.

Again, any time you must cause a feedrate motion when the spindle is stopped, you must use the feed per minute mode. One other example is when performing a light broaching operation.

Some programmers do prefer to use the feed per minute mode for operations when the feedrate will remain consistent in feed per minute for the entire machining operation (even though feed per revolution could be used). When drilling a hole, for example, once the feed per minute is calculated, it will work for the entire drilling operation.

An example of speed and feed usage

Figure 1.18 provides an illustration of what happens when you use the constant surface speed spindle mode with the per revolution feedrate mode to perform machining operations. Notice that spindle rpm changes with workpiece diameter. So does feedrate in inches per minute change with changes in spindle rpm.

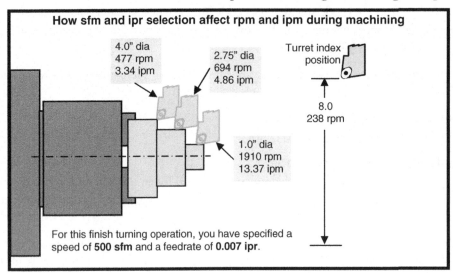

How sfm and ipr selection affect rpm and ipm during machining

4.0" dia
477 rpm
3.34 ipm

2.75" dia
694 rpm
4.86 ipm

Turret index position

8.0
238 rpm

1.0" dia
1910 rpm
13.37 ipm

For this finish turning operation, you have specified a speed of **500 sfm** and a feedrate of **0.007 ipr**.

Figure 1.18 ~ How spindle speeds and feedrates change as a workpiece is machined

As you study Figure 1.18, notice how many speed and feed calculations the machine is making – all based upon your spindle speed selection in surface feet per minute and feedrate selection in inches per revolution.

Lesson 5

Program Zero And The Rectangular Coordinate System

A programmer must be able to specify positions through which cutting tools will move as they machine a workpiece. The easiest way to do this is to specify each position relative to a common origin point called program zero.

You know that turning centers have two linear axes – X and Z. You also know these axes move and that they have a polarity (plus versus minus). In order to machine a workpiece in the desired manner, each axis must, of course, be moved in a controlled manner. One of the ways you must be able to control each axis is with precise *positioning control.*

In the early days of NC (before CNC, over forty years ago), a programmer was required to specify drive motor rotation in order to cause axis motion. This meant they had to know how many rotations of an axis drive motor equated to the desired amount of linear motion for the moving component of the axis (turret). As you can imagine, this was extremely difficult – it was not logical. There is no relationship between drive motor rotation and motion of the moving component. Today, thanks to program zero and the rectangular coordinate system, specifying positions through which cutting tools will move is much easier.

The rectangular coordinate system has an origin point that we'll be calling *program zero.* It allows you to specify all positions (we'll be calling *coordinates*) from this central location. As a programmer, *you* will be choosing the location for program zero – and if you choose it wisely, many of the coordinates you will use in the program will come directly from your workpiece drawing, meaning the number and difficulty of calculations required for your program can be reduced.

Graph analogy

We use a simple graph to help you understand the rectangular coordinate system as it applies to CNC. Since everyone has had to interpret a graph at one time or another, we should be able to easily relate what you already know to CNC coordinates. Figure 1.21 is a graph showing a company's productivity for last year.

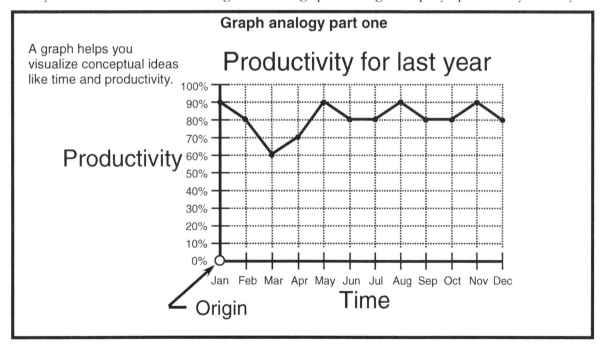

Figure 1.21   Graph example used to illustrate rectangular coordinate system

In Figure 1.21, the *horizontal base line* represents time. The *increment* of the time baseline is specified in months. One whole year is the *range*, given from January through December. The *vertical base line* represents productivity. The *increment* for this base line is specified in ten percent intervals and ranges from 0% to 100% productivity.

In order for a person to make this graph, they must have the productivity data for last year. They will plot a point along the vertical line corresponding to January (the vertical base line in this case) and along the horizontal line corresponding to the percentage of productivity (90% in our case). This plotting of points will be repeated for every month of the year. Once all of the points are plotted, a line or curve can be passed through each point to show anyone at a glance how the company did last year.

A graph is amazingly similar to the rectangular coordinate system used with CNC. Look at Figure 1.22.

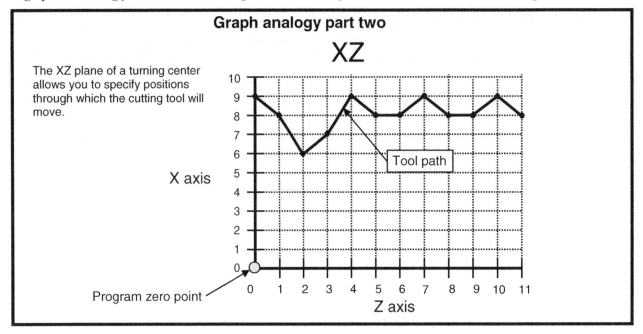

Figure 1.22 – The coordinate system of a turning center (XZ plane)

For CNC turning centers, the *horizontal base line* represents the Z axis. The *vertical base line* represents the X axis. The *increment* of each base line is given in linear measurement. If you work in the inch mode (which we use throughout this text), each increment is given in inches. The smallest increment is 0.0001 inch, meaning each axis has a very fine grid. If you work in the metric mode, each increment will be in millimeters. In the metric mode, the smallest increment is 0.001 mm. The *range* for each axis is the amount of travel in the axis (from one over-travel limit to the other).

A metric advantage – Again, since most people in the US are accustomed to the inch mode, we use it for all examples in this text. However, you should know that there is an accuracy advantage with the metric mode. The advantage has to do with the least input increment – or resolution – for each axis. As stated, in the metric mode the least input increment is 0.001 mm, which is less than half of 0.0001 inch (0.001 mm is actually 0.000039 in). Think of it this way: A ten inch long linear axis has 100,000 programmable positions in the inch mode. The same ten inch long linear axis has 254,000 programmable positions in the metric mode!

More about polarity

In the graph example shown in Figure 1.21, notice that all points are plotted after January and above 0% productivity. The area up and to the right of the two base lines is called a *quadrant*. This particular quadrant is quadrant number one. The person creating the productivity graph intentionally planned for coordinates to fall in quadrant number one in order to make it easy to read the graph.

From Lesson One, you know that each machine axis has a *polarity*. As the turret (and cutting tool) moves toward the workholding device in the Z axis, it is moving in the Z minus direction. As it moves toward the spindle center in the X axis, it is moving in the X minus direction (for almost all turning centers).

The rectangular coordinate system makes determining the polarity of coordinates used in a program *very* simple. From a programming standpoint, you will specify polarity for a given coordinate by determining whether a coordinate position is above or below program zero in X – or to the left or right of program zero in Z. Figure 1.23 shows how.

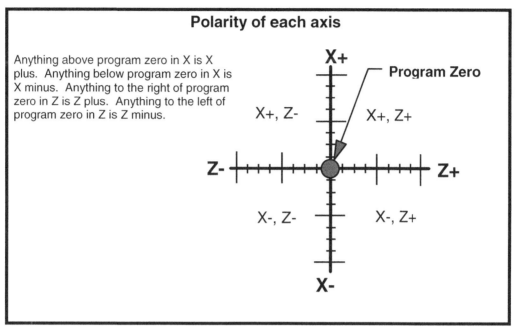

Figure 1.23 – How you determine polarity for coordinates in your program

It just so happens that all CNC machines will *assume* that a coordinate is positive (plus) unless a minus sign is specified. The CNC word: **X2.0**, for example, specifies a position along the X axis of *positive* two inches. Some controls will actually generate an alarm if the plus sign is included within the word, meaning you must let the control assume positive values. Only include a polarity sign if it is negative (-).

Remember an X coordinate specifies a diameter position. A diameter of 1.5" will be specified as **X1.5**.

Wisely choosing the program zero point location

As the programmer, *you* determine the program zero point location for every program you write. Frankly speaking, program zero could be placed in just about *any* location. As long as the coordinates used in your program are specified from the program zero point, the program will function properly. Though this is the case, the *wise* selection of the program zero point will make programming much easier. It may also make it easier for the setup person.

In X

Always place the X axis program zero point at the spindle/workpiece center in the X axis. This will allow you to specify all X coordinates as diameter positions.

In Z

For the most part, we recommend placing the Z axis program zero point at the end of the workpiece from which the workpiece is dimensioned (there is an exception to this recommendation that we will discuss in Lesson Six). Since most turned workpieces are dimensioned from the end you'll be machining, this means program zero in Z will be the right end of the finished workpiece.

Figure 1.24 shows an example of program zero placement based upon these recommendations.

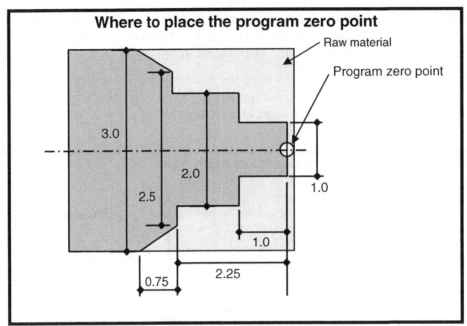

Figure 1.24 ~ An example of program zero placement

With this workpiece, notice that the Z related dimensions begin from the right side of the finished workpiece – so this is where we've placed the program zero point in Z. Since most coordinates used in the program will be to the left of the Z axis program zero point, the polarity for most Z coordinates will be negative.

For the X axis, again, always place the program zero point in the center of the workpiece. Based upon this placement of program zero in X and Z, Figure 1.25 shows the coordinates that will be needed to finish turn this workpiece (we're not showing any coordinates needed for rough turning).

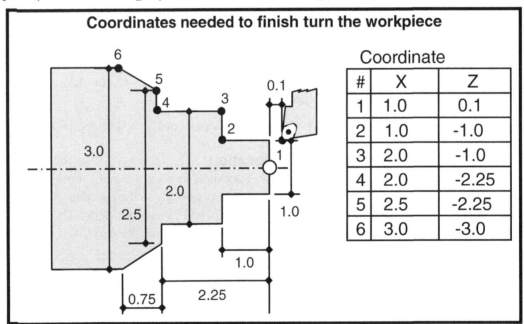

#	X	Z
1	1.0	0.1
2	1.0	-1.0
3	2.0	-1.0
4	2.0	-2.25
5	2.5	-2.25
6	3.0	-3.0

Figure 1.25 ~ Coordinates needed for finish turning

Notice that all coordinates for points one through six on the coordinate sheet are specified *from the program zero point*. For the Z coordinates, since points 2 through 6 are to the left of the program zero point, they are negative values. When program zero is placed at the right end of the finished workpiece in Z, the only time you'll have a positive Z coordinate (to the right of program zero) is when you specify an approach position (like point number one in Figure 1.25).

Fundamentals of CNC

With program zero placed at the center of the spindle almost all X coordinates will be positive – and again, the machine assumes a coordinate to be positive unless the minus sign is used. The only time you'll need a negative X coordinate is when the tool passes the spindle centerline – as is required when facing a workpiece to center. Figure 1.26 shows why.

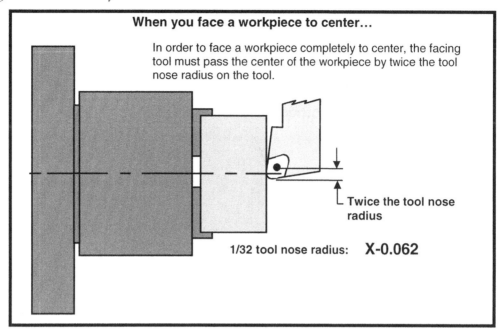

When you face a workpiece to center...

In order to face a workpiece completely to center, the facing tool must pass the center of the workpiece by twice the tool nose radius on the tool.

Twice the tool nose radius

1/32 tool nose radius: **X-0.062**

Figure 1.26 – Facing a workpiece completely to center requires a negative X coordinate

You will be programming the very tip of each cutting tool (in both axes). For this reason, you must send the extreme tip of a facing tool past center by twice the tool nose radius to ensure that the face gets completely machined. At a position of **X0**, there will still be material to be machined (in the form of a small point).

Once again, all X coordinates are specified in diameter. With the drawing shown in Figure 1.25, all X axis-related dimensions happen to be given directly in diameter, making it very easy to determine X coordinates needed in the program. It will not always be quite this simple. Consider, for example, the drawing shown in Figure 1.27.

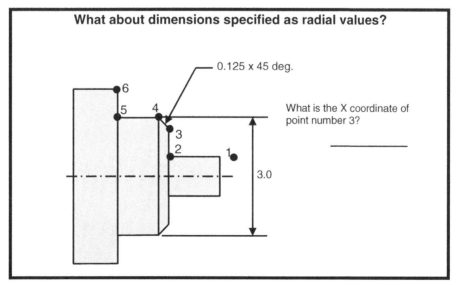

What about dimensions specified as radial values?

0.125 x 45 deg.

6
5   4
3
2   1

What is the X coordinate of point number 3?

3.0

Figure 1.27 – Drawing that has an X axis dimension specified as radial value

Notice that the design engineer has specified a chamfer on the middle diameter of this workpiece. The chamfer size is 0.125 (1/8") by 45 degrees. This means the distance from point 3 to point 4 along the X axis (on the side) is 0.125. A chamfer (or radius) is a *radial dimension*. The X coordinate for point 3, of course, must be the *diameter* of the workpiece at point number 3. To calculate this diameter, you must double the chamfer size (two times 0.125 is 0.25) and subtract the result from the 3.0 inch diameter. The X coordinate for point 3 is 2.75.

A common beginner's mistake is not to double the chamfer or radius size before adding or subtracting it from the closest diameter. For point 3 in Figure 1.27, they will incorrectly specify point number 3 as 2.875 instead of 2.75. This will result in the angle of the chamfer being considerably less than 45 degrees.

Figure 1.28 shows another example, this time for a fillet radius.

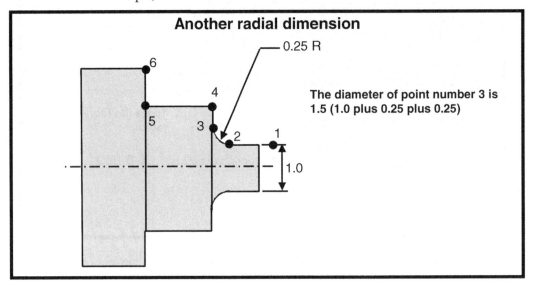

Figure 1.28 – another example of a radial dimension

The diameter of point 3 (X1.5) is determined by doubling the 0.25 radius (a radial dimension on the drawing) and adding the result to the 1.0 diameter.

Practice calculating coordinates

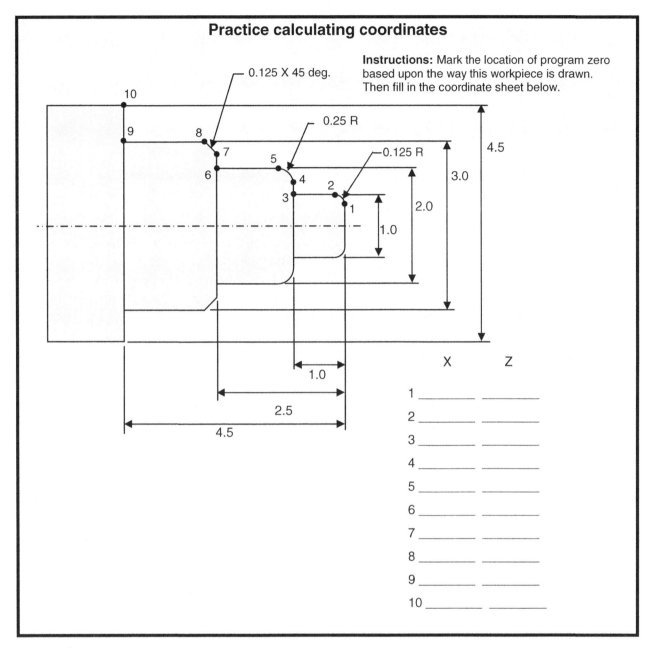

0.125 X 45 deg.

Instructions: Mark the location of program zero based upon the way this workpiece is drawn. Then fill in the coordinate sheet below.

0.25 R

0.125 R

4.5

3.0

2.0

1.0

1.0

2.5

4.5

	X	Z
1		
2		
3		
4		
5		
6		
7		
8		
9		
10		

Answers:

Program zero is at the center of the workpiece in X and at the right end of the finished workpiece in Z.

1: X0.75 Z0
2: X1.0 Z-0.125
3: X1.0 Z-1.0
4: X1. 5 Z-1.0
5: X2.0 Z-1.25
6: X2.0 Z-2.5
7: X2.75 Z-2.5
8: X3.0 Z-2.625
9: X3.0 Z-4.5
10: X4.5 Z-4.5

Lesson 6
Determining Program Zero Assignment Values

The programmer chooses the program zero point location. But the CNC turning center must also be told where program zero is located – it must be able to move cutting tools as instructed by the program. This means the setup person will probably be quite involved with program zero assignment.

You know from Lesson Five that the program zero point is the origin for your program. All coordinates specified in your program are taken from program zero (when using absolute positioning). You also know that the program zero point is determined based upon how the print is dimensioned. The program zero point is always placed at the center of the workpiece in the X axis (since X-related dimensions are specified on the print in *diameter*) and usually at the right end of the finished workpiece in the Z axis.

You must understand that just because you want the program zero point to be in a certain location, doesn't mean the CNC turning center is automatically going to know where it is. A conscious effort must be made to *assign* program zero. Program zero assignment is the task of telling the CNC turning center where the program zero point is located for each cutting tool. You can *think of this task as marrying your program to the workholding setup* that is made on the machine.

Much of what we present in Lessons Six and Seven is more related to setup than it is to programming. However, a CNC programmer must be able to *instruct and direct* setup people, providing instructions related to how a given setup must be made. This means they must understand as much about setups as setup people – this includes an understanding of how program zero is assigned.

The programmer must, of course, tell the setup person where the program zero point is located. This is commonly done in the setup documentation – on a *setup sheet* that provides directions for making the setup. For the purpose of this text, we'll say that the program zero point will always be placed at the center of the workpiece in the X axis and the right end of the finished workpiece in the Z axis – as we recommend in Lesson Five.

Program zero must be assigned independently for *each* cutting tool
A wide variety of cutting tools can be used on CNC turning centers. Figure 1.30 shows some of the most common cutting tools used in the normal turning mode (not considering live tools).

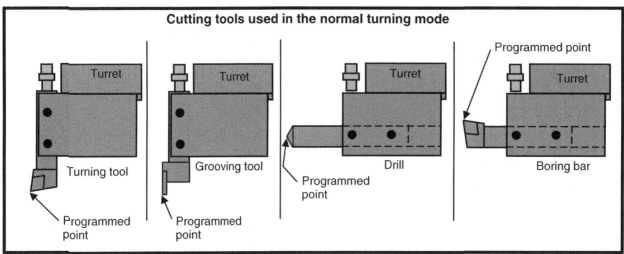

Figure 1.30 ~ Common cutting tools used in the normal turning mode

Additionally, machines that are equipped with live tooling can also use rotating tools. Figure 1.31 shows two end mills – one in an X axis tool holder and the other in a Z axis tool holder.

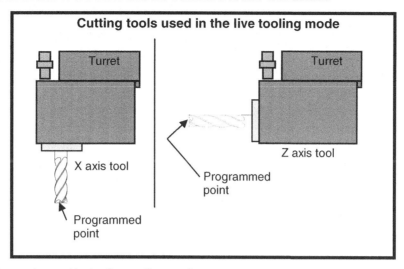

Figure 1.31 – Cutting tools used in the live tooling mode

As you look at Figures 1.30 and 1.31, notice that the point (or tip) of each cutting tool is in a different position. A program zero assignment for a turning tool will not be correct for a boring bar (or any other tool shown). Figure 1.32 more clearly illustrates this be super-imposing two cutting tools in the same turret station.

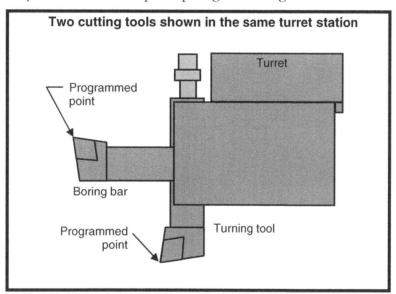

Figure 1.32 – The tool tip is in a different position for each style of cutting tool

Again, the programmed point for a turning tool is in a different location than it is for a boring bar – or a drill – or a grooving tool – or a threading tool – or any live tool. Yet the turning center must be able to correctly position any of these cutting tools to the coordinates you specify in a program – and of course these coordinates are specified relative to a central program zero point. This means the machine must be told – by one means or another – where each cutting tool is placed in the machine relative to the program zero point. *Each cutting tool will require its own program zero assignment.*

Even for very similar cutting tools a separate program zero assignment is necessary. Consider two identical turning tools being used in a job. One is being used for rough turning and the other is being used for finish turning. If the same program zero assignment is used for both tools, even a tiny deviation in the position of the programmed point (especially for the finishing tool) will cause the cutting tool to incorrectly machine the workpiece – probably resulting in a scrap workpiece. We'll discuss this in greater detail a little later. For now, the main point is: *each cutting tool requires a separate program zero assignment.*

Understanding program zero assignment values

Again, a CNC turning center must be told where each cutting tool is located relative to program zero. For the most part, *program zero assignment values specify the distances between the tool tip and the program zero point in each axis.* These distances are determined when the machine is positioned at the *zero return position* in each axis.

What is the zero return position?

The zero return position (also called the *home position* and *machine zero*), is a precise and consistent reference position in each axis. Many turning center manufacturers place the zero return position close to the plus over-travel limits in each axis. (An exception to this statement is with long-bed turning centers – with these machines the Z axis zero return position may be placed near the center of the Z axis.) Figure 1.33 shows an example.

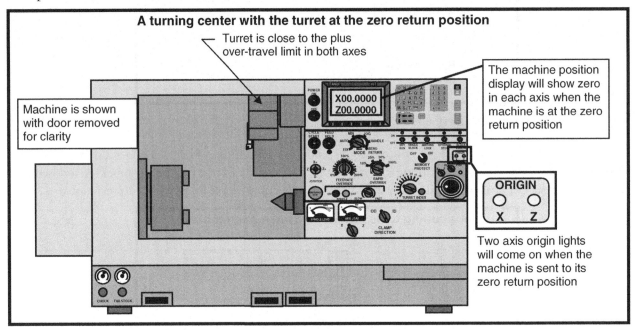

Figure 1.33 – A turning center with the turret resting at the zero return position

The actual procedure used to send each axis to its zero return position is shown in Lesson Twenty-Seven. For now, we simply want to make it clear that the zero return position is the point-of-reference for program zero assignment values.

Program zero assignment values

The X axis program zero assignment value is the diameter a cutting tool will machine while the turret (and selected cutting tool) is at the X axis zero return position. This is true for all cutting tools and regardless of which method is used to assign program zero.

The Z axis program zero assignment value will vary based upon the method of program zero assignment being used. With two of the methods we show (not our favorite methods), the Z axis program zero assignment value is the distance between the tool tip (with the turret at the Z axis zero return position) and the program zero point in the Z axis.

Figure 1.34 shows the program zero assignment values for this method of program zero assignment.

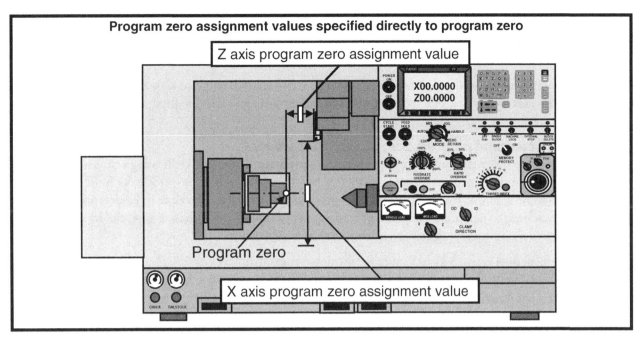

Figure 1.34 – Program zero assignment values specified directly to the program zero point in each axis

With the two other methods (our preferred methods), the Z axis program zero assignment value for each tool is the distance between the tool tip (while the turret is at the Z axis zero return position) and the *chuck face*. A special work shift value tells the machine how far it is from the chuck face to the program zero point. Figure 1.35 shows this.

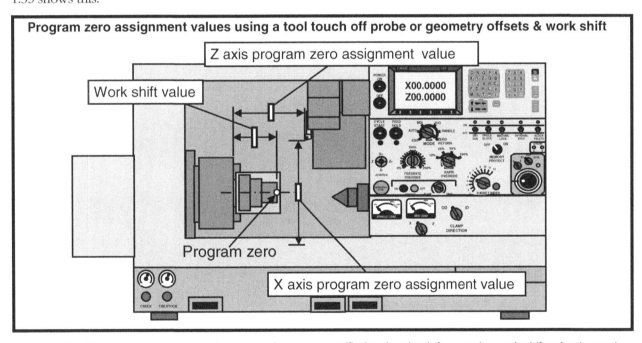

Figure 1.35 – Z axis program zero assignment values are specified to the chuck face and a work shift value is used.

We prefer the method shown in Figure 1.35 because it minimizes work from setup to setup. Any tool remaining in the turret from one job to the next will require but one program zero assignment – when it is used for the first time. That is, a program zero assignment done for a tool in one job will still be correct for it in the next job – and any upcoming job. The work shift value must be changed, of course, if the workpiece length changes from job to job.

With the program zero assignment values shown in Figure 1.34, if the workpiece length changes from one job to the next (as it almost always does), the Z axis program zero assignment value for *every* tool must be determined in *every* job. As you can imagine, this results in duplication of effort.

Regardless of which method of program zero assignment is being used, program zero assignment values must be determined – and they will be entered into the control during setup – into the program or into something called *geometry offsets*.

How do you determine program zero assignment values?

In Lesson Seven, we show the four most popular ways to assign program zero on turning centers. These methods vary based upon machine age and whether or not the turning center has a tool touch off probe. Every method of program zero assignment requires program zero assignment values to be determined. The actual techniques used to determine program zero assignment values also vary based upon the method used, and whether or not your turning center has a feature called *measure function*.

The **measure function** is an optional feature for turning centers that have *geometry offsets*. Most machine tool builders include it with their package of standard features – so if your machine has geometry offsets, it is likely to have the measure function. This function facilitates the calculation and entry of program zero assignment values – and is discussed during Lesson Seven.

Figure 1.36 shows one way to determine program zero assignment values. Frankly speaking, this method is rather crude. It is only used for older machines that *do not* have geometry offsets – when program zero must be assigned in the program. With it, we're simply using the turning center as a measuring device – and we're using the *relative position display screen* to help measure each program zero assignment value.

Again, this method is rather crude – and current model turning centers don't require this method. But it should nicely show what program zero assignment values represent. Note that this procedure can be easily modified for use with work shift. Instead of facing a workpiece to Z0 prior to setting the Z axis relative position display to zero (steps five and six in Figure 1.36), the tool tip can be brought flush with the face of the chuck.

Lesson 7

Assigning Program Zero

Once program zero assignment values are determined as shown in Lesson Six, you must assign program zero. The most popular way to do so is with geometry offsets.

Like Lesson Six, this lesson has more to do with setup than with programming. Again, programmers must know enough about making setups to instruct setup people. This includes knowing how to assign program zero.

You know that cutting tools used in turning centers vary from type to type. You know that each will require its own program zero assignment. You know that program zero assignment values are used to assign program zero. You know that program zero assignment values are the distances between the tool tip and the program zero point (or the face of the chuck if the work shift value is used). These values are determined while the machine is at its zero return position. And you have seen one rather crude way to determine program zero assignment values – manually measuring them at the machine during setup.

In this lesson we're going to show how to *assign* program zero. The most popular and efficient way to do so is with *geometry offsets*.

Understanding geometry offsets

Offsets are storage registers within the control. They are used to store numerical values (numbers) – and will not be used until they are *invoked* by the program. In general, offsets are used to separate certain *tooling related values* from the CNC program – keeping you from having to know them when a program is written.

Geometry offsets are the registers used to assign program zero. In them, you'll be placing program zero assignment values. Turning center controls come with at least sixteen sets of geometry offsets (most come with thirty-two). One set of geometry offsets is used per tool. Since most turning center turrets can hold no more than twelve cutting tools, there will always be more than enough geometry offsets.

Of the four ways we show to assign program zero, three of them use geometry offsets. And frankly speaking, you should use geometry offsets to assign program zero unless your machine does not have them – as will be the case with older machines (machines made before about 1987). Figure 1.39 shows the first display screen page of geometry offsets.

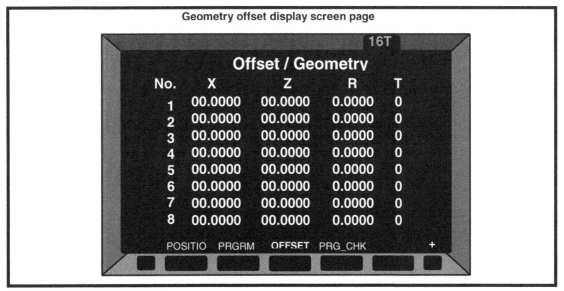

Geometry offset display screen page

16T

Offset / Geometry

No.	X	Z	R	T
1	00.0000	00.0000	0.0000	0
2	00.0000	00.0000	0.0000	0
3	00.0000	00.0000	0.0000	0
4	00.0000	00.0000	0.0000	0
5	00.0000	00.0000	0.0000	0
6	00.0000	00.0000	0.0000	0
7	00.0000	00.0000	0.0000	0
8	00.0000	00.0000	0.0000	0

POSITIO   PRGRM   **OFFSET**   PRG_CHK   +

Figure 1.39 – First page of geometry offsets

With this particular control model, the first eight geometry offsets are shown on page one. By pressing the page down button on the control panel, other pages of geometry offsets can be displayed. Notice that each geometry offset has four registers (X, Z, R, and T). Only the X and Z registers are used for program zero assignment. We'll discuss the R and T registers in Lesson Fourteen.

Say you have measured the program zero assignment values for every cutting tool to be used in a job using techniques shown in Figure 1.36 (Lesson Six). With the program zero assignment values at hand, they can be simply entered into geometry offset registers. However, geometry offsets do have a *polarity* (plus versus minus). They specify the distance *from* the tool tip (while the turret is at the zero return position) *to* the program zero point. If the zero return position is close to the plus over-travel limit for each axis (as it is with most turning centers), geometry offset values always will be negative (-). Figure 1.40 shows the geometry offsets page again, this time after program zero assignment values have been entered.

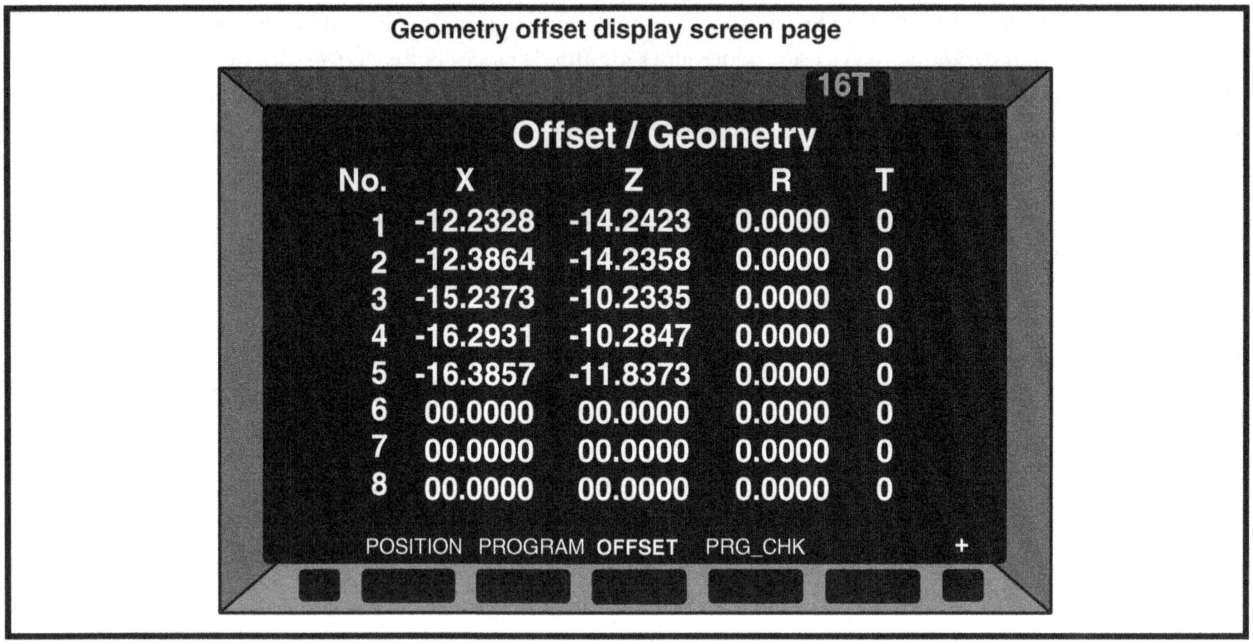

Geometry offset display screen page

No.	X	Z	R	T
			16T	
	Offset / Geometry			
1	-12.2328	-14.2423	0.0000	0
2	-12.3864	-14.2358	0.0000	0
3	-15.2373	-10.2335	0.0000	0
4	-16.2931	-10.2847	0.0000	0
5	-16.3857	-11.8373	0.0000	0
6	00.0000	00.0000	0.0000	0
7	00.0000	00.0000	0.0000	0
8	00.0000	00.0000	0.0000	0

POSITION PROGRAM **OFFSET** PRG_CHK   **+**

Figure 1.40 – Fixture offsets page after program zero assignment values are entered.

Again, the values placed in geometry offsets are simply the program zero assignment values needed for each cutting tool. In Figure 1.40, each X register represents the diameter of the tool tip while the machine is at the X axis zero return position (as negative values). And this will be true regardless of which method of program zero assignment is used. As you go from job to job, X axis geometry offset values will remain the same for any tools that remain in the turret.

The Z axis registers of Figure 1.40 represent the distances in Z from the tool tip while the machine is at the Z axis zero return position to the Z axis program zero point on the workpiece. When the next job must be run, all Z axis geometry offsets will change if the workpiece to be machined in the next job is longer or shorter than it is in the last job. As stated in Lesson Six, this can result in much duplicated effort. We'll be presenting two methods for program zero assignment that eliminate this duplication of effort – and in general – simplify the task of program zero assignment. These will be the first two methods shown in this lesson.

How geometry offsets are instated

Again, each cutting tool requires its own program zero assignment (geometry offset). From within a program, a four digit *T word* is used to instate geometry offsets. Actually the T word does three things.

- The first two digits of the T word specify the turret station and cause the turret to index
- The first two digits of the T word also specify and instate the geometry offset
- The second two digits of the T word specify and instate the wear offset

Consider the T word:

T0101

The first two digits command the turret to index to station one and instate geometry offset number one. The second two digits instate wear offset number one. Most programmers will keep things simple and use the same wear offset number as station number and geometry offset number (T0202, T0404, T0606, etc.).

When the machine executes the command

N030 T0505

it will index the turret to station five and instate geometry offset number five. It will also instate wear offset number five. The machine will look in geometry offset number five to find the related program zero assignment values. With this information, it can correctly position the cutting tool to the coordinates specified in the program and machine the workpiece properly. More is presented about programming with geometry offsets during Lesson Twelve.

The four most common ways to assign program zero

Actually, there are *many* ways to assign program zero with CNC turning centers. These methods vary based upon the age of the turning center as well as whether or not the turning center has a tool touch off probe (sometimes called a quick-setter or tool setter). Program zero assignment methods can even vary based upon personal preferences of setup people.

While you will have to be prepared for variations, we show the four most popular methods for program zero assignment in order from most preferable to least preferable:

- **With a tool touch off probe** (use this method if your machine has a tool touch off probe)

- **With geometry offsets using work shift** (use this method if your machine doesn't have a tool touch off probe but it does have the geometry offset and work shift features – this is the method we show in this text)

- **With geometry offsets without work shift** (use this method if your machine has neither a tool touch off probe nor the work shift function, yet does have the geometry offset feature)

- **With G50 in the program** (use this method if your machine doesn't have a tool touch off probe or geometry offsets – older machines)

Admittedly, this may get a little confusing. You'll have to check with an experienced person in your company to determine which method/s of program zero assignment your company uses. If you're unsure – or if you don't currently work with for a CNC-using company, please study the first two program zero assignment methods shown (with a tool touch off probe and with geometry offsets & work shift). They tend to be the most common methods used with *current* CNC turning centers.

If your company has a variety of different CNC turning centers ranging in age from very old to very new, you may find that all four of these methods are being used in your company (only one method for a given machine, of course). In this case, you'll eventually have to learn them all. But to get started, you might want to limit your studies to just one or two of these methods – we'd recommend starting with the first one shown (with the tool touch off probe) and then work your way through the list as you master each method.

You may even find that setup people are using a different method of program zero assignment all together. This often happens when setup people have no formal training – when they are left on their own to figure out how program zero assignment should be done. You may have to adapt to these methods – or try to convince them that methods shown in this text may be better. But if you can't convince them, you'll have to adapt to their methods. And if you truly understand what is presented in Lessons Six and Seven, you should be able to adapt to just about any method of program zero assignment.

Determining and entering the work shift value

Figure 1.39 shows the work shift display screen page for a popular Fanuc control model.

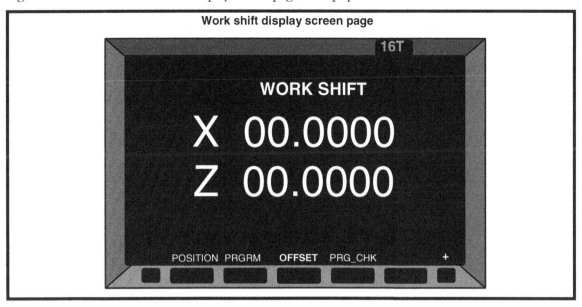

Figure 1.39 – Work shift page

The X and Z values on this page specify the distances from the center of the spindle (in X) and the face of the chuck (in Z) to the program zero point to be used by the program. In the X axis, the work shift will *always* be zero (program zero in X is always placed at the spindle center). In Z, the work shift value will be the length of the workpiece plus the jaw length (if program zero in Z is placed at the right end of the finished workpiece). With most machines, the polarity for this value is positive.

For almost all turned workpieces, both ends must be machined. For the first end (first operation), the work shift value is not terribly critical. In most cases, all that is important is that the raw material (stock) be balanced on both ends of the workpiece. That is, the stock on the front face will be the same as the stock on the back face. Figure 1.40 shows an example.

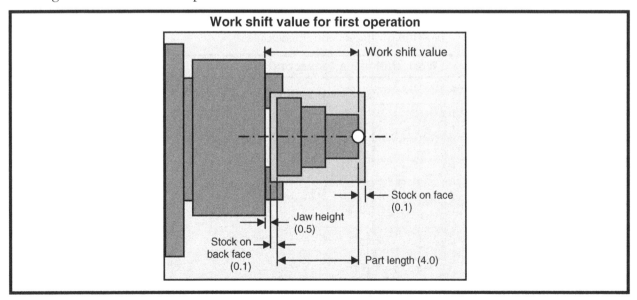

Figure 1.40 – The work shift value

Say, for example, the finished length of the finished workpiece is 4.0 inches. The raw material (bar) has been cut to a length of 4.2 inches. In this case, most setup people will machine 0.1 inch of material from the first operation – leaving another 0.1 inch of material to be machined from the other face in the second operation.

One way to determine the work shift value is to measure it. For this first operation, the setup person might use a caliper or scale (or even a tape measure) to measure the work shift value. Again, it is not very critical if they are simply balancing stock for first and second operations.

This value can also be calculated if you know the jaw height. The jaw height can, of course, be measured with a caliper or (more precisely) with a depth gauge. In Figure 1.40, we're showing that the jaw height is 0.5 inches. For the 4.0 inch workpiece that has been cut to 4.2 inches long, the setup person will add 4.1 to the 0.5 inch jaw height in order to come up with the Z axis work shift value. This entered as shown in Figure 1.41.

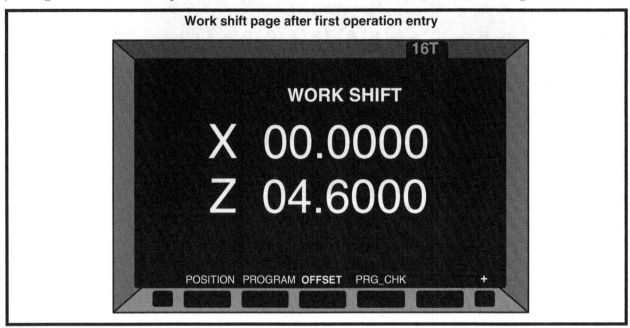

Figure 1.41 – Work shift entry for first operation

The Z axis work shift value is much more critical when it comes time to machine the *other* end of the workpiece (the second operation). This value will, of course, control the overall length of the finished workpiece – and depending upon the tolerance for the overall workpiece length, it might be pretty critical indeed.

Figure 1.42 shows the workholding setup for the second operation.

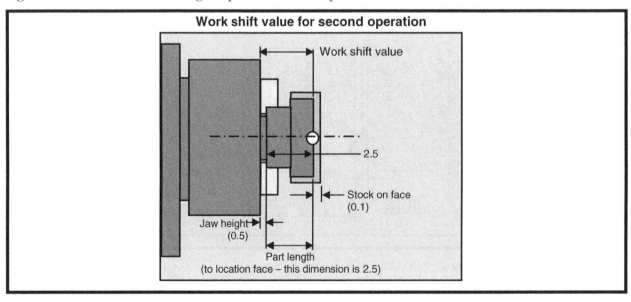

Figure 1.42 – Setup for the second operation

Notice that the jaws have been bored to hold on the middle diameter – and that they locate on the face of this diameter. The jaw height is still 0.5 inches (measured with a depth gauge). The distance from the locating face to the finished end of the workpiece is 2.5 (this dimension can be found on or calculated from the workpiece drawing). The work shift value for this operation will be Z3.0, as is shown in Figure 1.43.

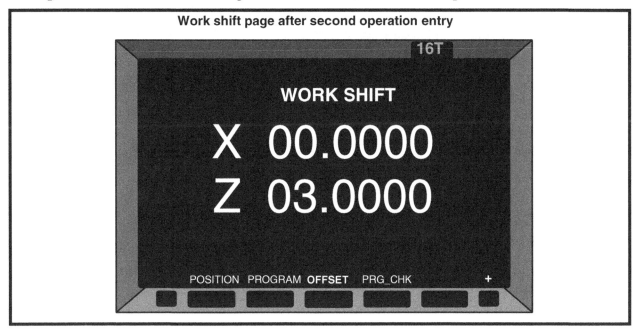

Figure 1.43 – Work shift value for second operation

A note about the polarity of the work shift value

We've been showing the polarity of the Z axis work shift value as *from* the chuck face *to* program zero. And of course, this will always result in a positive work shift value. You must be prepared, however, for control models that reverse this polarity. With these machines, the Z axis work shift value will be entered as a negative value. Ask an experienced setup person which polarity is used for work shift values.

Using geometry offsets with work shift to assign program zero

This is second-most preferred method. It should be used for turning centers that do not have a tool touch off probe, but do have geometry offsets and the work shift function. Advantages include:

- Eliminates the need for calculations
- Minimizes the possibility for entry mistakes (if measure function is used)
- Automatically clears the wear offset

Additionally, since the chuck face is used as the point of reference for geometry offset entries, and since the chuck face position does not change from job to job (in most companies), geometry offsets measured for a cutting tool used in one setup will still work in the next setup. That is, *cutting tools need only be measured once* – when they are first placed in the turret.

While tool touch off probes are becoming very popular, not all turning centers are equipped with them. Yet most current model Fanuc-controlled turning centers that do not have a tool touch off probe *are equipped with geometry offsets and the work shift value.*

These features are used in exactly the same way as they are on machines with tool touch off probes. In fact, the work shift value is determined and entered in exactly the same manner as it is when a tool touch off probe is used (see the presentation just given about the work shift value – we will not repeat it). The only difference is that more work is required to determine and enter geometry offset values.

Though geometry offsets and the work shift function are used in the same way as they are when using a tool touch off probe, remember that a tool touch off probe can also be calibrated to allow for tool pressure. And with a tool touch off probe, the tasks of measuring geometry offset values and entering them are automated. A tool touch off probe eliminates the need for trial machining.

This is not the case when the turning center does *not* have a tool touch off probe. With the methods we show from this point on, there is no (feasible) way to calibrate for tool pressure – and you *might* make a mistake with the measurement and/or entry of geometry offset values. So you must use trial machining techniques with cutting tools that must machine within small tolerances bands.

With this second method of program zero assignment, the **X axis geometry offset value** is the diameter of the tool tip while the X axis is at its zero return position. The **Z axis geometry offset value** is the distance from the tool tip at the Z axis zero return position to the chuck face. And the Z axis **work shift value** is the distance between the chuck face and the program zero point for the program in Z (the X axis work shift value is always zero). See the next illustration.

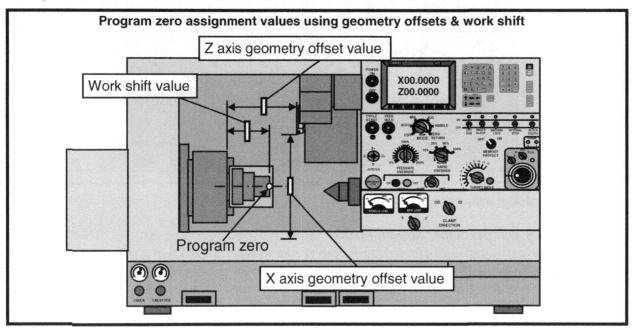

Instead of using a tool touch off probe to measure and enter geometry offset values, you will be doing so in a more manual manner. While most turning centers are equipped with a helpful *measure function* to simplify the task, you'll have to physically skim cut a sample workpiece and measure the machined surfaces during the task of program zero assignment.

Understanding the measure function

The measure function can be thought of as a poor man's tool touch off probe. With a tool touch off probe, the machine is calibrated to know the distance from each stylus position to the spindle center in X and the chuck face in Z. The machine also keeps track of axis position (in each axis) relative to the zero return position. So when a stylus position is contacted, the machine can automatically calculate and enter the related geometry offset value for the tool.

In similar function, the measure function will cause the machine to calculate and enter geometry offset values. But instead of contacting a stylus position, *you'll be telling the machine a cutting tool's current position* – relative to spindle center in X and the chuck face in Z. Based on this information, the machine will calculate and enter the related geometry offset values. Figure 1.44 shows an example of the procedure.

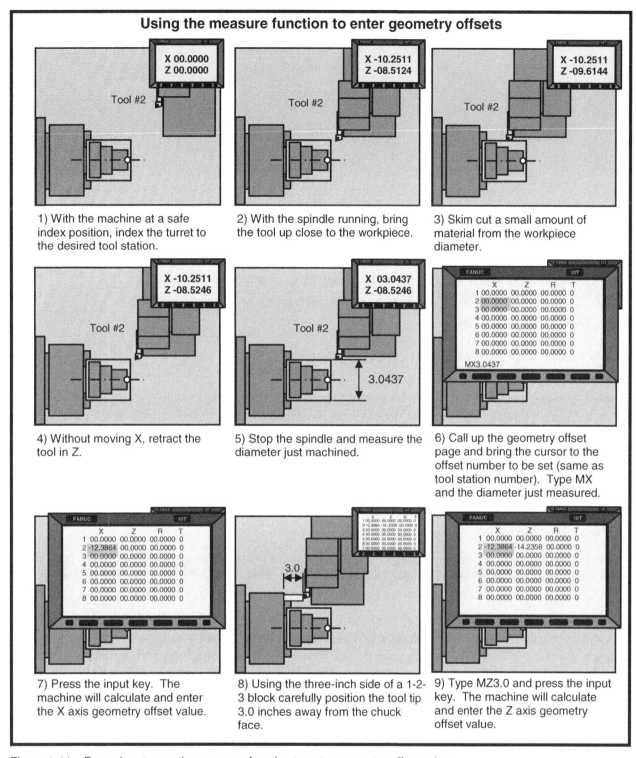

Figure 1.44 – Procedure to use the measure function to enter geometry offset values

In step five of Figure 1.44, notice that the tool is at a known diameter of 3.0437 (we've just machined this diameter and measured it). In step six and with the cursor on the related geometry offset number (number two in our case), you type **MX3.0437**. (MX stands for measure X.) Since the machine constantly knows the current X axis position relative to the zero return position, it can calculate and enter the X axis geometry value for this tool.

You Must Prepare To Write Programs

While this Key Concept does not involve any programming words or commands, it is among the most important of the Key Concepts. The better prepared you are to write a CNC program, the easier it will be to develop a workable program.

Key Concept Number Two is made up of but one lesson:

8: Preparation for programming

Frankly speaking, this Key Concept can be applied to any CNC related task, including setup and operation. Truly, the better prepared you are to perform a task, the easier it will be to correctly complete the task. When it comes to making setups, for example, if you have all of the needed components to make the setup at hand (jaws, cutting tools, cutting tool components, program, documentation, etc.), making the setup will be much easier and go faster. So *gathering needed components* is always a great preparation step to perform.

The same goes for completing a production run. If the CNC operator has all needed components (raw material, inserts for dull tool replacement, a place to store completed workpieces, gauging tools, etc.), they will be able to smoothly complete the production run.

In this Key Concept, however, we're going to limit our discussion to preparation steps that you can perform to get ready to write a CNC program. Remember, the actual task of *writing* a CNC program is but part of what you must do. Certain things must be done *before* you're adequately prepared to write the program.

Preparation for programming is especially important for entry-level programmers. For the first few programs you write, you will have trouble enough remembering the various CNC words – remembering how to structure the program correctly – and in general – you'll have trouble getting familiar with the entire programming process. The task of programming is infinitely more complicated if you are not truly ready to write the program in the first place.

Lesson 9

Preparation Steps For Programming

Any complex project can be simplified by breaking it down into small pieces. This can make seemingly insurmountable tasks much easier to handle. CNC turning center programming is no exception. Learning how to break up this complex task will be the primary focus of Lesson Nine.

As you now know, preparation will make programming easier, safer, and less error-prone. Now let's look at some specific steps you can perform to prepare to write CNC programs.

Doing the math

How dimensions are described on the part print will determine how much math is required for your program. Some companies have their design engineers supplying all drawings with datum surface dimensioning. When this technique is used, each dimension on the print will be taken from only one position in each axis. While this may reduce the need to do calculations, it does not completely eliminate the need for doing math. You'll see why in just a bit.

Of course, not all design engineers use datum surface dimensioning techniques – indeed most do not. You will probably be expected to do a great deal of math (commonly addition and subtraction) in order to determine X and Z coordinates required in the CNC program. Regardless of how much math is required to determine program coordinates, it is wise to perform these calculations *before* you write the program. This will keep you from having to break out of your train-of-thought to calculate coordinates as you write the program.

If you have been doing the practice exercises in this text, you've already done some coordinate calculations. You have also seen our method of using a *coordinate sheet* for documenting coordinates needed in a program. This method involves numbering each position on the print with a point (like connect the dots) and placing each point's X and Z values on a separate coordinate sheet.

However, our practice exercises have only required that you calculate positions that are *right on workpiece surfaces*. And we've provided all of the points for you to calculate. In reality, *not all coordinates* required in the program will actually be on the workpiece (for clearance positions and when finishing stock must be allowed with a roughing tool, for example). And you'll be on your own to determine which positions are required in the program. Once again, this brings up the need to understand basic machining practice. You must be able to plan each cutting tool's movement path (commonly called the tool path) in order to plot the points needed for your program.

Figure 2.4 shows the cross section of a workpiece that must be programmed. We'll use it to illustrate the coordinates that will be needed in the program.

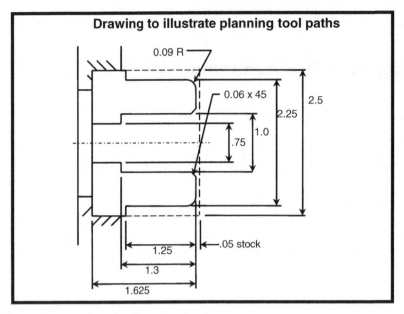

Figure 2.4 – Drawing to illustrate the planning of tool paths

Here is the process we will use to machine the workpiece.

Description	Tool	Stat.	Feedrate	Spindle speed
1) Rough face and turn	80 degree diamond tool	1	0.012 ipr	400 sfm
2) Drill through	3/4" drill	2	0.010 ipr	350 rpm
3) Rough bore	5/8" boring bar	3	0.007 ipr	400 sfm
4) Finish bore	5/8" boring bar	4	0.004 ipr	450 sfm
5) Finish face and turn	80 degree diamond tool	5	0.007 ipr	500 sfm

Figure 2.5 shows a series of tool paths needed to machine the workpiece. We're showing one tool path per tool. And – we're only showing the upper half of the workpiece in each example (above center), but you should be able to see what's going on.

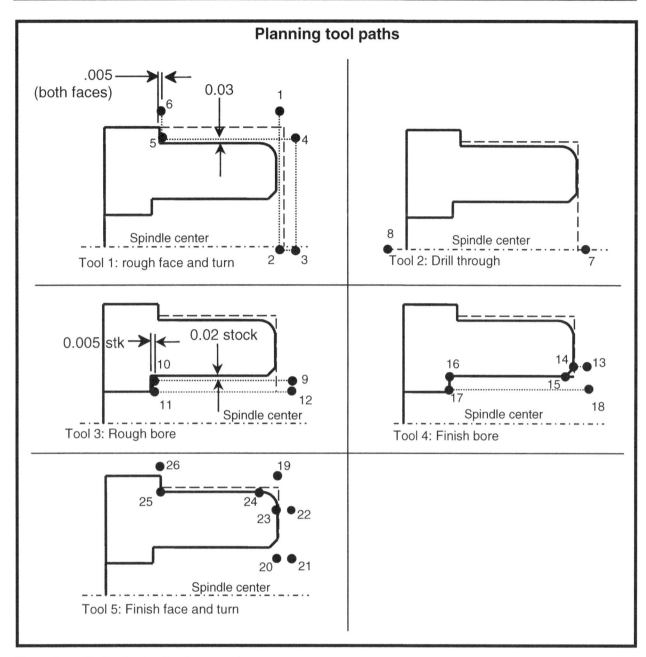

Planning tool paths

.005
(both faces)

0.03

Tool 1: rough face and turn

Spindle center

Tool 2: Drill through

Spindle center

0.005 stk

0.02 stock

Tool 3: Rough bore

Spindle center

Tool 4: Finish bore

Spindle center

Tool 5: Finish face and turn

Spindle center

Figure 2.5 – Tool path planning for example

For each cutting tool, we've planned the path it must take in order to properly machine the workpiece. And again, *you* will have to do this for every cutting tool you program. We've placed a dot next to every location that the cutting tool must pass through and numbered it. While it helps if the numbering is in order, it is not completely mandatory. At this point, we're simply coming up with *all* of the needed points.

Notice that we've also included things in each tool path drawing that are not on the print. For the roughing tools (tools one and three), for example, we've specified how much finishing stock will be left on each surface.

Also, there are some things in each tool path that are *not* labeled. Point number one, for example, appears to be above the rough stock diameter – but by how much? The same is true for points 3, 4, 7, 9, 11, 12, 13, 18, 19, 20, 21 and 22. Each of these points is at a *clearance position* (either approaching the workpiece or retracting from it). Most programmers will use an approach/retract distance of 0.1 inch (2.5 mm) – which is the approach/retract distance we use throughout this text.

Unfortunately, the drawings for many turned workpieces are quite small and cluttered. You may not be able to adequately draw your tool paths right on your working copy of the print. You may have to make a series of special sketches (like those shown above) to help with your tool path documentation. Beginners have a tendency to skip this step if it seems too time-consuming. But without a very clear understanding of the points needed in your program, you'll become confused with your tool paths when you try to write the program.

While it's a little off the subject of doing math, again, we'll mention again the programming features that minimize the need to come up with a special set of coordinates for roughing operations. *Multiple repetitive cycles*, discussed in Key Concept Six, help with roughing operations and truly minimize the number of points for which coordinates must be calculated. But as the programmer, you'll be responsible for determining which coordinates will be needed in your program – and for calculating them before you write the program.

Once all points are plotted, a simple coordinate sheet will help you document your math calculations. Here is a completed coordinate sheet for the previous example.

#	X	Z
1	2.7	0.005
2	-0.062	0.005
3	-0.062	0.1
4	2.31	0.1
5	2.31	-1.245
6	2.7	-1.245
7	0	0.1
8	0	-1.87
9	0.96	0.1
10	0.96	-1.295
11	0.75	-1.295
12	0.75	0.1
13	1.12	0.1
14	1.12	0
15	1.0	-0.06
16	1.0	-1.3
17	0.75	-1.3
18	0.75	0.1
19	2.45	0
20	0.8	0
21	0.8	0.1
22	2.07	0.1
23	2.07	0
24	2.25	-0.09
25	2.25	-1.25
26	2.7	-1.25

Completed coordinate sheet

Check the required tooling

The next preparation step is related to the cutting- and workholding- tools used by your program. Tooling problems can cause even a perfectly written CNC program to fail. You can avoid production delays by considering potential tooling problems while you prepare for programming. Again, there is no excuse for machine down-time for as avoidable a reason as poor preparation.

First, you will need to confirm that the various cutting tools to be used by your program are *available*. You may incorrectly assume, for instance, that common tools are always in your company's inventory. It is always wise to double-check, even with relatively common tools. And always be sure your company has lesser-used tools in stock.

If a given tool is not in stock, of course, it must be ordered. Or you may have to make do with what your company does have available. If this is the case, your entire machining process may have to be changed (another reason to check *before* you begin programming).

Second, you must confirm that cutting tools are capable of machining the workpiece attribute/s you intend. Every insert used in turning tools and boring bars has a certain *geometry* that controls what the tool can and cannot do. Figure 2.6 shows a potential problem.

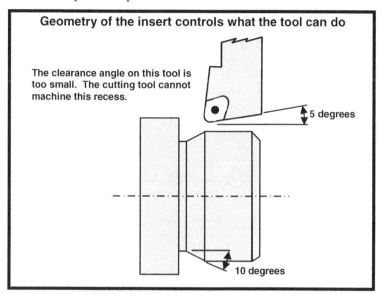

Figure 2.6 – A recess that cannot be machined with the intended turning tool

Notice that the cutting tool shown in Figure 2.6 has only a 5 degree clearance angle. It cannot machine a recess having more than a 5 degree angle.

Another tooling concern is the potential for *interference*. For the process given for the workpiece shown in Figure 2.4, for instance, the two boring bars must *fit* into the 3/4" drilled hole. We've selected 5/8" diameter bars, but we must confirm that they can machine the 1.3" deep face down to 0.75" in diameter without hitting on the back side of the boring bar. The only way to do this is to confirm (from the cutting tool drawing in the cutting tool manufacturer's catalog) that the distance from the tip of the boring bar to back of the boring bar is 0.75 inch or less, as is shown in Figure 2.7.

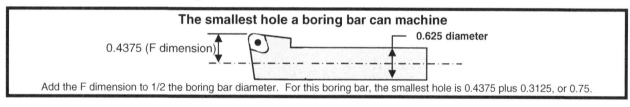

Figure 2.7 – How to determine the smallest hole a boring bar can machine

Yet another interference problem of which you must be constantly concerned has to do with adjacent turret stations. With many turning centers, you cannot place internal cutting tools (drills, boring bars, etc.) in turret stations that are adjacent to turning tools – or to other internal cutting tools. As a facing tool faces a workpiece to center, the adjacent drill or boring bar may collide with the chuck – or even the workpiece itself. The smaller the turret and the more tools it can hold – the more likely it will be that you must skip turret stations with internal cutting tools.

More practice calculating coordinates

Instructions: Fill in the coordinate sheet below. Use 0.1 for approach/retract clearance

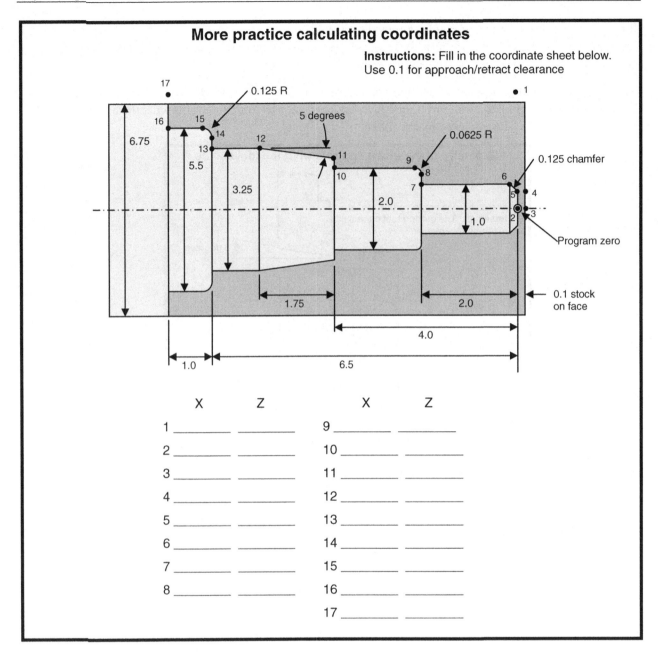

	X	Z		X	Z
1			9		
2			10		
3			11		
4			12		
5			13		
6			14		
7			15		
8			16		
			17		

Answers:

1: X6.95 Z0
2: X-0.0625 Z0
3: X-0.0625 Z0.1
4: X0.75 Z0.1
5: X0.75 Z0
6: X1.0 Z-0.125
7: X1.0 Z-2.0
8: X1.875 Z-2.0
9: X2.0 Z-2.0625

10: X2.0 Z-4.0
11: X2.9438 Z-4.0
12: X3.25 Z-5.75
13: X3.25 Z-6.5
14: X5.25 Z-6.5
15: X5.5 Z-6.625
16: X5.5 Z-7.5
17: X6.95 Z-7.5

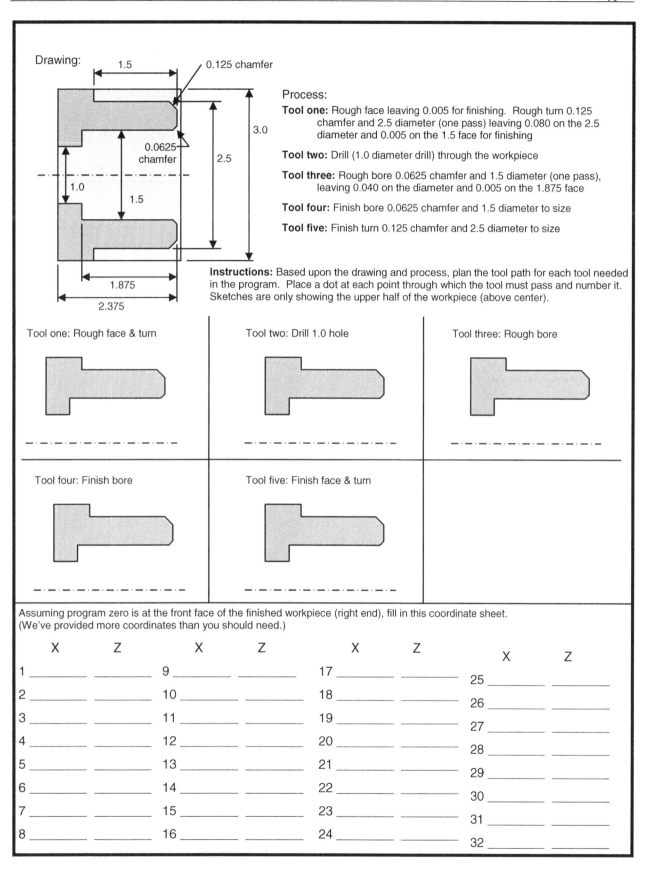

Drawing:

1.5

0.125 chamfer

0.0625 chamfer

3.0

2.5

1.5

1.0

1.875

2.375

Process:

Tool one: Rough face leaving 0.005 for finishing. Rough turn 0.125 chamfer and 2.5 diameter (one pass) leaving 0.080 on the 2.5 diameter and 0.005 on the 1.5 face for finishing

Tool two: Drill (1.0 diameter drill) through the workpiece

Tool three: Rough bore 0.0625 chamfer and 1.5 diameter (one pass), leaving 0.040 on the diameter and 0.005 on the 1.875 face

Tool four: Finish bore 0.0625 chamfer and 1.5 diameter to size

Tool five: Finish turn 0.125 chamfer and 2.5 diameter to size

Instructions: Based upon the drawing and process, plan the tool path for each tool needed in the program. Place a dot at each point through which the tool must pass and number it. Sketches are only showing the upper half of the workpiece (above center).

Tool one: Rough face & turn

Tool two: Drill 1.0 hole

Tool three: Rough bore

Tool four: Finish bore

Tool five: Finish face & turn

Assuming program zero is at the front face of the finished workpiece (right end), fill in this coordinate sheet. (We've provided more coordinates than you should need.)

	X	Z		X	Z		X	Z		X	Z
1			9			17			25		
2			10			18			26		
3			11			19			27		
4			12			20			28		
5			13			21			29		
6			14			22			30		
7			15			23			31		
8			16			24			32		

Can you plan the tool paths? (answers)

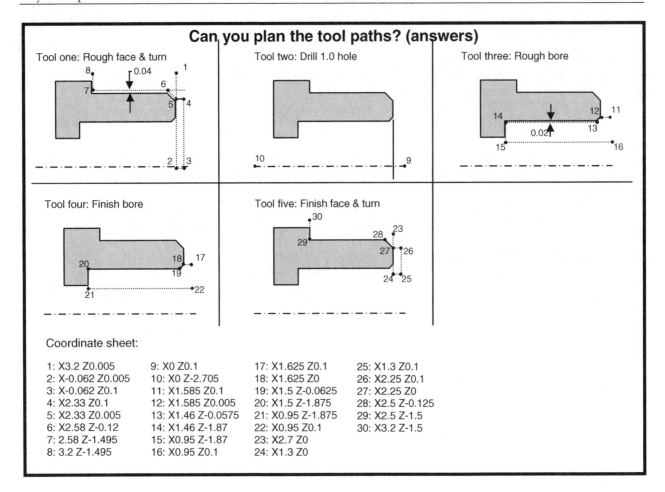

Tool one: Rough face & turn

Tool two: Drill 1.0 hole

Tool three: Rough bore

Tool four: Finish bore

Tool five: Finish face & turn

Coordinate sheet:

1: X3.2 Z0.005	9: X0 Z0.1	17: X1.625 Z0.1	25: X1.3 Z0.1
2: X-0.062 Z0.005	10: X0 Z-2.705	18: X1.625 Z0	26: X2.25 Z0.1
3: X-0.062 Z0.1	11: X1.585 Z0.1	19: X1.5 Z-0.0625	27: X2.25 Z0
4: X2.33 Z0.1	12: X1.585 Z0.005	20: X1.5 Z-1.875	28: X2.5 Z-0.125
5: X2.33 Z0.005	13: X1.46 Z-0.0575	21: X0.95 Z-1.875	29: X2.5 Z-1.5
6: X2.58 Z-0.12	14: X1.46 Z-1.87	22: X0.95 Z0.1	30: X3.2 Z-1.5
7: 2.58 Z-1.495	15: X0.95 Z-1.87	23: X2.7 Z0	
8: 3.2 Z-1.495	16: X0.95 Z0.1	24: X1.3 Z0	

Fundamentals of CNC

Understand The Motion Types

Motion control is at the heart of any CNC machine tool. CNC turning centers have at least three ways that motion can be commanded. Understanding the motion types you can use in a program will be the focus of Key Concept Number Three.

Key Concept Number Three is another one-lesson Key Concept:

10: Motion types

You know that all CNC turning centers have at least two axes (X and Z). You also know that you can specify positions (coordinates) along each axis relative to the program zero point. These coordinates will be positions through which your cutting tools will move. In Key Concept Number Three, we're going to discuss the ways to specify *how* cutting tools move from one position to another.

What is interpolation?

If a single linear axis is moving (X or Z), the motion will of course, be along a perfectly straight line. For example, look at Figure 3.1.

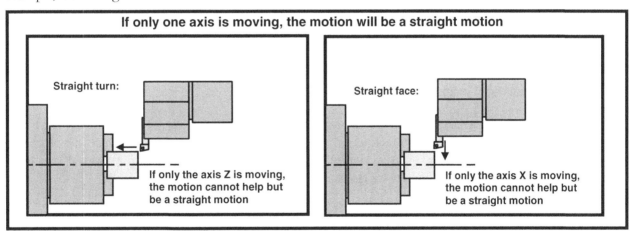

If only one axis is moving, the motion will be a straight motion

Straight turn:

If only the axis Z is moving, the motion cannot help but be a straight motion

Straight face:

If only the axis X is moving, the motion cannot help but be a straight motion

Figure 3.1 – A perfectly straight motion will occur if only one axis is moving

When turning a straight diameter on the workpiece (left view in Figure 3.1), only the Z axis is moving. And since the Z axis is a *linear* axis, this will force the motion to be perfectly straight – and parallel to the Z axis. In like fashion, when facing a workpiece (right view), only the X axis is moving – and again – this motion will be perfectly straight – parallel to the X axis.

But sometimes a straight surface on a workpiece is *tapered*. Additionally, there will be times when a cutting tool must machine a radius. This will require that *both* the X and Z axes move in a controlled manner, as shown in Figure 3.2.

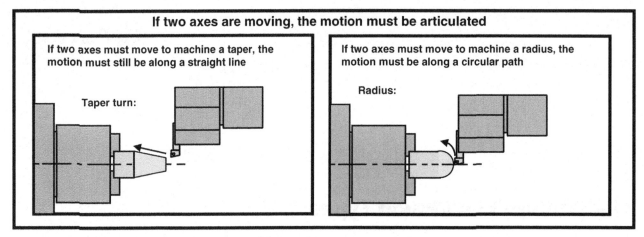

Figure 3.2 – Examples of when both the X and Z axes move in a controlled manner

The machine must *articulate* the motions shown in Figure 3.2. In CNC terms, this kind of articulation is called *interpolation*.

Look at Figure 3.3.

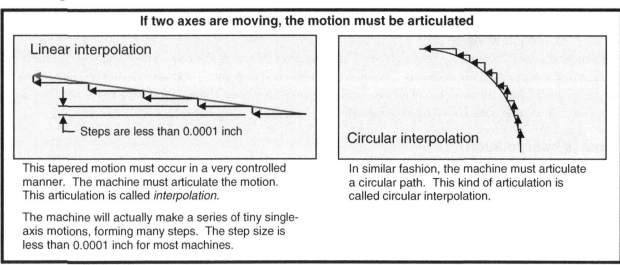

Figure 3.3 – The meaning of interpolation

As you can see, the machine will actually break these two-axis motions in to a series of very tiny single-axis steps. The step size will be under 0.0001 inch for most machines. These steps are so small that you will not be able to see them – or measure them – with most measuring devices. For all intents-and-purposes, all machined surfaces will appear to be perfectly straight (or circular) and without steps.

(By the way, the step size during interpolation is referred to as the machine's *resolution*. With many CNC machines, it is equal to the least-input-increment available in the measurement system you are using – 0.0001 inches in the inch mode or 0.001 millimeters in the metric mode. The smaller the step size, the finer will be machine's resolution – and the more precisely it will follow your commanded motions.)

Turning center manufacturers offer a variety of *interpolation types*, based upon what their customers will be doing with their machines. *All* offer at least the two interpolation types shown above – *linear interpolation* (also called straight-line motion) and *circular interpolation* (also called circular motion). Additionally, all turning centers come with a third motion type, called *rapid motion* – though we don't consider rapid motion to be a true form of interpolation since most machines do not articulate the tool's path during rapid motion.

Depending upon your company's specific needs, there may be other interpolation types available with your turning centers. However, these additional interpolation types will only be used for special applications. If, for example, your machine has live tooling, and if you'll be performing milling operations around the outside diameter of the workpiece (like milling flats), you'll need an interpolation type called *polar coordinate interpolation* (we discuss polar coordinate interpolation in the Appendix – after Lesson Twenty-Three). Again, special interpolation types are only used in special applications. The bulk of your programming will not require them.

Actually, it should be refreshing to know that there are *only* three common ways to cause axis motion – rapid, straight-line, and circular. Just about every motion a CNC turning center makes can be divided into one of these three categories. Once you master these three motion commands, you will be able to generate the movements required to machine a workpiece. Figure 3.4 illustrates them.

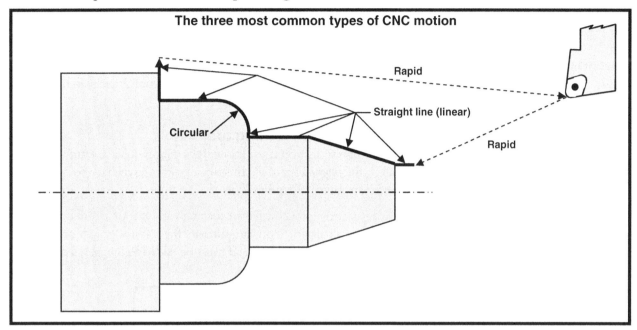

Figure 3.4 – The three kinds of motion available with all CNC turning centers

As you will see in Lesson Ten, it is actually quite easy to specify motion commands within a CNC program. In general, each motion will require you to specify the kind of motion (rapid, straight line, or circular) along with the motion's end point (coordinates at the end of the motion). Linear motion and circular motion additionally require that you specify the motion rate (feedrate) for the motion. Circular motion additionally requires that you specify the size of the arc being generated in the motion.

Lesson 10

Programming The Three Most Basic Motion Types

There are only three motion types used in CNC turning center programs on a regular basis – rapid, straight-line, and circular motion. You must understand how they are commanded.

While it helps to understand how the turning center will interpolate motion, it is not as important as *knowing how to specify motion commands in a program.* This is the focus of Lesson Ten. Let's begin our discussion by showing those things that all motion types share in common.

Understanding the programmed point of each cutting tool

In order to generate appropriate motions for cutting tools, you must know the location on each cutting tool that is being programmed – which we're calling the *programmed point.* In some cases, you will have to modify program coordinates to compensate for certain attributes of cutting tools.

For example, hole-machining tools like drills, taps, and reamers will have a certain amount of *lead* that must be compensated when Z axis (depth) coordinates are calculated. You can calculate the lead for a twist drill (118 degree point angle) by multiplying 0.3 times the drill diameter. This lead must be added to the hole's depth when plunging through holes. See Figure 3.5.

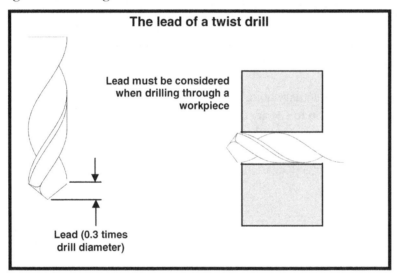

Figure 3.5 – The lead of a twist drill

Unfortunately, cutting tool manufacturers vary when it comes to *how much* lead there will be on certain hole-machining tools. A tap, for example, may have two, three, or four imperfect crests on the end of the tap. Reamer lead will vary with reamer size. You'll have to be quite familiar with the hole-machining tools your company uses in order to correctly compensate for lead when programming hole depths.

Single point tools, like turning tools, boring bars, and grooving tools will always have a small radius on the cutting edge. For the most part, you'll be programming the *extreme edges* of these tools in the X and Z axes. Figure 3.6 shows this.

Fundamentals of CNC

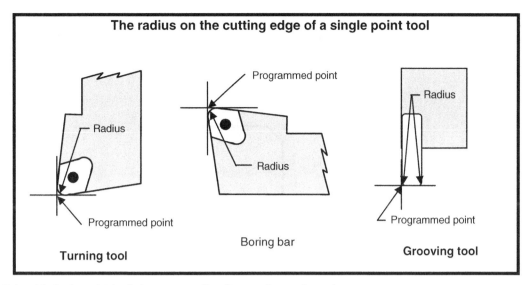

The radius on the cutting edge of a single point tool

Turning tool — Radius — Programmed point

Boring bar — Programmed point — Radius

Grooving tool — Radius — Programmed point

Figure 3.6 – All single point tools have a small radius on the cutting edge

There will be times when you'll have to compensate for a cutting tool's radius as you calculate coordinates for your program. One time is when facing a workpiece to center, as Figure 3.7 demonstrates.

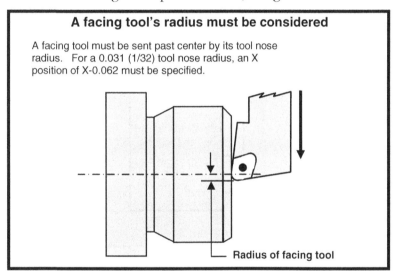

A facing tool's radius must be considered

A facing tool must be sent past center by its tool nose radius. For a 0.031 (1/32) tool nose radius, an X position of X-0.062 must be specified.

Radius of facing tool

Figure 3.7 – Compensating for a facing tool's radius

If you simply bring the programmed point of the facing tool to the workpiece center (**X0**), the tool will not completely machine the face. There will be a small nub left in the center of the workpiece. To eliminate the nub, you must program an X coordinate that is slightly past center. The facing tool must be sent to a negative X position that is twice the tool nose radius. If using a tool with a 0.031 (1/32) tool nose radius, the end point for the facing operation must be **X-0.062** (remember that X is specified in diameter).

Actually, there is a feature called *tool nose radius compensation* used to deal with many of the problems caused by the tool nose radius. Tool nose radius compensation will keep the cutting edge of the cutting tool flush-with (tangent-to) the work surface at all times. This dramatically minimizes the number of compensated coordinates you must calculate. Tool nose radius compensation is presented in Lesson Fourteen.

There is yet another time when you must calculate compensated coordinates based on tooling criteria. It has to do with grooving tools. Most programmers will program the *leading edge* of grooving tools (the side facing the chuck), as is shown in Figure 3.6. Again, you program the front (left) side of the groove and it machines the left side of the groove. If the groove is of the same width as the grooving tool, the *right* side of the grooving tool will machine the right side of the groove as the left side of the grooving tool machines the left side of the groove. Figure 3.8 shows the tool path needed for grooving tools.

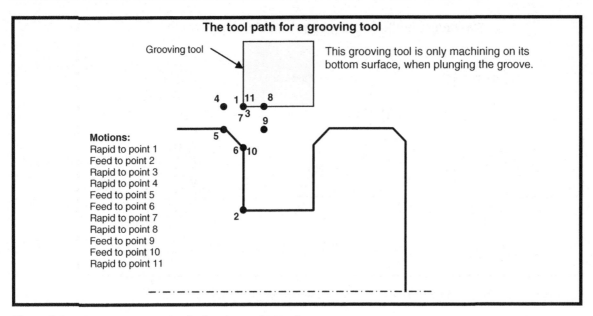

Figure 3.8 – The programmed point for a grooving tool

One last time we show that you must compensate for tooling during motion calculations has to do with threading. Though it is not a location on the cutting edge itself, most programmers program the *extreme leading edge* of the threading tool in the Z axis. Figure 3.9 shows this.

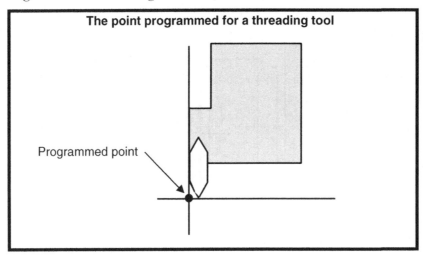

Figure 3.9 – The programmed point for a threading tool

Though some programmers do program the point of the threading tool insert, using the insert's leading edge as the programmed point will eliminate the possibility of interference when threading up to a shoulder. You can program a Z axis position quite close to the shoulder without fear of having the tool contact the shoulder.

On the other hand, if you program the point of the threading tool, calculating the Z end point will be more difficult and worrisome. You must know, of course, the distance between the tip of the threading tool and the leading edge of the threading tool. And if the setup person changes insert sizes (going to a bigger insert) a collision with the shoulder is likely to occur.

Figure 3.10 provides a summary, showing the programmed point for the most popular types of cutting tools used on turning centers.

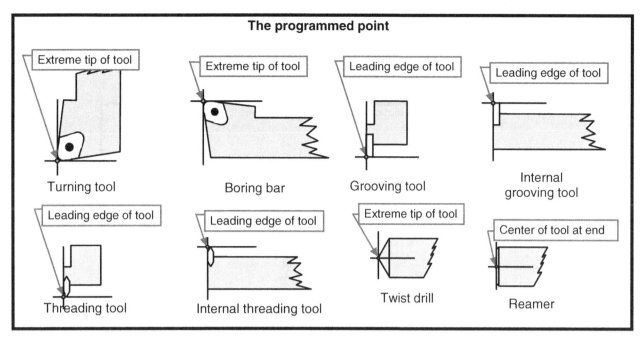

Figure 3.10 – The programmed point for the most common tools used on turning centers

G00 Rapid motion (also called positioning)

Rapid motion is commanded by G00 (or G0), and will cause the axes to move at the fastest possible rate. Rapid motion is used whenever the tool will not be machining during the motion. A good rule-of-thumb is: *if the tool won't be cutting, it should be moving at rapid.* Rapid motions are used for approaching the workpiece, making motions internal to the machining operation that are not cutting (like moving into position for another roughing pass), and when retracting the tool to the tool change position.

Current model turning centers have extremely fast rapid rates. Some can rapid at rates over 2,000 inches per minute. Even older turning centers have pretty fast rapid rates (at least 300 ipm). With these extremely fast rapid rates, you must be very careful with rapid movements. Turning center manufacturers offer two features that allow you to override the machine's rapid rate.

First, the *rapid override* function will allow you to slow the rapid rate at any time. However, machine tool builders vary with regard to how much control they provide with the rapid override switch. Some allow you to slow the machine in rather large increments. With one popular style, you'll have a four-position switch that allows 100%, 50%, 25%, and 10% of the rapid rate. While this is nice, 10% of rapid can still be quite fast. For example, if your turning center has a 1,200 ipm rapid rate, you'll only be able to slow the rapid rate to 120 ipm (10% of 1,200). This is still too fast to catch rapid motion mistakes.

Other turning center manufacturers provide more of a rheostat for rapid override control. With this kind of machine, your rapid override function will allow you to slow motion to a crawl.

Second, the *dry run* function of most machines allows you to use a special multi-position switch to override *all* motions the machine makes – including rapid motions. This function lets you slow motion to a crawl, but is not as handy as the second rapid override function mentioned above. We present more about how you take control of machine motions for the purpose of program verification in Lessons Twenty-Four and Twenty-Eight. For now, rest assured that you'll be able to keep the machine from moving at its very fast (and scary) rapid rate while you're getting familiar with the machine.

Again, the G code G00 (or G0) is used to specify rapid motion. Since all motion types are modal, you need only include a G00 in the first of consecutive rapid motion commands. You also include the motion's end point in X and/or Z in a G00 command.

If both axes will be moving in a rapid motion, it is important to know how the motion will be made. With most machines, each axis will move at its rapid rate until it reaches its destination. If the motions aren't equal distances, or if each axis has a different rapid rate (some turning centers have a slower rapid rate for the X axis), one axis will reach its destination before the other. This means that straight motion will not occur. The motion will appear as a *dog-leg* or *hockey stick*.

Figure 3.11 illustrates the points made about G00.

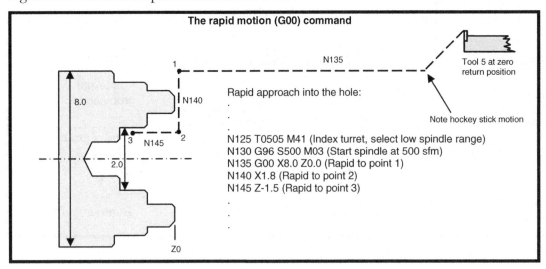

The rapid motion (G00) command

Tool 5 at zero return position

Note hockey stick motion

Rapid approach into the hole:

N125 T0505 M41 (Index turret, select low spindle range)
N130 G96 S500 M03 (Start spindle at 500 sfm)
N135 G00 X8.0 Z0.0 (Rapid to point 1)
N140 X1.8 (Rapid to point 2)
N145 Z-1.5 (Rapid to point 3)

Figure 3.11 - Rapid motion example

Notice that the first approach movement requires a two-axis motion (line N135). Since the X axis doesn't have to move as far as the Z axis, the X axis will reach its destination first. The Z axis will continue moving even after the X axis stops. Notice how this motion resembles a hockey stick (hence our odd name). From this original approach position, the internal grooving tool is brought to a diameter that is smaller than the current hole size. Then it's sent into the hole. These last two motions are single axis movements.

Since G00 is modal, there is no need to include a G00 in line N140 or N145. While it wouldn't hurt to do so, we recommend leaving out redundant words for two reasons. First, you can minimize program length, and given the need to conserve memory space even with current model controls, this is an important consideration. Second, and more importantly, including redundant words and commands will open the door to making mistakes as you write and/or type the program.

In line N140, since only X is moving, there is no need to include a Z value in this command. In line N145, only the Z axis is moving – so there is no need to include an X value in this command. Again, doing so wouldn't hurt anything, but we recommend leaving out redundant words.

G01 linear interpolation (straight line motion)

Linear interpolation, commanded by G01 (or G1), will cause the cutting tool to move along a straight path at a specified feedrate. Linear interpolation is needed for two purposes. First, it's necessary whenever a straight surface must be machined (turn, face, chamfer, taper, drilling a hole, etc.). Since most workpieces have many straight surfaces, there will be many G01 commands in your programs.

Second, there will be times when you'll want the machine to move along a straight path at a controlled feedrate even though you're not cutting anything. A bar feeding turning center will require this kind of motion whenever advancing bar stock.

All turning centers have a limitation when it comes to *maximum feedrate*. Though it's not heavily publicized, the maximum feedrate for most turning centers is *about half the rapid rate*. If the machine can rapid at 800 ipm, for example, its maximum feedrate will be about 400 ipm. For the most part, the turning center's maximum feedrate will be sufficient for machining with optimum cutting conditions. About the only exception to this statement might be when machining very coarse threads at high spindle speeds on older machines.

Feedrate (the F word) is also modal. So once a feedrate is selected, it will remain in effect until it is changed. Many operations can be completed with just one feedrate. For example, even though a rough turning operation may require several roughing passes, every pass can be made at the same feedrate as long as feedrate is specified in per revolution fashion (**G99**). For most tools, you'll need but one feedrate word, regardless of the motion types required for the tool (linear or circular) and how many cutting motions the tool must make.

Remember from Lesson Two that there are *two* feedrate modes, per revolution (**G99**) and per minute (**G98**). If in the inch mode (**G20**), this equates to inches per revolution or inches per minute. If in metric mode (**G21**), this equates to millimeters per revolution or millimeters per minute. Feed per revolution (**G99**) is the best feedrate mode for almost all turning center machining operations, especially if you're using constant surface speed spindle speed mode (**G96**). The bulk of your feedrate specifications will be in per revolution fashion.

As with the rapid motion, there is a way to override the feedrate for all cutting motions, including straight line cutting motions. On most machines, the *feedrate override switch* provides control of feedrates from 0% though about 200% in 10% increments, meaning you can halt the motion – and you can double the programmed feedrate – or you can select any feedrate in between. If programmed feedrates are correct, the feedrate override switch will be left in the 100% position for the entire program.

Again, the G code **G01** (or **G1**) is used to specify the linear interpolation mode. And since all motion types are modal, you need only include a **G01** in the first of consecutive linear motion commands. Also included in the **G01** command will be the X and/or Z end points for the motion. The first **G01** command for each tool must also include a feedrate word (F word).

As with **G00**, you need only include the moving axes in a **G01** command. If the machine will be moving in only the X axis, for example, (possibly facing a workpiece), you need not to include a Z word in the command.

Figure 3.12 shows an example of linear interpolation

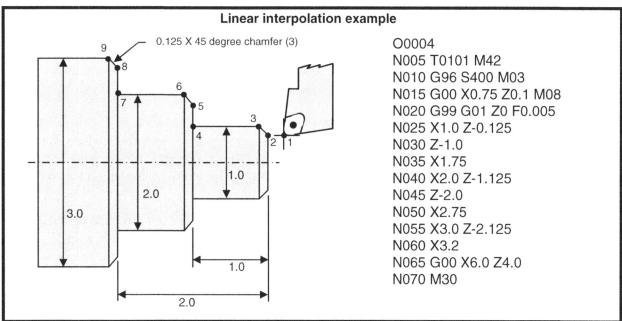

Linear interpolation example

```
O0004
N005 T0101 M42
N010 G96 S400 M03
N015 G00 X0.75 Z0.1 M08
N020 G99 G01 Z0 F0.005
N025 X1.0 Z-0.125
N030 Z-1.0
N035 X1.75
N040 X2.0 Z-1.125
N045 Z-2.0
N050 X2.75
N055 X3.0 Z-2.125
N060 X3.2
N065 G00 X6.0 Z4.0
N070 M30
```

Figure 3.12 – Linear interpolation example

The series of linear motions from point one to point nine finish turns the workpiece. The tool approaches to point one and retracts from point nine with rapid motion commands.

Program with comments:

O0004 (Program number)

N005 T0101 M42 (Index to station number one and invoke wear offset one, select high spindle range)

N010 G96 S400 M03 (Turn the spindle on cw at 400 SFM)

N015 G00 X0.75 Z0.1 M08 (Rapid to point 1, turn on the coolant)

N020 **G99 G01** Z0 **F0.005** (Feed to point 2)

N025 X1.0 Z-0.125 (Feed to point 3)

N030 Z-1.0 (Feed to point 4)

N035 X1.75 (Feed to point 5)

N040 X2.0 Z-1.125 (Feed to point 6)

N045 Z-2.0 (Feed to point 7)

N050 X2.75 (Feed to point 8)

N055 X3.0 Z-2.125 (Feed to point 9)

N060 X3.2 (Feed off diameter)

N065 G00 X6.0 Z4.0 (Rapid away from the workpiece)

N070 M30 (End of program)

After indexing the turret, starting the spindle, and approaching to point one in lines N005, N010, and N015, line N020 begins the series of linear motions. Again, G01 is modal, so it is only required in the first of this series of straight motions (line N020).

The feedrate is also modal, so it is only required in this first cutting command (again, line N020). Every cutting motion will be done at 0.005 ipr, and F0.005 in line N020 specifies this feedrate.

We've also included a G99 in line N020 even though feed per revolution mode is initialized (automatically instated at power-up). This ensures that the feedrate in line N020 will be taken as 0.005 inches per revolution, *not* 0.005 inches *per minute*.

In line N065, the rapid mode is re-selected for the retract motion to the turret index position. Line N070 ends the program. M30 will stop anything that's still running (spindle and coolant in our case) and rewind the program.

G02 and G03 Circular motion commands

Circular interpolation will cause the cutting tool to move along a circular path at a specified feedrate. Circular interpolation is used to machine circular surfaces on a workpiece. While not nearly as often required as linear interpolation, almost all turned workpieces have at least one radius that must be machined, meaning you'll use circular motion in almost every program.

Again, the feedrate (the F word) is modal. So once a feedrate is specified, it will remain in effect until it is changed. This is true even as you switch from linear to circular motions. As stated previously, many machining operations can be completed with but one feedrate word. And as stated during the presentation of G01, all points made during Lesson Two about feed per revolution (G99) and feed per minute (G98) still apply – as do all points made the feedrate override function.

There are two G codes used to command circular motion. G02 (or G2) is used to specify *clockwise* motion and G03 (or G3) is used to specify *counter clockwise* motion. You evaluate which kind of motion is needed by looking at the way the tool will be moving through the motion. In almost all cases, this simply means looking at the print from above. Figure 3.13 shows the difference between clockwise and counter clockwise motion.

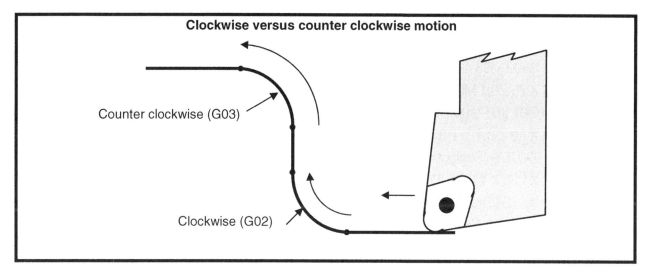

Figure 3.13 – Clockwise versus counter clockwise motion

By the nature of circular motion, both axes will be moving during a circular motion. You'll almost always specify an X and Z coordinate in circular motion commands. As with all other motion types, these coordinates specify the *end point* for the motion.

There is one more specification that must be made within every circular motion command, and it has to do with the *size* of the arc being machined. There are actually two ways to specify arc size. The easiest is to use an R word. We recommend that beginning programmers use this method. The older, more difficult way to specify arc size is to use *directional vectors*. Letter addresses I and K are used for this purpose.

Specifying a circular motion with the radius word

In Figure 3.14, the series of motions from point one to point eleven will finish turn the workpiece. We approach to point one and retract from point eleven with rapid motion commands. Figure 3.5 includes a complete program that will finish turn the workpiece. Notice that this program requires the use of both linear and circular motion. We're using the R word to specify the size of each circular arc.

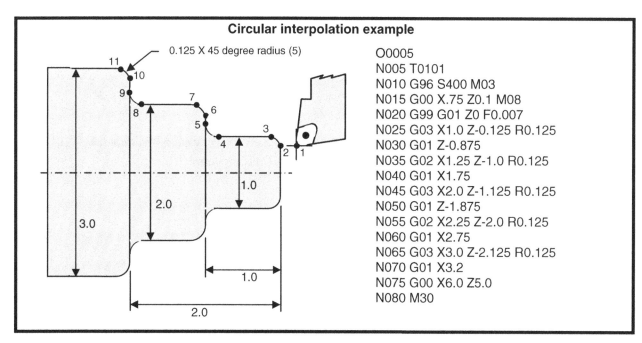

Figure 3.14 - Circular interpolation example

Program with comments:

O0005 (Program number)

N005 T0101 M42 (Index to station number one, select high spindle range)

N010 G96 S400 M03 (Start spindle fwd at 400 SFM)

N015 G00 X.75 Z0.1 M08 (Rapid to point 1, turn coolant on)

N020 G99 G01 Z0 F0.007 (Feed to point 2)

N025 **G03 X1.0 Z-0.125 R0.125** (CCW circular motion to point 3)

N030 **G01** Z-0.875 (Straight move to point 4)

N035 **G02 X1.25 Z-1.0 R0.125** (CW circular motion to point 5)

N040 **G01** X1.75 (Straight move to point 6)

N045 **G03 X2.0 Z-1.125 R0.125** (CCW circular motion to point 7)

N050 **G01** Z-1.875 (Straight move to point 8)

N055 **G02 X2.25 Z-2.0 R0.125** (CW circular motion to point 9)

N060 **G01** X2.75 (Straight move to point 10)

N065 **G03 X3.0 Z-2.125 R0.125** (CCW circular motion to point 11)

N070 **G01** X3.2 (Straight move off workpiece)

N075 G00 X6.0 Z5.0 (Rapid away from workpiece)

N080 M30 (End of program)

After indexing the turret, starting the spindle, and approaching the workpiece in lines **N005** through **N015**, line **N020** uses **G01** to contact the workpiece. The motion from point two to three is counter clockwise, so a **G03** is used (line **N025**). This command includes the end point in X and Z, as well as the arc size with the R word. The value of the R word is specified right on the print as 0.125 inch. (As you can see, it is relatively simple to specify circular motion using the R word.)

The motion from point three to four is linear, so we *must* to include a **G01** in line **N030** (again, all motion types are modal). The motion from point four to point five is clockwise, so a **G02** is used in line **N035**. Again the end point is specified with X and Z and the radius is specified with R. word. The balance of this series of motions alternates between linear and circular motion using the same techniques just shown.

Note the importance of including a **G01** or **G02/G03** as you switch motion types. Since these words are modal, the control will not interpret the program correctly if one is left out. For example, if in line **N040**, you leave out the **G01** during the motion from point five to point six, the control will retain the clockwise motion (**G02**) from line **N035**. In this case, it's likely that an alarm will be generated, since only one axis is commanded and no R word is included in line **N040**.

Important point: *The R word is not modal.* In our example, even though each of the radii has a 0.125 radius, the **R0.125** must be included in *every* circular command of this program.

Working on your first program

Instructions: Using the coordinates you calculated for this practice exercise in Lesson Nine, fill in the blanks for the program that follows. Note that point 1 is now up close to the 1.0 inch diameter (X of point 1 will be X1.2).

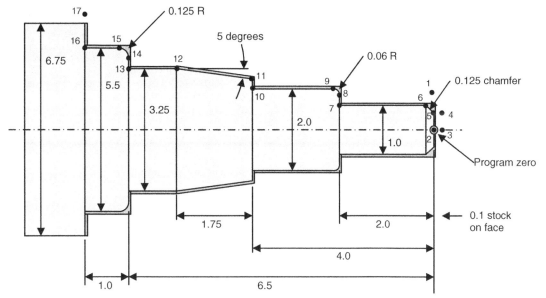

This program will finish face and finish turn the workpiece (it will not do any roughing). Use a feedrate of 0.007 ipr:

O0001

N005 T0202 M42 (Index turret, select high spindle range)

N010 G96 S500 M03 (Start spindle fwd at 500 sfm)

N015 ____ _____ _____ M08 (Rapid to point 1, turn on coolant)

N020 ____ _____ _____ (Face to point 2)

N025 ____ _____ (Rapid away to point 3)

N030 _____ (Rapid up to point 4)

N035 ____ _____ (Feed to point 5)

N040 _____ _____ (Chamfer to point 6)

N045 _____ (Turn 1.0 diameter to point 7)

N050 _____ (Feed up face to point 8)

N055 ____ _____ _____ _____ (Turn radius to point 9)

N060 ____ _____ (Turn 2.0 diameter to point 10)

N065 _____ (Feed up face to point 11)

N070 _____ _____ (Turn taper to point 12)

N075 _____ (Turn 3.25 diameter to point 13)

N080 _____ (Feed up face to point 14)

N085 ____ _____ _____ _____ (Turn radius to point 15)

N090 ____ _____ (Turn 5.5 diameter to point 16)

N095 _____ (Feed up face to point 17)

N100 G00 X8.0 Z6.0 (Rapid to tool change position)

N105 M30 (End of program)

Answer program for exercise on previous page

O0001
N005 T0202 M42 (Index turret, select high spindle range)
N010 G96 S500 M03 (Start spindle fwd at 500 sfm)
N015 **G00 X1.2 Z0** M08 (Rapid to point 1, turn on coolant)
N020 **G01 X-0.062 F0.007** (Face to point 2)
N025 **G00 Z0.1** (Rapid away to point 3)
N030 **X0.75** (Rapid up to point 4)
N035 **G01 Z0** (Feed to point 5)
N040 **X1.0 Z-0.125** (Chamfer to point 6)
N045 **Z-2.0** (Turn 1.0 diameter to point 7)
N050 **X1.875** (Feed up face to point 8)
N055 **G03 X2.0 Z-2.0625 R0.0625** (Turn radius to point 9)

N060 **G01 Z-4.0** (Turn 2.0 diameter to point 10)
N065 **X2.9438** (Feed up face to point 11)
N070 **X3.25 Z-5.75** (Turn taper to point 12)
N075 **Z-6.5** (Turn 3.25 diameter to point 13)
N080 **X5.25** (Feed up face to point 14)
N085 **G03 X5.5 Z-6.625 R0.125** (Turn radius to point 15)
N090 **G01 Z-7.5** (Turn 5.5 diameter to point 16)
N095 **X6.95**(Feed up face to point 17)
N100 G00 X8.0 Z6.0 (Rapid to tool change position)
N105 M30 (End of program)

Key Concept

4

Know The Compensation Types

CNC turning centers provide three kinds of compensation to help you deal with tooling related problems. In essence, each compensation type allows you to create your CNC program without having to know every detail about your tooling. The setup person will enter certain tooling information into the machine separately from the program.

Key Concept Four is made up of four lessons:

      11: Introduction to compensation
      12: Geometry offsets
      13: Wear offsets
      14: Tool nose radius compensation

The fourth Key Concept is that you must know the three kinds of compensation designed to let you ignore certain tooling problems as you develop CNC programs. In Lesson Eleven, we'll introduce you to compensation, showing the reasons why compensation is needed on CNC turning centers.

Two of the compensation types are related to cutting tools: wear offsets and tool nose radius compensation. We discuss them in Lessons Thirteen and Fourteen. The third compensation type, geometry offsets, is more related to work holding devices – and we'll discuss geometry offsets in Lesson Twelve. While we do introduce geometry offsets in Key Concept Number One, we present more about them in Lesson Twelve.

Frankly speaking, not all of the material presented in Key Concept Four may be of immediate importance to you. For example, you may not need to know any more about geometry offsets than what is shown in Lessons Six and Seven. Depending upon your needs, you may want to simply skim this material to gain an understanding of what's included – then move on. You can always come back and dig in should your needs change.

On the other hand, wear offsets and tool nose radius compensation are very important features - features used in *every tool of every program you write*. You'll need to completely master these features.

Lesson 11
Introduction To Compensation

An airplane pilot must compensate for wind direction and velocity when setting a heading. A race-car driver must compensate for track conditions as they negotiate a turn. A marksman must compensate for the distance to the target when firing a rifle. And a CNC programmer must compensate for certain tooling-related problems as programs are written.

Understanding offsets

All three compensation types use *offsets*. Offsets are storage registers for numerical values. They are very much like memories in an electronic calculator. With a calculator, if a value is needed several times during your calculations, you can store the value in one of the memories. When the value is needed, you simply type one or two keys and the value returns. In similar fashion, the setup person or operator can enter important tooling-related values into offsets. When they are needed by the program, a command within the program will *invoke* the value of the offset. And by the way, just as a calculator's memory value has no meaning to the calculator until it is invoked during a calculation, neither does a CNC offset have any meaning to the CNC machine until it is invoked by a CNC program.

Like the memories of most calculators, offsets are designated with *offset numbers*. Geometry offset number one's X register may have a value of -12.5439 (each geometry offset has four registers). Geometry offset number two's X register may have a value of -11.2957. Almost all offsets used on turning centers are related to cutting tools (the only exception being the work shift offset discussed in Lesson Seven), and offset numbers are made to correspond to *tool station numbers*. For example, program zero assignment values for tool number one are entered in geometry offset number one.

Unlike the memories of an electronic calculator that will be lost when the calculator's power is turned off, CNC offsets are more permanent. They will be retained even after the machine's power is turned off – and until an operator or setup person changes them.

Offsets are used with each compensation type to tell the control important information about tooling. From the marksman analogy, you can think of offset values as being like the amount of sight adjustment a marksman must make prior to firing a shot. Tooling related information entered into offsets includes each cutting tool's program zero assignment values (in geometry offsets), the tool nose radius for single point tools (also in geometry offsets), and the amount of wear a tool experiences during its life (in wear offsets).

Offset organization

First of all, rest assured that you will always have enough offsets to handle your applications. Most machine tool builders supply many more offsets than are required even in the most elaborate applications. Most turning centers have at least two distinct groups of offsets, one for cutting tools and another for program zero assignment.

Offset pages on the display screen

These display screen pages are introduced in Lessons Six and Seven. Figure 4.1 shows the geometry offset page, used for two purposes: to enter program zero assignment values and to enter tool nose radius compensation values.

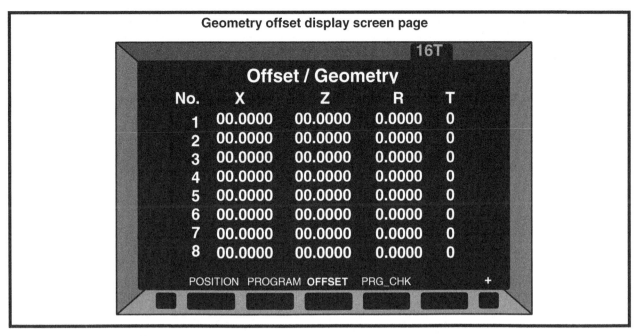

Figure 4.1 – Geometry offset display screen page

Actually, Figure 4.1 shows the *first page* of geometry offsets for a popular Fanuc control model. Only the first eight geometry offsets are shown. All Fanuc controls have at least sixteen sets of geometry offsets (most have thirty-two sets). Since most turning centers have turrets that can hold no more than twelve tools, you'll surely have more than enough geometry offsets.

The R and T registers are used to specify tool nose radius compensation values. These registers will be discussed in Lesson Fourteen.

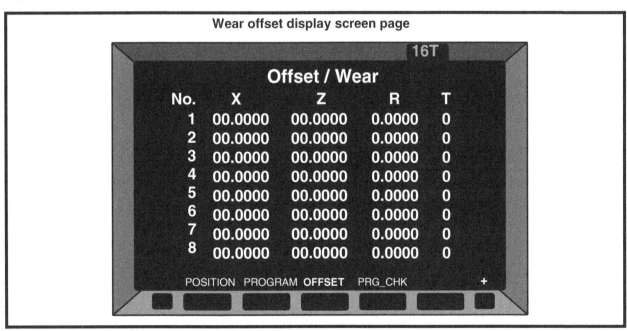

Figure 4.2 – Wear offset display screen page

First of all, notice how similar the wear offset display screen page is to the geometry offset display screen page. While some control models do make the difference a little more distinct, you'll always want to confirm that you're looking at the correct page before you make an entry.

The X and Z registers of the wear offset page are used to enter sizing adjustments that are needed as a cutting tool wears. We discuss wear offsets in Lesson Thirteen.

Do notice that the wear offset page also has a set of R and T registers. Though you can use either the geometry offset page or the wear offset page to enter tool nose radius compensation values, we recommend doing so on the geometry offset page. We explain why in Lesson Fourteen.

Figure 4.3 shows the work shift display screen page.

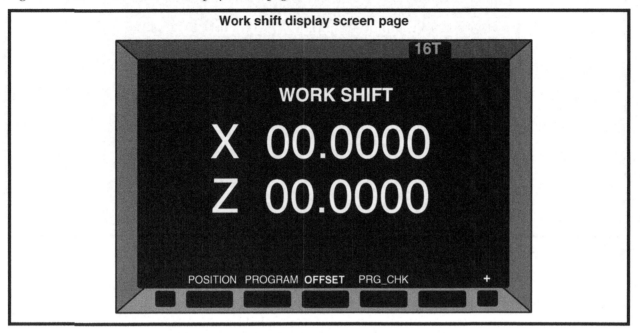

Work shift display screen page

16T

WORK SHIFT

X 00.0000

Z 00.0000

POSITION PROGRAM **OFFSET** PRG_CHK +

Figure 4.3 – Work shift display screen page of one popular control model

As is presented in Lesson Seven, the work shift page is used to help specify the program zero position in the Z axis. We discuss it in more detail during Lesson Twelve.

How offsets are instated

Again, offsets have no meaning to the control until they are *instated* by a program command. Geometry offsets and wear offsets are instated by the T word alone. Tool nose radius compensation additionally requires a special command in the program to completely instate them. A **G41** or **G42** is required to specify the relationship between the cutting tool and the surface/s being machined. These G codes are presented in Lesson Fourteen.

The T word for most turning centers has a four digit format. The first two digits cause the machine to index the turret to the specified station and instate the geometry offset. The second two digits of the T word instate the wear offset. Since the R and T registers are included in each offset table, The T word will cause the machine to know the tool nose radius compensation values for the time when a **G41** or **G42** is executed. For example, consider this command.

N160 T0505

The first two digits of the T word (05) will cause the machine to index the turret to station number five and instate geometry offset number five. This means the machine will know the program zero assignment values (previously placed in the X and Z registers of geometry offset number five). If the tool requires tool nose radius compensation, and if the setup person has entered the cutting tool's tool nose radius value and tool type in the R and T registers of the geometry offset, the machine will also know the tool nose radius and tool type.

The second two digits of the T word (also 05) will cause the machine to instate the wear offset.

Most programmers will always make the geometry offset number the same as the wear offset number (**T0505**, **T0202**, **T0404**, etc.).

Lesson 13

Wear offsets

This compensation type allows the setup person or operator to deal with minor size variations as workpieces are machined. While many CNC people (somewhat inappropriately) use wear offsets to compensate for minor setup imperfections and tool pressure, the major application for wear offsets is to compensate for tool wear during a tool's life.

You know that the tolerances commonly held on CNC turning centers are quite small. It is not unusual to see at least one overall tolerance of under 0.001 inch (0.254 mm) on turned workpieces.

You also know that each cutting tool has its own program zero assignment – and that there are several ways to assign program zero. And you know that unless you are using a properly calibrated tool touch off probe, mistakes with program zero assignment – even minor ones – as well as the effects of tool pressure, make it difficult to *perfectly* assign program zero. That is, even after program zero is assigned, there is no guarantee that every tool will machine the workpiece perfectly – or even within specified tolerances. The tighter the tolerances you must hold, the greater the potential there will be for sizing problems.

Wear offsets provide a way to make minor adjustments when machined surfaces are not within their tolerance bands – or when they're close to a tolerance limit.

There are at least four times when a typical CNC setup person or operator will use wear offsets:

- **During setup and after mounting a cutting tool in the turret** - After machining with the new tool, if the setup person discovers that the cutting tool has not machined a surface within the tolerance band, or if the surface is close to a tolerance limit, they can change a wear offset to make the needed adjustment.

- **When trial machining** – Trial machining is done when the setup person or operator is worried that a cutting tool (that has just been placed in the turret) will not machine within the tolerance band. Wear offsets are commonly used with which to make trial machining adjustments (Remember, if you are using a properly calibrated tool touch off probe, you shouldn't need to trial machine.)

- **When compensating for tool wear** – As a cutting tool wears, it will cause the surfaces it machines to grow or shrink in size. Wear offsets are used to keep cutting tools machining on-size for their entire lives.

- **After a dull tool is replaced** – Again, during a cutting tool's life, tool wear commonly causes the need for sizing adjustments in wear offsets. When a dull tool is replaced with a new one, the wear offset must be set back to its initial value – otherwise the new tool will machine too much material from the workpiece.

In the introduction to Key Concept Four, there is a lengthy presentation about *tolerance interpretation*. From this presentation, you know that *every* dimension has a tolerance. You know that each tolerance will have a high limit (largest acceptable dimension), a low limit (smallest acceptable dimension), and a mean value (the dimension right in the middle of the tolerance band.

You also know that each dimension to be machined on a workpiece will have a *target value* – this is the dimension you shoot for when an adjustment must be made. Many CNC people use the mean value of the tolerance band as the target value. That is, when an adjustment must be made, they target the mean value.

While there are times when this may not be appropriate (large lot sizes with small tolerances), we'll use this technique throughout Lesson Thirteen. That is, for each surface needing an adjustment, we'll be targeting the mean value.

You know that the *deviation* is the amount of needed adjustment. It is the difference between the measured value and the target value. In all cases, there will be a polarity to the deviation (plus or minus). The current wear offset value must be either increased or decreased by the amount of the deviation. The polarity is determined by judging which way the cutting tool must move (plus or minus) in order to bring the dimension back to the target value. We'll discuss more abut the deviation and its polarity later in this lesson.

Which dimension do you choose for sizing?

It is very common for a finishing tool to machine several surfaces. One finish turning tool may, for example, finish three external diameters and faces. Each surface may have its own tolerance specification. And of course, you program the mean value for each tolerance. In almost all cases, only one wear offset will be used to control *all* surfaces machined by each tool, meaning one adjustment will handle all surfaces machined by the tool.

For example, say a finish turning tool machines three external diameters: a 1.5 inch diameter, a 2.0 inch diameter, and a 3.0 inch diameter. Unless there is a substantial tool pressure problem, when the 1.5 inch diameter is coming out correctly (based on a wear offset adjustment in X), so will the 2.0 and 3.0 inch diameters.

When it comes to making sizing adjustments, you should use the two surfaces (one for diameters and one for lengths) that have the smallest tolerances. Use them to make sizing adjustment decisions.

How wear offsets are programmed

As you know, wear offsets are invoked by the second two digits of the T word. The command

N140 T0303

for example, indexes the turret to station number three, instates geometry offset number three (the first two digits) and instates wear offset number three (the second two digits). In almost all applications, you'll be making the wear offset number the same as the tool station number (tool one: wear offset one, tool two: wear offset two, and so on).

What actually happens when a wear offset is instated will vary based upon the CNC control (model and age). With newer controls (and especially if geometry offsets are used to assign program zero), the T word appears to simply index the turret. The geometry and wear offsets will not actually be instated (made active) *until the next motion command*. But with some older controls – and especially when program zero must be assigned in the program – the T word indexes the turret and the wear offset will be immediately instated. This will cause the turret to actually move by the amount of the wear offset.

What if my machine doesn't have geometry offsets?

It's much better if the turret remains in position during a turret index, and you can rest assured it will if you're assigning program zero with geometry offsets.

If you're assigning program zero in the program with G50 commands, you may notice that the turret *jumps* at each turret index. This motion can be troublesome for two reasons. First, the additional motions will add to cycle time. The turret may have to move in the plus direction when instating the wear offset. Then it will have to move in the negative direction when approaching the workpiece. Second, and more importantly, if the machine's starting position is close to a plus over travel limit in either axis (and the zero return position usually is), any large plus wear offset will cause an axis over-travel as soon as the turret index command is executed.

If you find that your machine's turret moves by the amount of the wear offset during turret indexes, here's a tip that will eliminate the associated motion problems. As you index the turret in your program, do not (yet) instate the wear offset. The command

T0300

for example, indexes the turret to station number three but will not instate a wear offset. In the first approach motion for the tool, instate the offset. Here is a program segment that eliminates unwanted motion for a machine that will move by the amount of the wear offset as soon as the offset is instated.

```
O0003 (Program number)
N005 G28 U0 W0
N010 G50 X8.3432 Z10.2383 (Assign program zero)
N015 T0100 M41 (Index to station one, no wear offset, and select the low range)
N020 G96 S600 M03 (Start the spindle CW at 600 sfm)
N025 G00 X2.2 Z0.005 T0101 M08 (Instate wear offset during the approach movement)
  .
  .
  .
```

In line N015, the turret is indexed to station one but no wear offset is being instated (yet). During the first approach movement in line N025, notice the T0101. Since the turret is already indexed to station one, it will not index – but the wear offset will be instated. The machine will actually consider the current values in wear offset number one as it makes its approach movement. In essence, the machine will modify the commanded end point in X and Z by the amounts stored in wear offset number one, and then rapid to the modified position. Note that this technique will only be required older machines when program zero assignment is done in the program – and only if the machine moves as soon as a wear offset is instated.

What about wear offset cancellation?

Canceling a wear offset has the inverse effect of instating it. Programmed positions will not be affected by the values stored in wear offsets once the wear offset is cancelled.

If you're using geometry offsets to assign program zero, *you don't need to cancel wear offsets* (but doing so will have no adverse affect). If you are assigning program zero in the program with G50 commands, you must cancel the active wear offset at the end of every tool. The wear offset is canceled during the tool's return to the tool change position.

One way to cancel a tool's offset is to repeat the tool station number in the T command but make the last two digits zero. The command

```
N240 T0500
```

for example, will cancel tool station number five's wear offset. However, there's an easier way that will cancel *any* tool's wear offset. The command

```
T0
```

will do so.

Here is a program segment that simply faces a workpiece. The technique shown in this program is not required if you are using geometry offsets to assign program zero. We're using G50 to assign program zero (it's an older machine), so the tool offset must be canceled at the end of the tool.

Sample program that uses G50 to assign program zero:

```
O0003 (Program number, example when G50 is used to assign program zero)
N003 G28 U0 W0
N005 G50 X8.3432 Z10.2383 (Assign program zero)
N010 T0100 M41 (Index to station one, no wear offset, and select the low range)
N015 G96 S600 M03 (Start the spindle fwd at 600 sfm)
N020 G00 X2.2 Z0 T0101 M08 (Invoke offset during the approach movement)
N025 G01 X-0.062 F0.010 (Face workpiece)
```

N030 G00 Z0.1 (Move away in Z)
N035 G00 X8.3432 Z10.2383 **T0** (Move back to tool change position, cancel offset)
N040 M01 (Optional stop)
N045

.
. (Program continues)
.

In line **N035**, we're sending the tool back to where it started and canceling the offset with **T0**.

Once again, remember that programming wear offsets is simpler when geometry offsets are used to assign program zero. Here is the same program segment modified to use *geometry offsets* to assign program zero.

O00003 (Program number, example when geometry offsets are used to assign program zero)
N005 T0101 M41 (Index to station one, select wear offset one, and select the low range)
N010 G96 S600 M03 (Start the spindle fwd at 600 sfm)
N015 G00 X2.2 Z0 M08 (Move into approach position, turn on the coolant)
N020 G01 X-0.062 F0.010 (Face workpiece)
N025 G00 Z0.1 (Move away in Z)
N030 G00 X7.0 Z5.0 (Move back to tool change position)
N035 M01 (Optional stop)
N040

.
. (Program continues)
.

Again, geometry offsets make things easier. Notice that program zero assignment values are not in the program (they're in the geometry offsets). In line **N005**, the wear offset can be easily instated right in the turret index command. And at the end of the tool in line **N030**, there's no need to cancel the wear offset (though doing so won't hurt anything).

How wear offsets are entered

Figure 4.6 shows the wear offset display screen page.

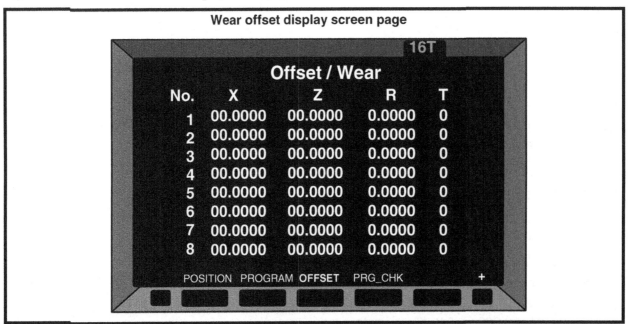

Figure 4.6 – The wear offset display screen page

Fundamentals of CNC   © CNC Concepts, Inc.

Wear offsets use only the X and Z registers on this page. The X registers allow adjustments to diameters (adjustments related to the X axis). The Z registers allow adjustments to lengths (adjustments related to the Z axis). The R and T registers are related to tool nose radius compensation, and are discussed in Lesson Fourteen. Notice once again that wear offsets are organized by number (wear offsets one through eight are shown in Figure 4.6). As you know, you'll be making the primary wear offset number for a cutting tool the same as the tool station number. That is, tool number five will use wear offset number five. Tool number ten will use wear offset number ten. And so on.

The wear offset page shown in Figure 4.6 is blank. That is, none of the wear offsets have been used – they're all set to zero. Let's look at how wear offsets can be entered.

There are two common ways to actually enter a wear offset, with INPUT and with +INPUT. As you begin typing the value of a wear offset, the soft keys at the bottom of the screen will change. Look at Figure 4.7.

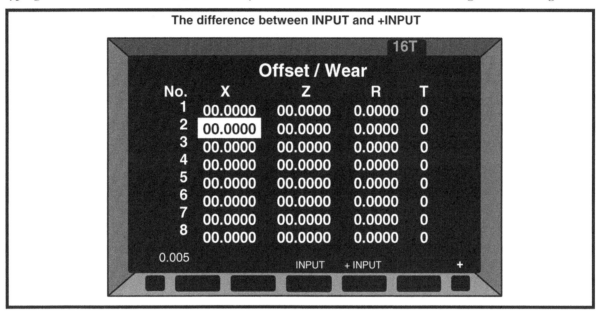

Figure 4.7 – INPUT versus +INPUT

Notice that the cursor is currently on the X register of wear offset number two. The operator has just typed 0.005 from the keyboard on the control panel (the entry shows up in the lower-left corner of the display screen. As the operator types the first digit (zero in our case), the soft keys change. While Fanuc control models vary with regard to what the soft keys show, two of the soft keys will be INPUT and +INPUT. These two keys are used to enter offset adjustments.

INPUT – This soft key (or the INPUT key on the control panel) will cause the machine to *replace* the value that is currently in the offset register with the entered value. For the example in Figure 4.7, if the operator presses the INPUT soft key, the X register of offset number two will be set to 0.005.

+INPUT – This soft key will cause the machine to *modify* the value that is currently in the offset register by the entered value. For the example in Figure 4.7, if the operator presses the +INPUT soft key, the X register of offset number two will be set to 0.005.

Since the initial value of the offset register is zero, INPUT and +INPUT happen to have the same result in this example (0.005 being placed in the register). But consider what happens after an initial entry. Figure 4.8 illustrates.

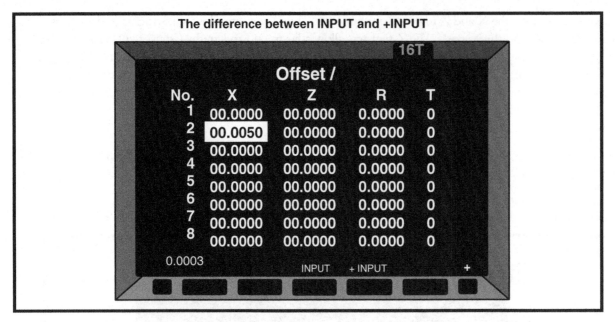

Figure 4.8 – Better example to stress the difference between INPUT and +INPUT

This time, the current value in the offset register is 0.005. The operator has typed the value 0.0003. If they press the **INPUT** soft key, the register's value will change to 0.0003. If they press the **+INPUT** soft key, the register's value will change to 0.0053.

Which is better, INPUT or +INPUT?

In almost all cases, a CNC setup person or operator will need to *modify* the current value in the offset register by the amount of the offset entry, meaning the **+INPUT** soft key will be the offset entry soft key. Think about it. Whenever a wear offset adjustment is required, you will have calculated the deviation – which is the difference between the target value and the dimension you have measured on the workpiece. You must modify the offset's current setting by the amount of this deviation. The **+INPUT** soft key will keep you from having to calculate the resulting offset value – the machine will make this calculation for you.

Remember that the deviation has a polarity. Sometimes it will be negative. The **+INPUT** soft key is still used. Say for example, the current value in the offset register is 0.005 (like it is in Figure 4.8). You've determined the deviation – and it is -0.0007. To make the adjustment, you will type -0.0007, and then press the **+INPUT** key. After you do, the register will show 0.0043 (when you add 0.005 to -0.0007, the result is 0.0043). Again, the machine will do the calculation for you when you use the **+INPUT** soft key.

What if my machine doesn't have a +INPUT soft key?

Older Fanuc controls don't have a soft key labeled **+INPUT**. Indeed, they don't have soft keys. If you have one of these controls, rest assured that there is still a way to enter the amount of deviation and let the control do the needed calculation. If your machine doesn't have a **+INPUT** soft key, you will use letter addresses X and Z to cause the control to overwrite the offset (like the **INPUT** soft key) and U and W to enter modify the offset (like the **+INPUT** soft key). In either case, the **INPUT** key will be used to enter the value.

For example, say the current value of the X offset register is 0.005. If you type X0.0007 and then press the **INPUT** key (this **INPUT** key is on the keyboard), its new value will be 0.0007. But if you type U-0.0007 and then press the **INPUT** key, its new value will be 0.0043.

While operation techniques are different, the result is the same. You can enter the amount of deviation and cause the machine to do the calculation if you use U (for X axis registers) and W (for Z axis registers) when making offset entries.

Sizing in a tool after it has just been placed in the turret

This example shows the first of four times when wear offsets can be used. While this example doesn't show the method we recommend, it should help you understand how wear offsets work. Look at Figure 4.8.

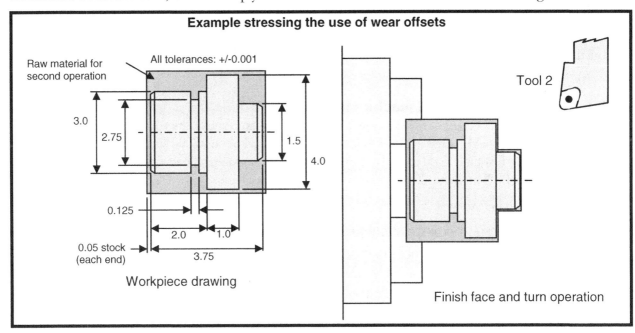

Figure 4.8 – Example to help stress the use of wear offsets

From the drawing on the left side of Figure 4.8, notice that all dimensions have a tolerance specified as +/-0.001 inch (0.002 overall). On the right side, the finish face and turning operation is about to take place and is done by tool number two. The roughing operation has already been done.

We'll say that this is the first workpiece being machined in the job, and during setup, tool number two (the finish face and turn tool) has just been placed in the turret. The setup person has done their best to measure program zero assignment values and enter them into geometry offsets. The X and Z wear offset registers for this tool are currently set to zero.

Notice that tool number two will be machining the 1.5 inch diameter – and it is this diameter we'll concentrate on for this example.

After tool number two runs – machining the 1.5 diameter – the setup person stops the cycle and measures the 1.5 inch diameter. Say they find that this diameter is currently 1.5024. It is, of course, oversize – larger than the high limit for the tolerance band (1.501). An adjustment must be made.

Since we're targeting the mean value, the adjustment must bring the diameter to 1.5 inches. The amount of the adjustment – the deviation – will be the difference between the target dimension and our measured dimension of 1.5024 (this comes out to 0.0024). Since the measured diameter is *larger* than the target diameter, the deviation will be *negative*. And since this is a diameter deviation, we must modify the X register of the offset (not the Z register).

So we must modify the current value of wear offset number two's X axis register by -0.0024. After bringing the cursor to the X register of offset number two, we type -0.0024 and press the soft key under **+INPUT**. After doing so, the value in this register will be -0.0024 (remember that it started at zero). The next time tool number two machines a workpiece, this diameter will come out 0.0024 smaller – and right at our target value of 1.5 inches.

In the scenario just given, we got lucky. Since the 1.5 inch external diameter is oversize, there is still a little stock to be removed. After making the offset adjustment, the setup person can re-run tool number two on the same workpiece. After tool number two runs the second time, the 1.5 diameter will be within its tolerance band (though it may not be exactly 1.5 since the amount of material being removed now is only 0.0024).

Consider the other possibility. After machining with tool number two for first time, maybe the setup person measures the 1.5 inch diameter and finds it to be 1.4975. Now the diameter is *undersize* – smaller than the low limit. Since there is no (feasible) way t put material back on this diameter, the workpiece is scrap. The deviation in this case is (positive) 0.0025, and the X register of wear offset number two must be modified by this amount before the *next* workpiece can be machined.

We call the technique just shown *sizing after the fact*. While it may be acceptable for some cutting tools (like roughing tools), it is not recommended when the cutting tool must machine surfaces having tight tolerances.

Sizing in a new tool with trial machining

This example stresses the second of four times when wear offsets can be held. As you have just seen, when tight tolerances must be held, there is a fifty-fifty chance that the workpiece will be scrapped when a cutting tool machines for the first time. You should use trial machining techniques whenever there is *any* doubt about whether a cutting tool will machine all surfaces within their tolerance bands.

Trial machining involves five steps:

- Recognizing a workpiece attribute with a tight tolerance
- Making an initial adjustment to machine the surface with some excess stock
- Letting the cutting tool machine under the influence of the trial machining offset
- Measuring the surface and adjusting accordingly
- Making the cutting tool run a second time (this time the surface will be within its tolerance band)

Consider these five steps to trial machining for the example just shown. In the first step, we know there could be a problem holding the +/-0.001 tolerance on the 1.5 inch diameter. We doubt whether the initial geometry offset setting for tool number two is accurate enough to make the tool machine this diameter within its tolerance band.

In the second step, the trial machining adjustment will be made in the X register of offset number two. To make this tool leave more material on the diameter, the adjustment must be positive. And usually an amount of about 0.010 inch (0.25 mm) is an appropriate amount of excess material to leave. So the X register of wear offset number two must be increased by 0.01 inch (the setup person types 0.01 and presses the soft key under **+INPUT**).

In the third step, the setup person lets tool number two run. The 1.5 inch diameter is machined with excess material.

In the fourth step, the setup person measures the 1.5 inch diameter. It will, of course, be oversize. Say they find it to be 1.5082. The needed adjustment is -0.0082 to bring the diameter back to the target value of 1.5 inch. So the X register of wear offset number two is modified by -0.0082 (setup person types -0.0082 and presses the soft key under **+INPUT**). While it may not be important at this time, the X value of wear offset number two will end up as (positive) 0.0018.

In the fifth step, the setup person reruns tool number two. This time it will machine the 1.5 inch diameter within its tolerance band – and for the next workpiece to be machined, the 1.5 inch diameter should come out to precisely to its target value, 1.5 inches.

By the way, in the trial machining example just given, trial machining saved us from machining a scrap workpiece. After trial machining, the diameter came out to only 0.0082"oversize. Yet we had initially increased the offset by 0.01". If we had not used trial machining techniques, the 1.5 inch diameter would have come out to 1.4982, which is smaller than the low limit – again, this diameter would have been undersize.

What causes the initial deviation?

Before we show an example of the third time when wear offsets can be used, we want to make an important point. This point will only apply if geometry offsets are being used to assign program zero.

In each example just given, when the (new) finish turning tool machines for the first time, there is a substantial deviation. That is, the difference between the target value and the measured value is quite large – large enough to cause the 1.5 inch diameter to be out of its tolerance band. And by the way, unless you have a properly calibrated tool touch off probe, these are realistic examples. It is not unusual for the initial deviation to be large enough to cause the surface being machined to be out of its tolerance band.

What is causing such a large deviation? As you know from Lesson Eleven, it could be one of two things, or a combination of both. First, it is possible that the setup person has made a mistake with the measurement and/or entry of the program zero assignment values for the finish turning tool. And second, even if these values are perfectly measured and entered, the difference in tool pressure from when the program zero assignment values are measured to the actual machining of the workpiece may be causing the initial deviation.

Either way, *this initial deviation is more related to incorrect program zero assignment than it is to tool wear.* Keep in mind that you can just as easily make the initial adjustment (including trial machining adjustments) in *geometry offsets* as you can in wear offsets. The machine will behave in *exactly* the same way. The **+INPUT** key can even be used to modify the current value of the (geometry) offset.

If you make the initial adjustment in the geometry offset, the wear offset values for the (new) cutting tool will start out at zero. The benefit of having the wear offset begin at zero with a new cutting edge will become clear as this presentation continues.

Dealing with deviations caused by tool wear

This is the third time when wear offsets can be used. As you know, when certain cutting tools wear, the surfaces they machine will change. Single point turning tools and boring bars are prime examples. When a finish turning tool wears, a small amount of material will be removed from its cutting edge. When this happens, any external diameter machined by the tool will *grow*. That is, external diameters will get larger as turning tools wear. At this point, the cutting tool is not worn out – it is just showing signs of wear.

It is not unusual for as much as 0.002 inch (0.050 mm) or more material to be removed from an insert before it must be replaced. This means a diameter machined by the finish turning tool will grow by 0.004 inch during the cutting tool's life.

The same occurs with boring bars. But an internal diameter will *shrink* as the boring bar insert wears.

Whether or not this will present a problem holding size depends upon two factors: lot size and the smallest tolerance machined by the tool.

With small lots, it is likely that the production run will be completed before the cutting tools wear enough to require an adjustment. And with wide open tolerances, it is likely that a new cutting tool can completely wear out without causing the surfaces being machined to approach tolerance limits.

But with large lots, cutting tools will eventually wear out and will need to be replaced. And if tolerances are small, surfaces will eventually grow (or shrink) out of their tolerance bands long before the tool is completely dull. This means adjustments must be made during the tool's life in order to keep the cutting tool machining all surfaces within tolerance bands.

In the drawing shown on the left side of Figure 4.8, notice once again that the 1.5 inch diameter has a tolerance of +/-0.001. The high limit is, of course, 1.501 and the low limit is 1.499. During setup, we'll say the setup person uses the trial machining techniques shown in the second example to size in the 1.5 inch diameter – they use wear offsets to do so. When the production run begins, a value of 0.0018 is in the X register of wear offset number two and the 1.5 inch diameter is coming out to *precisely* 1.5 inches.

For this discussion, we'll say there are thousands of workpieces to produce in this production run. So maybe the job is turned over to a CNC operator at this point to run out the job. As they begin running workpieces, the 1.5 inch diameter will continue coming out very close to 1.5 inches. But after fifty workpieces are machined, the operator notices that the 1.5 inch diameter is 1.5002. The finish turning tool has worn by 0.0001 inch. The diameter is still well within the tolerance band, so the operator simply continues running workpieces.

After fifty more workpieces, the 1.5 inch diameter comes out to 1.5005. The growth trend is continuing. But the diameter is still within the tolerance band – and nothing needs to be done (yet).

After fifty more workpieces (one-hundred-fifty total), the 1.5 inch diameter comes out to 1.5008. While this diameter is still within its tolerance band, it is getting dangerously close to the high limit (1.501). If this trend continues much longer, the diameter will be larger than the high limit (oversize). So at this point, the operator decides to make a sizing adjustment.

Since the target value is the mean value of the tolerance band, the operator will target 1.5. Currently the workpiece is 1.5008. So the operator must modify the current X value of wear offset number two by -0.0008.

To do so, they position the cursor to the X register of wear offset number two (which currently happens to have a value of 0.0018). Then they type -0.0008 and press the soft key under **+INPUT**. The machine will subtract 0.0008 from the current value of the offset.

With the next workpiece machined, the 1.5 inch diameter will come out to precisely 1.5 inches – which is back at the target value. The operator will be able to run another one-hundred-fifty workpieces or so before another offset adjustment will be required.

After running one-hundred-forty more workpieces, say they find the 1.5 inch diameter has grown to 1.5008 again. So they must make another adjustment. To do so, they position the cursor to the X register of wear offset number two (which currently happens to have a value of 0.0010). Then they type -0.0008 and press the soft key under **+INPUT**. The machine will subtract 0.0008 from the current value of the offset.

The operator may have to make several such adjustments during a given cutting tool's life – the smaller the tolerance, the more adjustments that will have to be made. Eventually, the cutting tool will be completely dull and in need of replacement.

After a dull tool is replaced

This is the fourth time when wear offsets can be used. While different types of cutting tools vary when it comes to what must be done to replace them, most single point turning tools used on turning centers (like turning tools and boring bars) incorporate *inserts*. With these cutting tools, the task of replacing a dull cutting tool simply means indexing or replacing the insert.

One advantage of most inserts is that they have more than one cutting edge (some have as many as eight cutting edges). When one edge of the insert is dull, the operator will remove the insert from the cutting tool holder, rotate it to an unused cutting edge, and replace it in the cutting tool holder. This task is known as *indexing* the insert. When all cutting edges on the insert have been used, of course, the insert must be replaced.

The key to insert indexing and replacement is *consistency*. If you can index or replace the insert in such a way that the new cutting edge is in exactly the same location as the last cutting edge, the new cutting edge will machine in exactly the same way as the last cutting edge *when it was new*. The only problem you'll have to deal with is related to the amount of offset change you made during the last cutting edge's life.

In the previous example, after sizing the first workpiece in during setup with a new cutting edge, the X value of wear offset number two started out at a value of 0.0018. During this tool's life, several adjustments are made. Eventually this cutting edge will be dull and the insert must be indexed or replaced. As long as it can be perfectly replaced (the new cutting edge is in the same location as the old cutting edge), the operator will set the X axis register for wear offset number two back to the initial value of 0.0018. This will cause the new cutting edge to machine the 1.5 diameter precisely to 1.5, just as the last cutting edge did when it was new.

This is one time when the **INPUT** soft key will be used. Since you know the value you want in the offset (0.0018 in our case), it will be easier to use than the **+INPUT** soft key – which would first require a calculation to be made.

By the way, this is the reason why we recommend making the initial adjustment (including trial machining adjustments) in geometry offsets. If you do, the wear offset will *begin at zero*. When cutting tools are replaced, you won't have to remember the initial wear offset value (0.0018 in our example). Instead, you can simply set the wear offset value back to zero.

Consistently replacing inserts

Unfortunately, inserts used for turning center applications may not be very consistent. That is, the insert itself may have a rather large tolerance. If inserts vary in size – even by a little bit – it will be impossible to *perfectly* replace them.

Consider this specification for a very common 80 degree diamond-shaped insert:

CNMG-432

Each letter and number of an insert's specification has a special (and universal) meaning. It just so happens that the third letter in the specification (M in our case), specifies the tolerance for the insert.

- A- Included circle = plus or minus .0002, thickness = plus or minus .001
- B- Included circle = plus or minus .0002, thickness = plus or minus .005
- C- Included circle = plus or minus .0005, thickness = plus or minus .001
- D- Included circle = plus or minus .0005, thickness = plus or minus .005
- E- Included circle = plus or minus .001, thickness = plus or minus .001
- G- Included circle = plus or minus .001, thickness = plus or minus .005
- **M Included circle = plus or minus .002, thickness = plus or minus .005**
- U- Included circle = plus or minus .005, thickness = plus or minus .005

The M specification in **CNMG-432** specifies a tolerance for the included circle of 0.004 (+/-0.002). The tolerance for the insert's thickness is +/-0.005. While the thickness of an insert isn't very critical, the size of its included circle controls the position of the cutting edge, as Figure 4.9 shows.

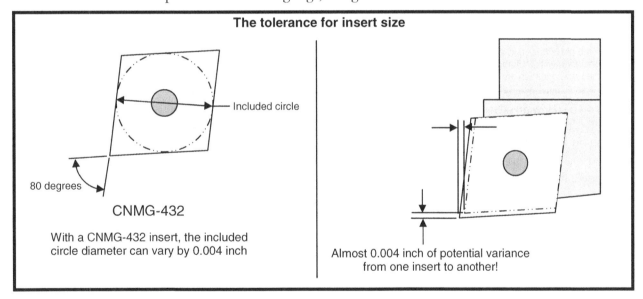

The tolerance for insert size

Included circle

80 degrees

CNMG-432

With a CNMG-432 insert, the included circle diameter can vary by 0.004 inch

Almost 0.004 inch of potential variance from one insert to another!

Figure 4.9 – Included circle of an 80 degree insert

A position variance of almost 0.004 inch is possible with a **CNMG-432** insert. This can result in as much as 0.008 inch variance when machining a diameter. With this much variation, of course, you will not be able to change inserts precisely enough to eliminate trial machining or using a tool touch off probe when inserts are replaced.

Note that the tolerances specified above are *maximum* variations. In practice, you'll probably find them to be much less from insert to insert – especially with inserts coming out of the same package. You can easily measure the amount of variation from insert to insert with a micrometer. If you find the variance to be under about 0.0005 inch from insert to insert, you may be able to replace inserts precisely enough to eliminate the need for trial machining or the need to use the tool touch off probe after insert replacement.

Consistently indexing inserts

As stated, almost all inserts have more than one cutting edge. The CNMG-432, for example, has *four* cutting edges. This means after a new insert is placed in the tool holder, it can be *indexed* three times.

The variation among new inserts will not apply to an insert when it is indexed. If an insert is consistently placed in a tool holder, there will be no variation from one cutting edge of an insert to another. But the key word is *consistently*. In order to eliminate the need for trial machining or the tool touch off probe, you must be able to index inserts perfectly – the next cutting edge must be in precisely the same location as the last cutting edge. While it may take a little practice, you can save much time and effort if you can master the ability to consistently index inserts.

Many inserts used on turning centers incorporate an eccentric pin for clamping and location. This pin can be turned in either direction to clamp the insert. When it is turned in one direction, the pin will press the insert against one of the location surfaces of the tool holder. When it is turned in the other direction, the eccentric pin will press the insert against the other location surface of the tool holder. Figure 4.10 shows this.

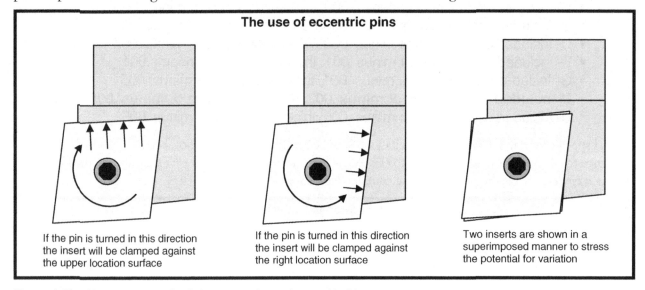

The use of eccentric pins

If the pin is turned in this direction the insert will be clamped against the upper location surface

If the pin is turned in this direction the insert will be clamped against the right location surface

Two inserts are shown in a superimposed manner to stress the potential for variation

Figure 4.10 – How an eccentric pin locates an insert in a tool holder

When you turn the pin in the direction shown in the left-most illustration of Figure 4.10, the insert will be pressed against the upper location surface of the tool holder. When you turn the pin in the direction shown in the middle illustration, the insert will be pressed against the right location surface of the tool holder. The right-most illustration super-imposes two inserts that are clamped in opposite directions. If the seat in the tool holder is not perfect, you can see how inconsistently clamped inserts will vary in position.

To determine which way the pin should be turned, consider what the cutting tool is doing. For the tool shown in Figure 4.10, if the tool will be predominantly facing, turn the eccentric pin in the direction shown in the left-most illustration. If it will be predominantly turning, turn the eccentric pin in the direction shown in the middle drawing. Once you determine which way the eccentric pin should be turned, *use the same method every time an insert must be indexed or changed.*

The method by which inserts are located in the tool holder affects more than just the position of the cutting edge. It also affects basic machining practice – especially for tools that perform powerful machining operations. If an insert is not properly located in the tool holder, it will prematurely fail.

Lesson 14
Tool Nose Radius Compensation

This compensation type allows you to deal with problems caused by the small radius that is on the cutting edge of all single point turning tools and boring bars. With the addition of two G codes per tool in your program – and with the entry of two simple values in the tool's offset – the machine will automatically keep the cutting edge of cutting tools flush with the surfaces they machine.

You know from Lesson Ten that you must sometimes compensate for tooling attributes when you calculate the coordinates for your programs. With a twist drill, for example, you must compensate for the lead of the drill when calculating hole depth. In similar fashion, you must compensate for the lead of a reamer – or the number of imperfect threads on a tap.

In Lesson Ten, we mention one other important time when you must compensate for attributes of cutting tools. It has to do with the small radius that is on the cutting edge of any single point cutting tool – like a turning tool or a boring bar. Figure 4.12 shows this radius.

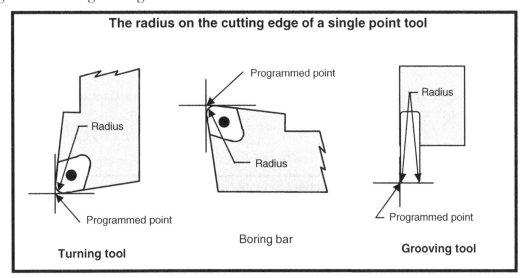

Figure 4.12 – All single point tools have a small radius on the cutting edge

For cutting tools used in the United States, the actual size of the radius will be specified in inches – and there are four standard tool nose radius sizes for turning and boring inserts:

- 1/64 inch (0.0156)
- 1/32 inch (0.0316)
- 3/64 inch (0.0468)
- 1/16 inch (0.0625)

Though you may consider these radii to be quite small, this small nose radius on the edge of the cutting tool will be sufficient to cause a small deviation from the programmed shape of your workpiece – at least when angular and circular surfaces must be machined.

Remember that you are programming the extreme edges of the cutting tool in each axis. This is illustrated in Figure 4.12 (specified as *programmed point*). Notice the small gap between the programmed point and the actual cutting edge.

This small gap will not affect the turning of diameters (parallel to the Z axis) and the machining of faces (parallel to the X axis). Figure 4.13 shows this.

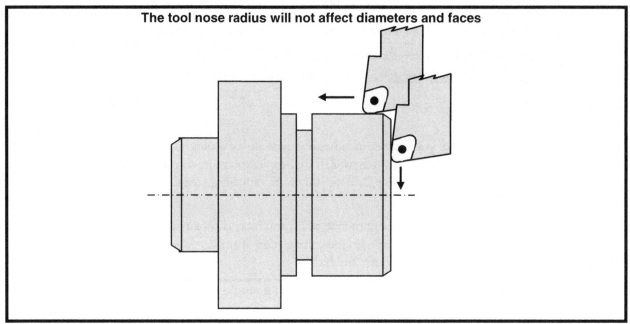

Figure 4.13 – The tool nose radius will not affect the machining of straight turns and faces

When a cutting tool is turning diameters and machining faces, the cutting edge is in direct contact with the workpiece surface being machined.

But when angular (tapered) surfaces and circular surfaces must be machined, the gap between the programmed point and the cutting edge will affect machining – as Figure 4.14 shows.

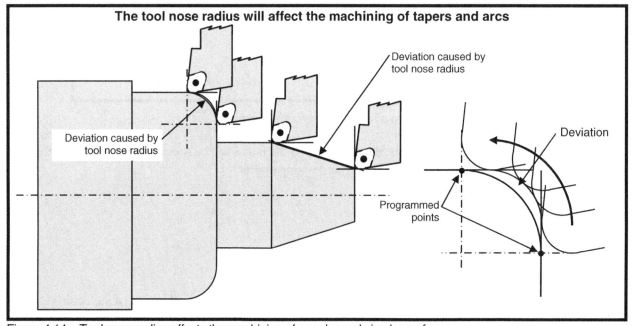

Figure 4.14 – Tool nose radius affects the machining of angular and circular surfaces

How much deviation are we talking about?
The deviation is at its worst at a forty-five degree angle – when the cutting tool is half-way through a ninety-degree arc – or when it is machining a forty-five degree chamfer. Figure 4.15 shows this.

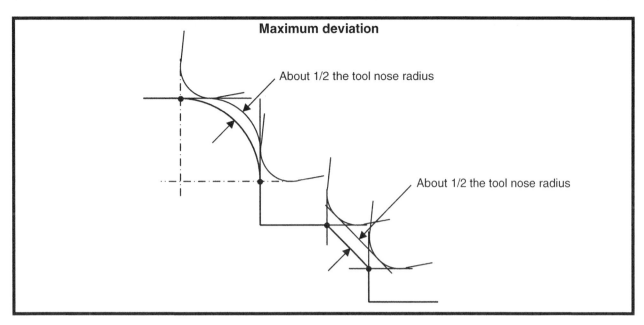

Figure 4.15 – Maximum deviation caused by tool nose radius

If you are using a cutting tool with a 1/32 inch (0.0312) nose radius, the deviation at forty-five degrees will be about 0.0155 inch.

In some cases, this may not be enough to cause problems. Maybe you are machining a chamfer or radius for the purpose of breaking sharp corners. On the other hand, there are many applications when even a tiny deviation will cause a scrap workpiece. Consider, for example, machining a Morse taper. The taper angle and position will be critical to the function of the workpiece.

Keeping the cutting edge flush with the work surface at all times

One way to handle the deviations is to manually compensate for the tool nose radius with your programmed coordinates. Note that this is *not* the method we recommend. We're only showing it to help you gain an understanding of how tool nose radius compensation works – and what it's doing for you. Look at figure 4.16.

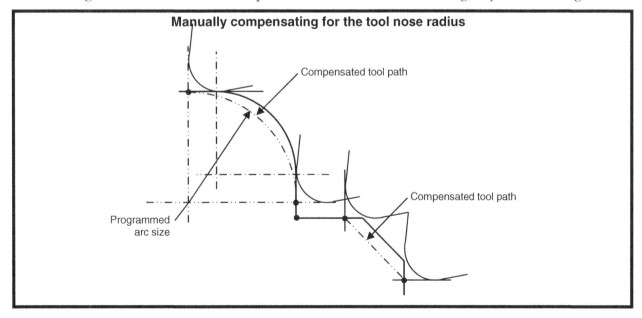

Figure 4.16 – Coordinates that compensate for the tool nose radius

Notice that the programmed points are no longer related to workpiece coordinates. For the chamfer, the start point is at a smaller diameter than the chamfer's beginning position. The end point is more negative in Z.

With these programmed points (which we're calling the compensated tool path), the cutting edge of the tool will remain in contact with the workpiece chamfer during the motion.

We've done the same for the radius motion. The start point is smaller in X and the end point is more negative in Z. And the size of the radius must be increased by the tool nose radius. With this compensated tool path, the cutting edge will remain flush with the workpiece radius during the motion – and the radius on the workpiece will be machined properly.

Again, *this is not the method we recommend* for dealing with the deviations caused by the tool nose radius. We're showing it for two reasons. First, this is exactly what the feature *tool nose radius compensation* will (automatically) do for you. Tool nose radius compensation allows you to program the work surface path (all coordinates right on the workpiece). It will automatically create the compensated path – and keep the cutting edge flush with all workpiece surfaces.

Second, some computer aided manufacturing (CAM) systems will create a CNC program with compensated tool paths. During programming, the programmer tells the CAM system the size of the tool nose radius and the CAM system outputs a CNC program with compensated tool paths. This eliminates the need for CNC-based tool nose radius compensation, which in turn eliminates some of the work a setup person must do during setup. If your company uses a CAM system, you'll want to determine whether it is generating compensated tool paths – or whether it is generating programs with CNC-based tool nose radius compensation commands (G41 and G42).

When to use tool nose radius compensation

As stated, tool nose radius compensation is used with single point cutting tools, like turning tools and boring bars. And it should only be used for *finishing* operations. We don't recommend using tool nose radius compensation for roughing operations. And it is never required for center-cutting operations like drilling, reaming, and tapping.

Steps to programming tool nose radius compensation

Programming tool nose radius compensation is relatively easy. Here are the three programming steps:

- Instate tool nose radius compensation

- Program the motions to machine the workpiece

- Cancel tool nose radius compensation

Instating tool nose radius compensation

To instate tool nose radius compensation, you must first be able to determine how the tool will be related to the work surface during the machining operation. Look in the direction the tool will be moving during the operation (rotate the print if it's necessary). Looking in this direction, ask the question, *"What side of the work surface is the cutting tool on, the left side or the right side?"* If the cutting tool is on the left side of the surface to be machined, you will use a G41 to instate tool nose radius compensation. If the cutting tool is on the right side of the surface to be machined, you will use a G42 to instate tool nose radius compensation. Look at Figure 4.17 to see some examples of how this is done.

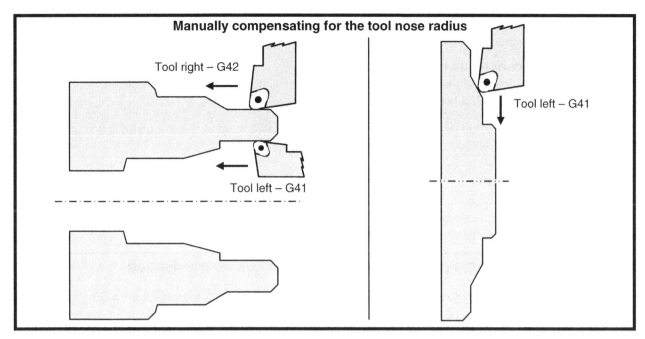

Figure 4.17 – Drawing illustrates how to decide between G41 and G42

Notice that **G42** is always used for turning toward the chuck and **G41** is always used for facing (toward the spindle center) and boring operations.

Once you know which side of the surface to be machined the tool is on (left or right), you simply include the appropriate G code (**G41** or **G42**) in the cutting tool's approach movement.

Programming motion commands to machine the workpiece

Once you've instated tool nose radius compensation, you simply program the movements to finish turn or bore the workpiece using work surface coordinates. Again, these programmed coordinates are right on the work surface. The control will automatically keep the cutting edge radius on the specified side of the workpiece (left or right), and tangent to surfaces being machined.

You must be sure that the machine has the ability to keep the tool radius tangent to one surface without violating an upcoming surface. The most common problem in this regard has to do with machining narrow recesses. The recess must be wide enough to allow the tool nose radius into the recess. If it's not, the tool will actually violate one side of the recess before the depth of the recess is reached. Before this will be allowed to happen, an alarm will be sounded. Figure 4.18 demonstrates this possible problem.

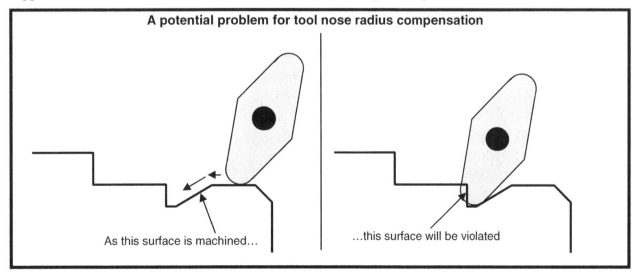

Figure 4.18 – Narrow recesses can present problems for tool nose radius compensation

Once tool nose radius compensation is instated, the machine will simply keep the tool on the left or right side of all programmed surfaces until tool nose radius compensation is canceled.

Canceling tool nose radius compensation

You must remember to cancel tool nose radius compensation. If it is not canceled, the machine will remain under its influence even with subsequent tools.

Canceling tool nose radius compensation is easy. Just include a **G40** word in the command that sends the tool back to the tool change position.

An example program

Figure 4.19 shows the workpiece to be used for our first example program. It is the drawing used for the exercise you worked on in Lesson Ten. Again, we're only finishing this workpiece – the roughing has already been done by another tool.

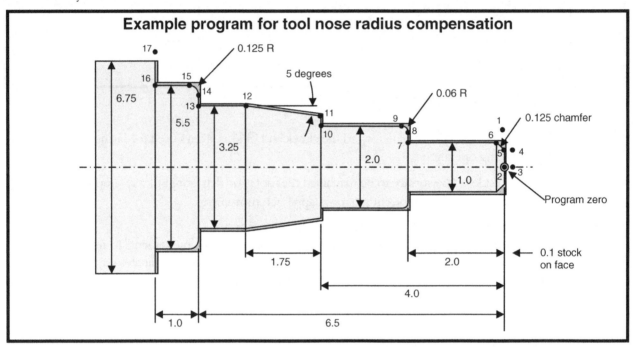

Figure 4.19 – Drawing for tool nose radius compensation example program

Here is the program. Notice that it is identical to the one shown in Lesson Ten except for the addition of two G codes.

```
O0001
N005 T0202 M42 (Index turret, select high spindle range)
N010 G96 S500 M03 (Start spindle fwd at 500 sfm)
N015 G00 X1.2 Z0 M08 (Rapid to point 1, turn on coolant)
N020 G01 X-0.062 F0.007 (Face to point 2)
N025 G00 Z0.1 (Rapid away to point 3)
N030 G42 X0.75 (Instate tool nose radius compensation, rapid up to point 4)
N035 G01 Z0 (Feed to point 5)
N040 X1.0 Z-0.125 (Chamfer to point 6)
N045 Z-2.0 (Turn 1.0 diameter to point 7)
N050 X1.875 (Feed up face to point 8)
N055 G03 X2.0 Z-2.0625 R0.0625 (Turn radius to point 9)
N060 G01 Z-4.0 (Turn 2.0 diameter to point 10)
```

N065 X2.9438 (Feed up face to point 11)

N070 X3.25 Z-5.75 (Turn taper to point 12)

N075 Z-6.5 (Turn 3.25 diameter to point 13)

N080 X5.25 (Feed up face to point 14)

N085 G03 X5.5 Z-6.625 R0.125 (Turn radius to point 15)

N090 G01 Z-7.5 (Turn 5.5 diameter to point 16)

N095 X6.95 (Feed up face to point 17)

N100 G00 **G40** X8.0 Z6.0 (Rapid to tool change position, cancel tool nose radius compensation)

N105 M30 (End of program)

In line **N030**, notice the addition of the **G42** word. Since this is a finish turning tool – and it will be on the right side of the work surfaces being machined, a **G42** is being used to instate tool nose radius compensation.

Do notice that we waited until after the finish facing has been done before instating tool nose radius compensation. Since the face being machined is parallel to the X axis, there is no need for tool nose radius compensation. (If you must face surfaces that include angular and circular movements, tool nose radius compensation must be used – and a **G41** will be used to instate it.)

In line **N020**, we do manually compensate for the tool nose radius. We send the facing tool past the workpiece centerline to a diameter that is twice the tool nose radius. This ensures that the face will be completely machined.

From lines **N035** through **N095**, the programmed coordinates are directly on the work surface. The **G42** in line **N030** will ensure that the cutting tool will remain on the right side of all programmed surfaces. That is, the cutting edge radius will remain tangent to (flush with) all surfaces being machined.

In line **N100**, the **G40** cancels tool nose radius compensation on the tool's motion to the tool change position.

Tool nose radius compensation from a setup person's point of view

As you've seen, programming tool nose radius compensation is relatively easy. You simply instate tool nose radius compensation during the tool's approach movement and cancel it during its return to the tool change position. But there is one more thing you must know – and it is related to the cutting tool's offset.

By one means or another, two offset registers must be entered for each tool that is using tool nose radius compensation. In many companies, the setup person manually enters this information during setup. Figure 4.20 shows the geometry offset display screen page.

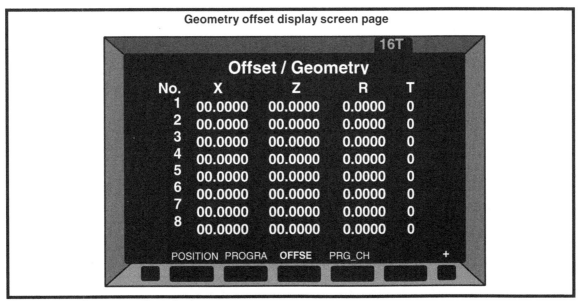

Figure 4.20 – Geometry offset display screen page

Notice the R and T registers. These registers are used with tool nose radius compensation. The R register is used to specify the tool nose radius. As you can see, this register requires a decimal value. For a 1/32 inch tool nose radius, a value of 0.0312 must be entered.

The T register is used to specify the cutting tool *type*. It is a code number. Figure 4.21 provides a chart, showing the various T register values.

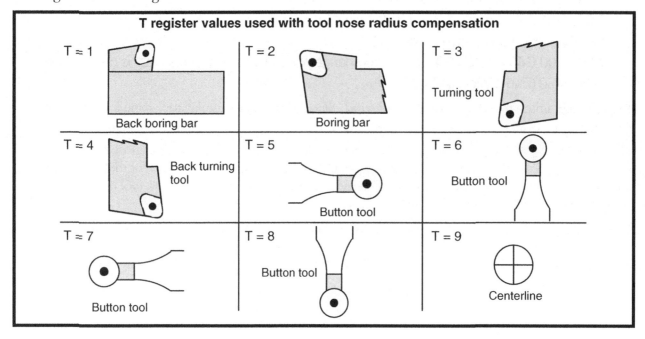

Figure 4.21 ~ Types of cutting tools used with tool nose radius compensation and the related T values

Of the tools shown on the chart in Figure 4.21, by far the two most popular types are the boring bar (type number two) and the turning tool (type number three). You should try to remember these two types.

In order for the example program for the workpiece shown in Figure 4.19 to work, geometry offset number two's R and T registers must be entered. Figure 4.22 shows this. Note that tool number two is a finish turning tool and we'll say it has a 0.0312 nose radius.

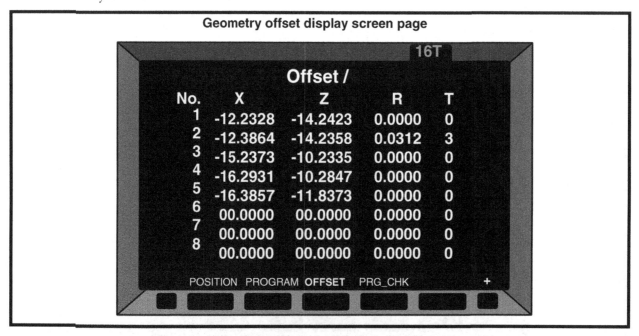

Figure 4.22 ~ How the R and T registers must be set for the example program

While the X and Z program-zero-assignment registers are set for several tools (there must be other tools currently in the turret), we're only concerned with offset number two. Notice that the R register is set to the cutting tool's nose radius (0.0312) and the T register is set to the tool type (3, which specifies a turning tool).

Add tool nose radius compensation to this program

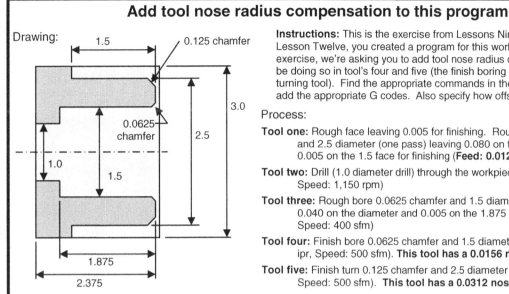

Drawing:

Instructions: This is the exercise from Lessons Nine and Twelve. After Lesson Twelve, you created a program for this workpiece. In this new exercise, we're asking you to add tool nose radius compensation. You will be doing so in tool's four and five (the finish boring bar and the finish turning tool). Find the appropriate commands in the program below and add the appropriate G codes. Also specify how offsets must be set.

Process:

Tool one: Rough face leaving 0.005 for finishing. Rough turn 0.125 chamfer and 2.5 diameter (one pass) leaving 0.080 on the 2.5 diameter and 0.005 on the 1.5 face for finishing (**Feed: 0.012 ipr, Speed: 500 sfm**)

Tool two: Drill (1.0 diameter drill) through the workpiece (Feed: 0.008 ipr, Speed: 1,150 rpm)

Tool three: Rough bore 0.0625 chamfer and 1.5 diameter (one pass), leaving 0.040 on the diameter and 0.005 on the 1.875 face (Feed: 0.007 ipr, Speed: 400 sfm)

Tool four: Finish bore 0.0625 chamfer and 1.5 diameter to size (Feed: 0.005 ipr, Speed: 500 sfm). **This tool has a 0.0156 nose radius.**

Tool five: Finish turn 0.125 chamfer and 2.5 diameter to size (Feed: 0.005 ipr, Speed: 500 sfm). **This tool has a 0.0312 nose radius.**

O0002 (Program number)
(ROUGH FACE AND TURN)
N005 T0101 M41 (Index turret, select low spindle range)
N010 G96 S500 M03 (Start spindle fwd at 500 sfm)
N015 G00 X3.2 Z0.005 M08 (Rapid to point 1, turn on coolant)
N020 G99 G01 X-0.062 F0.012 (Face to point 2 at 0.012 ipr)
N025 G00 Z0.1 (Rapid away to point 3)
N030 X2.33 (Rapid up to point 4)
N035 G01 Z0 (Feed up to point 5)
N040 X2.58 Z-0.12 (Chamfer to point 6)
N045 Z-1.495 (Turn 2.5 diameter to point 7)
N050 X3.2 (Feed up face and off part to point 8)
N055 G00 X8.0 Z7.0 (Rapid to tool change position)
N060 M01 (Optional stop)

(DRILL 1.0" HOLE)
N065 T0202 M41 (Index turret, select low spindle range)
N070 G97 S1150 M03 (Start spindle at 1,150 rpm)
N075 G00 X0 Z0.1 M08 (Rapid to point 9, turn on coolant)
N080 G01 Z-2.705 F0.008 (Feed to point 10 at 0.008 ipr)
N085 G00 Z0.1 (Rapid back to point 9)
N090 X8.0 Z7.0 (Rapid to tool change position)
N095 M01 (Optional stop)

(ROUGH BORE)
N100 T0303 M41 (Index turret, select low spindle range)
N103 G96 S400 M03 (Start spindle at 400 sfm)
N105 G00 X1.585 Z0.1 M08 (Rapid to point 11, start coolant)
N110 G01 Z0 F0.007 (Feed to point 12 at 0.007 ipr)
N115 X1.46 Z-0.0575 (Chamfer to point 13)
N120 Z-1.87 (Rough bore 1.5 to point 14)
N125 X0.95 (Feed down face to point 15)
N130 G00 Z0.1 (Rapid to point 16)
N135 X8.0 Z7.0 (Rapid to tool change position)
N140 M01 (Optional stop)

Specify the tool nose radius compensation offset entries for this job:

#	R	T
4	_____	____
5	_____	____

(FINISH BORE)
N145 T0404 M42 (Index turret, select high spindle range)
N150 G96 S500 M03 (Start spindle at 500 sfm)
N155 G00 X1.625 Z0.1 M08 (Rapid to point 17, turn on coolant)
N160 G01 Z0 F0.005 (Feed to point 18 at 0.005 ipr)
N165 X1.5 Z-0.0625 (Chamfer to point 19)
N170 Z-1.875 (Bore 1.5 to point 20)
N175 X0.95 (Feed down face to point 21)
N180 G00 Z0.1 (Rapid to point 22)
N185 X8.0 Z7.0 (Rapid to tool change position)
N190 M01 (Optional stop)

(FINISH FACE AND TURN)
N195 T0505 M42 (Index turret, select high spindle range)
N200 G96 S500 M03 (Start spindle at 500 sfm)
N205 G00 X2.7 Z0 M08 (Rapid to point 23, turn on coolant)
N210 G01 X1.3 F0.005(Face to point 24 at 0.005 ipr)
N215 G00 Z0.1 (Rapid to point 25)
N220 X2.25 (Rapid to point 26)
N225 G01 Z0 (Feed to point 27)
N230 X2.5 Z-0.125 (Chamfer to point 28)
N235 Z-1.5 (Turn 2.5 to point 29)
N240 X3.2 (Feed up face to point 30)
N245 G00 X8.0 Z7.0 (Rapid to tool change position)
N250 M30 (End of program)

Offset entry answers:

#	R	T
4	0.0156	2
5	0.0312	3

Program answers:
In line N155, add G41
In line N185, add G40
In line N220, add G42
In line N245, add G40

Fundamentals of CNC

Key Concept

5

You Must Provide Structure To Your CNC Programs

While there are many ways to write programs, you must ensure that your programs are safe and easy to use, yet as efficient as possible. This can be a real challenge since safety and efficiency usually conflict with one another.

Key Concept Number Five contains two lessons:

> 15: Introduction to program structure
> 16: Four types of program format

The fifth Key Concept is you must structure your CNC programs using a strict format – while incorporating a design that accomplishes the objectives you intend. Though it is important to create *efficient* programs, *safety and ease-of-use* must take priority – at least until you gain proficiency. In Key Concept Number Five, we'll be showing techniques that stress *safety as the top priority.*

To this point in the text, we have been presenting the building blocks of CNC programming – providing the needed individual tools. Machine components, axes of motion, program zero, and programmable functions are presented in Key Concept One. Preparation steps in Key Concept Two. Motion types in Key Concept Three. Compensation types in Key Concept Four. In Key Concept Five, we're going to draw all of these topics together, showing you what it takes to write CNC turning center programs completely on your own – making you a *self-sufficient programmer.*

Lesson Fifteen will introduce you to a CNC program's structure, showing you the reasons *why* programs must be strictly formatted. We'll also review some the program-structure-related points we've made to this point – and present a few new ones. And we'll address some variations related to how certain machine functions are handled.

In Lesson Sixteen, we'll show the four types of program format as they are applied to CNC turning centers. These formats can be used as a crutch to help you write your first few programs.

You may be surprised at how much you already know about a program's structure, especially if you have been doing the exercises included in this text and/or those in the workbook. We have introduced many of the CNC words used in programming, and we have been following the structure-related suggestions that we will be recommending here in Key Concept Five.

Lesson 15

Introduction To Program Structure

Structuring a CNC program is the act of writing a program in a way that the CNC machine can recognize and execute safely, efficiently, and with a high degree of operator-friendliness.

You know that CNC programs are made up of commands, that each command is made up of words, and that each word is made up of a letter address and a numerical value.

You also know that programs are executed sequentially – command by command. The machine will read, interpret, and execute the first command in the program. Then it will move on to the next command. Read – interpret – execute. It will continue doing so until the entire program has been executed.

You have seen several complete programs so far in this text – you have even worked on a few if you have done the exercises in this text and in the workbook. You have probably noticed that there is quite bit of consistency and *structure* in the CNC programs we have shown.

Our focus in Lesson Fifteen will be to help you understand more about the structure that is used in CNC programming.

Objectives of your program structure

CNC machines have come a long way. In the early days of NC (before computers), a program had to be written *just so*. If *anything* was out-of-place, the machine will go into an alarm state – failing to execute the program. While today's CNC machines are *much* more forgiving, you must still write CNC programs in a rather strict manner.

There are many ways to write a workable program – and the methods you use in structuring your programs will have an impact on the three most important objectives:

- Safety
- Efficiency
- Ease-of-use (operator friendliness)

It may be impossible to come up with a perfect balance among these objectives. Generally speaking, what you do to improve one objective will negatively affect the other two. When faced with a choice, a beginning programmer's priorities should always lean toward *safety* and *ease-of-use*. Our recommended programming structure stresses these two objectives. We will, however, show some of the efficiency-related short-comings of our recommended methods – so you can improve efficiency as you gain proficiency.

We're going to be assuming that *you have control* of the structure you use to write programs. Your company may, however, already have a programmer that is writing programs with a different structure. As long as these programs are working – and satisfying the company's objectives – you're going to have to adapt to the established structure. If you understand the reasons for formatting, and if you understand one successful method for structuring programs, it shouldn't be too difficult to adapt.

Machine variations that affect program structure

As mentioned, we provide two sets of format in Lesson Sixteen – one for use when geometry offsets are used to assign program zero and the other for use when program zero is assigned in the program. These formats can be applied to just about any of the machine configurations shown in Lesson One. But even within a given machine type, there are lots variations among machine tool builders.

Our given formats are aimed at basic turning centers – those without a lot of bells and whistles. About the only programmable functions we address are turret indexing, spindle, coolant, and feedrate. If your machine has other accessories (like a tailstock, steady rest, or bar feeder), you must consult your machine tool builder's programming manual to learn how they are programmed. As is discussed in Lesson One, most accessories are handled with M codes – so look first for the machine's list of M codes. (Note that we also provide information about certain accessories in the Appendix that follows Lesson Twenty-Three.)

Even within our limited coverage of accessories, there are variations. Here we list them.

M code differences

M code number selection is left completely to the discretion of your machine tool builder/s, and no two builders seem to be able to agree on how all M codes should be numbered. For *very* common machine functions like spindle on/off (M03, M04, M05) and flood coolant on/off (M08, M09), machine tool builders have standardized. But for less common functions like tailstock activation, chuck open/close, bar feeder activation, chip conveyer activation, door open/close, and high pressure coolant systems, you must find the list of M codes in your machine tool builder's programming manual.

Here is a list of very common M codes you will need for our given program formats. Fortunately, most machine tool builders do utilize these M code numbers just as we show.

M00 - Program stop (halts the program's execution until the operator reactivates the cycle)

M01 - Optional stop (used to cause the machine to stop between tools during program verification)

M03 - Spindle on forward (for right hand tools)

M04 - Spindle on reverse (for left hand tools)

M05 - Spindle off (not normally needed with our formats since the spindle need not stop when the turret is indexed)

M08 - Flood coolant on (cools and lubricates the machining operation)

M09 - Coolant off (not normally needed with our formats since the coolant can continue flowing when the turret is indexed)

M41 - Low spindle range selection

M42 - High spindle range selection

M30 - End of program (on many machines, M02 can also be used)

Here are some other functions commonly handled with M codes, but that are not consistently numbered by machine tool builders. We've left a blank next to each M code number. After checking your machine tool builder's programming manual, fill in the blanks for those machine functions that are equipped on your CNC turning center/s.

_____ - Coolant on and spindle on forward (together)

_____ - Coolant on and spindle on reverse (together)

_____ - Open chuck jaws

_____ - Close chuck jaws

_____ - Open machine door

_____ - Open machine door

_____ - Tailstock body forward

_____ - Tailstock body back

_____ - Tailstock quill forward

_____ - Tailstock quill back

_____ - Activate bar feeder

_____ - Deactivate bar feeder

_____ - Advance part catcher

_____ - Retract part catcher

_____ - Activate tool touch-off probe

_____ - Deactivate tool touch-off probe

_____ - Select normal turning mode

_____ - Select live tooling mode

_____ - Turn on chip conveyor

_____ - Turn off chip conveyor

G code numbering differences

Machine tool builders even vary when it comes to certain G code numbers. Even if you have a Fanuc or Fanuc-compatible control, you must be aware of this potential for variation. Generally speaking, imported machines (Japanese, Korean, Taiwanese, etc.) use what Fanuc calls the *standard G code specifications*. We use this set of G code specifications throughout this text. On the other hand, American machine tool builders tend to use the *special G code specifications*. Rest assured that the usage of these G codes remains exactly the same, only the invoking G code *number* will vary. Again, there are just a few differences. Here is a list of the G codes that vary based upon being standard or special.

Standard	Special	Description
G50	G92	Program zero assignment and maximum rpm designator
G90	G77	One-pass turning cycle
G92	G78	One pass threading cycle
G94	G79	One pass facing cycle
G98	G94	Feed per minute
G99	G95	Feed per revolution
X/Z	G90	Absolute positioning
U/W	G91	Incremental positioning

Again, most of these differences simply require that you translate a G code number (G92 instead of G50, for example) if your machine/s require the special G code specifications. However, notice how incremental and absolute motions are commanded. With the standard G code specification (used in this text), X and Z specify absolute positions. U and W specify incremental motions. With the special G code specification, G90 specifies incremental mode and G91 specifies incremental mode. With both special G codes, X and Z are used to specify motion. The command

        N040 G90 G00 X1.0 Z1.0

tells the control to rapid to a position (relative to program zero) of one inch in X and one inch in Z. On the other hand, the command

        N040 G91 G00 X1.0 Z1.0

tells the control to rapid from its current position one inch larger (positive) in diameter, and one inch in the Z plus direction.

The only other functions affected by the G code numbering are feedrate modes and simple canned cycles (discussed in Key Concept Six).

Turret variations

The design of a turning center's turret is left completely to the machine tool builder. Some incorporate external tool holders that are mounted on the outside of the turret. Others hold cutting tools directly in the turret itself. Some make it very easy to direct coolant lines to the cutting tool tip. Others make it more difficult. Some are large enough to eliminate interference when a long boring bar or drill is placed in a station that is adjacent to a face and turn tool. With others, you must be very careful with cutting tool placement in order to avoid interference problems.

As you know, a T word is used to specify turret index. This T word also instates the geometry and wear offsets. The command

N240 T0404

for example, indexes the turret to station number four, instates geometry offset number four, and instates wear offset number four.

Most machines have a *bi-directional* turret, meaning the turret can index in either direction. To select *which* direction, most turning centers are designed to automatically select the direction that provides the shortest rotational distance to the specified tool.

There are some turning centers, however, that let you specify the direction of turret rotation. Three M codes are usually involved. One M code specifies that the machine will select the shortest rotational distance (this is commonly the initialized M code). A second M code will force the turret to index in a clockwise direction. And a third M code will force the turret to index in the counter clockwise direction. You'll have to reference the machine tool builder's programming manual to find out whether your machine has this ability – and if so, which M codes are involved.

Choosing the appropriate spindle range

Some current model turning centers have but one spindle range. In this case, you will not be selecting spindle ranges. It's likely the machine has a high torque spindle drive motor that can drive the spindle with adequate force.

Most current model turning centers do have two or more spindle ranges. You can think of these ranges as like the transmission of an automobile. You get power (but limited speed) in the low range/s (gears) and you get speed (but limited power) in the high range/s (gears).

One rule of thumb is to perform roughing operations in the low range and finishing operations in the high range. While this is a good rule-of-thumb, there are times when following this rule can be somewhat inefficient. For your first few programs, use this rule-of thumb. In Lesson Sixteen, we do show how to more appropriately determine which spindle range you should select.

Which direction do you run the spindle?

You must, of course, have the spindle running in the appropriate direction in order for machining to occur. While this might sound like a very basic statement, turning center tooling is available in both right-hand and left-hand versions. Spindle direction is directly related to which hand of tooling you're using. In most cases, right hand tools require that the spindle be running in a forward (M03) direction and that left hand tools require the reverse (M04) direction. It is important that you specify the hand of tooling you're using on the setup sheet to eliminate the possibly for confusion in this regard. It's also quite important to confirm that the spindle is running in the right direction for each tool during the program's verification.

How do you check what each tool has done?

The optional stop command, programmed with M01, gives the setup person or operator the ability to stop the machine by turning on an on/off switch (labeled M01 or optional stop) on the operation panel. If the optional stop switch is currently on when the machine executes an M01 in the program, the machine will stop. For most machines, the spindle, coolant, and anything else currently activated will be stopped. If the switch is turned off, the machine will ignore M01 commands in the program (not stop the machine).

Our formats recommend that you include an M01 at the end of every tool. This will allow the setup person or operator to stop the machine after each tool by simply turning on the optional stop switch. This will be necessary during the program's verification and whenever trial machining.

Fundamentals of CNC © CNC Concepts, Inc.

Lesson 16

Four Types Of Program Format

The formats shown in this lesson will keep you from having to memorize most of the words and commands needed in CNC programming. As you'll see, a large percentage of most programs is related to structure.

You know the reasons why programs must be formatted using a strict structure. In Lesson Sixteen, we're going to show the actual formats. We will show two sets of format, one for use when geometry offsets are used to assign program zero – the other for use when program zero is assigned in the program. We will also explain every word in each format in detail. And we'll show example programs that stress the use of our given formats. Finally, we'll show some efficiency-related limitations of our given formats.

There are four types of program format:

> Program startup format
>
> Tool ending format
>
> Tool startup format
>
> Program ending format

Any time you begin writing a new program, follow the program startup format. You can copy this *structure* to begin your program. The actual values of some words will change based on what you wish to do in your own program, but the structure will remain the same every time you begin writing a new program.

After writing the program startup format, you write the motions for the first tool's machining operations. When finished with the first tool motions, you follow the tool ending format. You then follow the tool startup format for the second tool and write the second tool's machining motion commands. From this point, you toggle among tool ending format, tool startup format, and machining motion commands until you are finished machining with the last tool. You then follow the program ending format.

One of the most important benefits of using these formats is that you *will not have to memorize anything.* You simply copy the structure of the format.

Remember that you should use the formats related to assigning program zero with geometry offsets unless your machine does not have geometry offsets.

Format for assigning program zero with geometry offsets

Once again, this should be your format set of choice. The only reason not to use this format is if your turning center does not have geometry offsets (as is the case with older machines).

Program startup format (using geometry offsets)

If you've been doing the programming exercises in this text and/or in the workbook, you may be surprised at how many of these words and commands you are already familiar with. There really won't be all that many new words to learn.

O**0001** (Program number)

N005 G99 G20 G23 (Select feed per revolution feedrate mode, select inch mode, cancel stored stroke limit)

N010 G50 S**4000** (Limit spindle speed to 4,000 rpm)

N015 **T0101** M**41** (Index to first tool, instate geometry and wear offset number one, select spindle speed range)

N020 G**96** S**350** M03 (Select spindle mode, speed, and activate spindle in the forward direction)

N025 G00 X**3.0** Z**0.1** M08 (Move to approach position, and turn on the coolant)

N030 G01 X... Z... F**0.015** (Select feed per revolution mode, first cutting motion must include feedrate)

This is the format you'll use whenever you begin writing a new program. The values shown in bold will vary from program to program. The program number (O**0001** in our case) will change from program to program. So will the spindle speed limiting value (S**4000** in line N010). In line N015, is possible that your program will start with a tool station other than number one – and you may need a different spindle range. In line N020, you may need a different speed mode, a different speed, and you may sometimes need to start the spindle in the reverse direction. In line N025, you'll likely send the tool to a different approach position. And in line N030, your first machining command, you'll probably need a different feedrate.

While there will surely be variations from program to program, the basic *structure* will remain the same for every program you write. Again, you'll follow this program startup format whenever you begin writing a new program. Use it as a crutch until you have it memorized.

After the program number (specified by the O word), line N005 is the safety command that ensures that important initialized modes are still active.

N010 specifies spindle limiting. Since most workpieces don't require spindle limiting, you normally specify the machine's maximum rpm (4,000 rpm in our case).

N015 specifies the turret index and selects the desired spindle range. The T word selects the appropriate turret station (again, your first tool may not always be station number one) and instates the geometry and wear offsets. We've assumed this tool will be run in the low spindle range, but you may want to begin with the high spindle range (M42 with most turning centers). We're also assuming that your machine has more than one spindle range. If it does not, you must omit the M41 (or M42) from this command.

N020 selects the spindle mode (css or rpm – G96 or G97), the appropriate speed with the S word, and turns the spindle on in the appropriate direction. We're assuming that you're using right hand tools. If you use a left hand tool, the spindle must be started in the reverse direction (M04).

N025 makes the rapid approach movement to within a small distance (usually 0.1 inch) from the surface being machined. We're assuming you want to run the tool with flood coolant. If you do not, leave the M08 word out of this command.

N030 performs the first machining command. The feedrate word (F) is part of the program startup format and must be included in the tool's first machining command.

Tool ending format (using geometry offsets)

N075 G00 X8.0 Z6.0 (Move to turret index position)

N080 M01 (Optional stop)

These commands will be used to end every tool (but the last one) in your program.

N075 rapids the tool back to a safe turret index position. There is a presentation in Lesson Fifteen that shows how to select a safe turret index position.

N080 is the optional stop (M01) that gives the setup person or operator the ability to stop the machine (with the optional stop on/off switch) to see what this tool has done. This is very helpful during the program's verification – and whenever trial machining must be done.

Tool startup format (using geometry offsets)

N140 T0202 M42 (Index to next tool station, instate geometry and wear offset, select spindle speed range)

N145 G97 S600 M03 (Select spindle mode, speed, and turn on spindle in forward direction)

N150 G00 X0 Z0.1 M08 (Move to first position, turn on the coolant)

N155 G01 Z... F0.010 (In first cutting movement, include a feedrate word)

This is the format you'll follow whenever beginning a tool in your program. Notice how similar the tool startup format is to program startup format. We've simply eliminated the program number, the safety command, and the spindle limiting command.

Program ending format (using geometry offsets)

N210 G00 X8.0 Z6.0 (Move to tool change position)

N215 M30 (End of program)

Program ending format is almost identical to tool ending format. The only difference is that you end with M30 (end of program) instead of M01. M30 will turn off anything that's still running (spindle & coolant), rewind the program to the beginning for the next workpiece, and stop the cycle.

Example programs showing format for turning centers

Figure 5.2 is the drawing we use to stress turning center format.

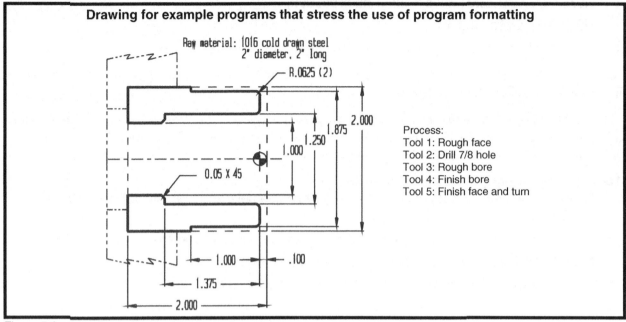

Drawing for example programs that stress the use of program formatting

Raw material: 1016 cold drawn steel
2" diameter, 2" long

R.0625 (2)

2.000
1.875
1.250
1.000

0.05 X 45

1.000 .100
1.375
2.000

Process:
Tool 1: Rough face
Tool 2: Drill 7/8 hole
Tool 3: Rough bore
Tool 4: Finish bore
Tool 5: Finish face and turn

Figure 5.2 – Drawing to be used for example programs

Example when assigning program zero with geometry offsets

Program:

O0002 (Program number)

(ROUGH FACING TOOL)

N002 G99 G20 G23 (Ensure that initialized modes are still in effect)

N004 G50 S5000 (Limit spindle speed to machine's maximum)

N005 T0101 M41 (Index turret, select spindle range)

N010 G96 S400 M03 (Start spindle in forward direction at 400 sfm)

N015 G00 X2.2 Z0.005 M08 (1) (Rapid to starting position, start coolant)

N020 G01 X-0.062 **F0.012** (2) (Select per revolution feedrate mode, face workpiece at 0.012 ipr)

N025 G00 Z0.1 (3) (Rapid away)

N030 X6.0 Z5.0 (Rapid to tool change position)

N035 M01 (Optional stop)

(7/8 DRILL)

N040 T0202 M41 (Index turret, select spindle range)

N045 G97 S354 M03 (Start spindle forward at 354 rpm)

N050 G00 X0 Z0.1 M08 (4) (Rapid into position, start coolant)

N055 G01 Z-2.2 **F0.008** (5) (Drill hole at 0.008 ipr)

N060 G00 Z0.1 (4) (Rapid out of hole)

N065 X6.0 Z5.0 (Rapid to tool change position)

N070 M01 (Optional stop)

(3/4 ROUGH BORING BAR)

N075 T0303 M42 (Index turret, select spindle range)

N080 G96 S350 M03 (Start spindle forward at 350 sfm)

N085 G00 X1.19 Z0.1 M08 (6) (Rapid into position, start coolant)

N090 G01 Z-1.37 **F0.007** (7) (Begin boring operation at 0.007 ipr)

N095 X0.875 (8)

N100 G00 Z0.1 (9) (Rapid out of hole)

N105 X5.0 Z6.0 (Rapid to tool change position)

N110 M01 (Optional stop)

(3/4 FINISH BORING BAR)

N115 T0404 M42 (Index turret, select spindle range)

N120 G96 S400 M03 (Start spindle forward at 400 sfm)

N125 G00 X1.375 Z0.1 M08 (10) (Rapid into position, start coolant)

N130 G01 Z0 **F0.005** (11) (Begin finish boring at 0.005 ipr)

N135 G02 X1.25 Z-0.0625 R0.0625 (12)

N140 G01 Z-1.375 (13)

N145 X1.1 (14)

N150 X1.0 Z-1.425 (15)

N155 Z-2.0 (16)

N160 G00 X0.8 (17)

N165 Z0.1 (18) (Rapid out of hole)

N170 X6.0 Z5.0 (Rapid to tool change position)

N175 M01 (Optional stop)

(FINISH FACE AND TURN TOOL)

N180 T0505 M42 (Index turret, select spindle range)

N185 G96 S450 M03 (Start spindle forward at 450 sfm)

N190 G00 X2.075 Z0 M08 (19) (Rapid into position, start coolant)

N195 G01 X1.05 **F0.006** (20) (Start finish facing and turning at 0.006 ipr)

N200 G00 Z0.1 (21)

N205 X1.75 (22)

N210 G01 Z0 (23)

N215 G03 X1.875 Z-0.0625 R0.0625 (24)

N220 G01 Z-1.0 (25)

N225 X2.2 (26)

N230 G00 X6.0 Z5.0 (Rapid back to tool change position)

N235 M30 (End of program)

All of the commands and words shown in bold are related to format. The structure for these commands can be copied from our given formats. You'll only be on your own to develop the cutting commands for each tool. Truly, a great percentage of most programs is related to format.

Where are the restart commands?

We've discussed re-running tools several times. And you know that the program must be properly structured in order to re-run tools (which programs written with our given formats are). In order to re-run a cutting tool, you must know the *restart command* for the tool (which is also called the *pickup block*). You will scan to this command before activating the cycle.

To re-run the first tool, of course, you can simply run the program from the beginning. For other tools, finding the restart command with our format is still pretty simple. It is the command that includes the turret index (T word). In the program above, the restart commands are N040 for tool two, N075 for tool three, N115 for tool four, and N180 for tool five.

Write your first complete program

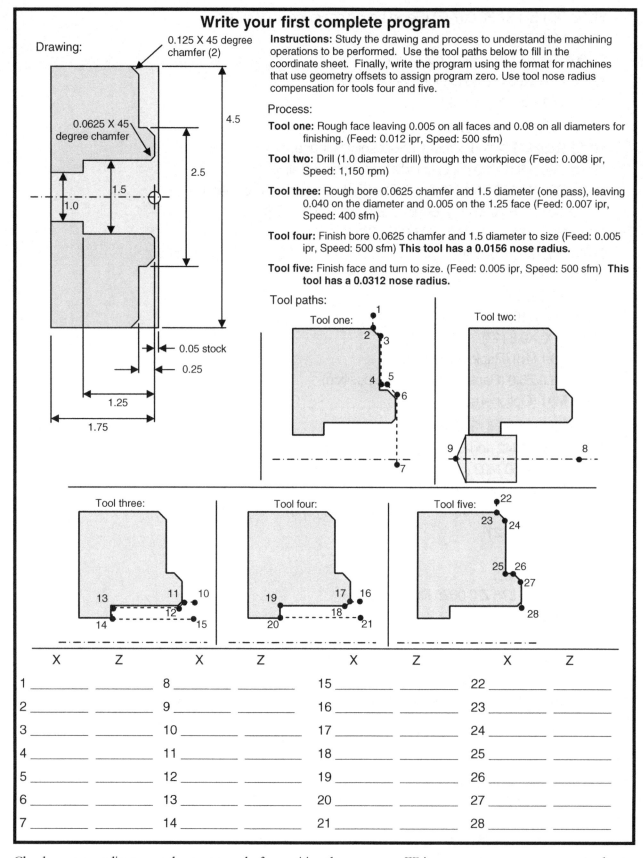

Drawing:

0.125 X 45 degree chamfer (2)

0.0625 X 45 degree chamfer

4.5

2.5

1.5

1.0

0.05 stock

0.25

1.25

1.75

Instructions: Study the drawing and process to understand the machining operations to be performed. Use the tool paths below to fill in the coordinate sheet. Finally, write the program using the format for machines that use geometry offsets to assign program zero. Use tool nose radius compensation for tools four and five.

Process:

Tool one: Rough face leaving 0.005 on all faces and 0.08 on all diameters for finishing. (Feed: 0.012 ipr, Speed: 500 sfm)

Tool two: Drill (1.0 diameter drill) through the workpiece (Feed: 0.008 ipr, Speed: 1,150 rpm)

Tool three: Rough bore 0.0625 chamfer and 1.5 diameter (one pass), leaving 0.040 on the diameter and 0.005 on the 1.25 face (Feed: 0.007 ipr, Speed: 400 sfm)

Tool four: Finish bore 0.0625 chamfer and 1.5 diameter to size (Feed: 0.005 ipr, Speed: 500 sfm) **This tool has a 0.0156 nose radius.**

Tool five: Finish face and turn to size. (Feed: 0.005 ipr, Speed: 500 sfm) **This tool has a 0.0312 nose radius.**

Tool paths:

Tool one:

Tool two:

Tool three:

Tool four:

Tool five:

	X	Z		X	Z		X	Z		X	Z
1			8			15			22		
2			9			16			23		
3			10			17			24		
4			11			18			25		
5			12			19			26		
6			13			20			27		
7			14			21			28		

Check your coordinates on the next page before writing the program. Write your program on a separate sheet of paper.

Coordinates for your program

1: X4.7 Z-0.37
2: X4.5 Z-0.37
3: X4.33 Z-0.245
4: X2.58 Z-0.245
5: X2.58 Z-0.12
6: X2.33 Z0.005
7: X-0.062 Z0.005

8: X0 Z0.1
9: X0 Z-2.08
10: X1.585 Z0.1
11: X1.585 Z0.005
12: X1.46 Z-0.0575
13: X1.46 Z-1.245
14: X0.95 Z-1.245

15: X0.95 Z0.1
16: X1.625 Z0.1
17: X1.625 Z0
18: X1.5 Z-0.0625
19: X1.5 Z-1.25
20: X0.95 Z-1.25
21: X0.95 Z0.1

22: X4.7 Z-0.375
23: X4.5 Z-0.375
24: X4.25 Z-0.25
25: X2.5 Z-0.25
26: X2.5 Z-0.125
27: X2.25 Z0
28: X1.5 Z0

Answer program

O0010
N002 G99 G20 G23
N004 G50 S5000

(ROUGH FACING TOOL)
N005 T0101 M41
N010 G96 S500 M03
N015 G00 X4.7 Z-0.37 M08 (1)
N020 G01 X4.5 F0.012 (2)
N025 X4.33 Z-0.245 (3)
N030 X2.58 (4)
N035 Z-0.12 (5)
N040 X2.33 Z0.005 (6)
N045 X-0.062 (7)
N050 X6.0 Z5.0
N055 M01

(1.0" DRILL)
N060 T0202 M41
N065 G97 S1150 M03
N070 G00 X0 Z0.1 M08 (8)
N075 G01 Z-2.2 F0.008 (9)
N080 G00 Z0.1 (8)
N085 X6.0 Z5.0
N090 M01

(ROUGH BORING BAR)
N095 T0303 M42
N100 G96 S400 M03
N105 G00 X1.585 Z0.1 M08 (10)
N115 G01 Z0 F0.007 (11)
N120 X1.46 Z-0.0575 (12)
N125 Z-1.245 (13)
N130 X0.95 (14)
N135 G00 Z0.1 (15)
N140 X5.0 Z6.0
N145 M01

(FINISH BORING BAR)
N150 T0404 M42
N155 G96 S500 M03
N160 G41 G00 X1.625 Z0.1 M08 (16)
N165 G01 Z0 F0.005 (17)
N170 X1.5 Z-0.0625 (18)
N175 Z-1.25 (19)
N180 X0.95 (20)
N185 G00 Z0.1 (21)
N190 X6.0 Z5.0
N195 M01

(FINISH FACING TOOL)
N200 T0505 M42
N205 G96 S500 M03
N210 G41 G00 X4.7 Z-0.375 M08 (22)
N215 G01 X4.5 F0.005 (23)
N220 X4.25 Z-0.25 (24)
N225 X2.5 (25)
N230 Z-0.125 (26)
N235 X2.25 Z0 (27)
N240 X1.5 (28)
N245 G00 X6.0 Z5.0
N250 M30

Fundamentals of CNC

Key Concept

6

Special Features That Help With Programming

You now have the basic tools you need to write CNC programs. However, writing programs with only the CNC features we have shown will be very tedious. In Key Concept Number Six, we'll be showing features that make programming easier, shorten the program's length, and in general, facilitate your ability to write programs.

Key Concept Number Six contains seven lessons and an appendix:

This key concept should be of great interest to you. While you have learned how to write simple CNC programs, you may be wondering if there are easier ways to get things done. As you'll see in this key concept, there are.

You know, for example, that programming any kind of roughing operation (rough turning, rough facing, or rough boring) can be quite tedious. With only **G00**, **G01**, **G02**, and **G03** to work with, every roughing pass will require at least four commands. Consider rough turning a workpiece down from 8.0 inches to 2.0 inches in diameter, taking a depth-of-cut of 0.125 inches. This will require twenty-four passes – at least ninety-six commands. And if a roughing pass intersects a chamfer, taper, or radius, it will be difficult to calculate the end point for the roughing pass. You will learn in Lesson Eighteen that as long as you can program the finish pass, the machine will completely rough turn or bore the workpiece for you – based on one simple command – regardless of how many passes are required. In Lesson Nineteen, you'll see that there is a similar command for rough facing.

In Lesson Twenty, we'll present a simple way to machine threads – one command will machine the entire thread regardless of how many passes it takes. In Lesson Twenty-One, you'll learn how to use a program-shortening feature called *sub-programming*. In Lesson Twenty-Three, we'll present several special features that can help you create programs. While they won't all be of immediate use, you'll surely find at least some of them to be very helpful.

We also include a programming appendix in Key Concept Six. In this appendix, we'll be showing how to handle special machines and machine accessories. For instance, live tooling, bar feeders, and tailstocks will be presented in the appendix.

Almost every CNC function presented thus far is *essential* to your ability to program CNC turning centers. You will be using the features and techniques presented in Key Concepts One through Five in every program you write. Think of what you have learned so far as the rudimentary tools of CNC programming.

You may be wondering why we waited so long to show these features, since they do make programming *so* much easier. Do not underestimate the importance of what you have learned so far. Just as it helps to understand how arithmetic calculations are done manually before you use a calculator, so does it help to understand the long (more difficult) way of programming machining operations before you move on to the more advanced methods. At the very least, the programming activities you have done so far should give you a good appreciation for what is presented in Key Concept Six. More importantly, you should now have a *very* good foundation in CNC that will let you easily grasp the importance of the features we show in this key concept.

Some of the features shown in Key Concept Six are considered by Fanuc to be *options*. But most machine tool builders will include them with the machines they sell. If you come across one of these features that your machine does not have, remember that all of them are *field-installable*, meaning if you have a need for them, they can be added to your control at any time (for an additional price).

As stated, you will not have immediate need for some of the features discussed in this key concept. In fact, some you may never need. But we urge you to study this information if for no other reason than to gain a better understanding of what is possible in CNC programming. You can always come back and review a needed feature when the need arises, *if you know it is available*.

Control series differences

Some of these special programming features (especially multiple repetitive cycles and subprograms) require slightly different programming words within commands based upon the Fanuc control model you are programming. The basic functions of each cycle are exactly the same from one control model to the next. Only the programming format changes slightly from one control to another.

To handle this minor format problem during this text, we will first look at how these functions are programmed for the majority of Fanuc controls. These include the Fanuc 0TC, the 6T, 6TB, 10T, 11T, 15T, 16T, and 20T, and 21T. Most controls that claim to be Fanuc-compatible also use this format. Then we will discuss the minor format differences for Fanuc 0TA, 0TB, 3T, and 18T. These minor program formatting differences exist because the 0TA, some 0TBs, 3T, and 18T controls do not have all of the letter addresses needed to program certain cycles in one command.

If you happen to have a 0TA, 0TB, 3T, or 18T, you must first follow all presentations. You will be exposed to the various functions and features of each cycle and see how they work. When we discuss control model differences (in Lesson Twenty-Two), we'll summarize the command differences.

Lesson 18

G71 And G70 – Rough Turning And Boring Followed By Finishing

These cycles are extremely helpful. They allow you to completely rough turn or rough bore a workpiece based upon a finish pass definition and one special command. In this command you specify important information like depth-of-cut and the amount of stock you want to leave for finishing. Based upon this simple information, the machine will completely rough the workpiece – regardless of the number of passes required.

You know that one-pass canned cycles can be used to minimize the number of commands needed to machine a workpiece. But as you have seen, they are somewhat cumbersome to use. By comparison, the multiple repetitive cycles we discuss in this lesson are *much* more helpful. They make it relatively simple to program complex rough turning and rough boring operations for even the most difficult workpieces. As you'll see, they rival what can be done with a computer aided manufacturing (CAM) system.

As mentioned earlier, first we will show how these multiple repetitive cycles are programmed for the majority of Fanuc turning center controls. Then we will discuss the slight differences required for 0TA, 0TB, 3T, and 18T Fanuc controls.

G71 - Rough turning and boring

If you've been doing the programming exercises in this text and in the workbook, you've seen that rough turning and rough boring operations can be very tedious and error prone to program with only the three basic motion types (rapid, straight line, and circular). Every rough turning or boring pass requires at least four commands. And in our programming exercises so far, we've kept things as simple as possible. In reality, roughing a workpiece is usually much more difficult.

For example, all of the roughing passes you've programmed so far end at a known position (at a face or diameter). If a roughing pass ends at a chamfer, taper, or radius, calculating the end point for the pass will require more difficult math calculations (including trigonometry), even for relatively simple workpieces. Figure 6.6 shows an example workpiece for which programming rough turning will be more difficult.

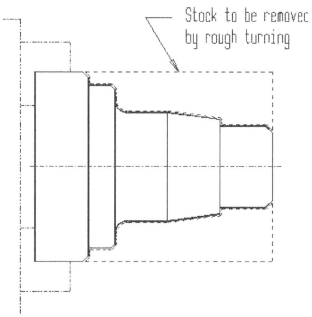

Stock to be removed by rough turning

Figure 6.6 – Drawing that stresses the difficulty of programming roughing passes long-hand

Though this is a relatively simple workpiece, notice how much stock must be removed by the rough turning operation. To program this long hand (without any special cycles), a programmer will have to plan and program several roughing passes based on a previously determined *depth-of-cut*. This creates a real problem for manual programmers. And consider the additional difficulty if one or more of the roughing passes ends in the middle of the large taper or fillet radius. Trigonometry will be required in both cases.

Knowing how difficult it can be to program individual turning or boring passes, Fanuc has designed three very helpful multiple repetitive cycles for roughing:

- **G71** - Rough turning and rough boring
- **G72** - Rough facing
- **G73** - Pattern repeating

In this lesson, we limit our discussions to **G71**, which is used for rough turning and boring. In the Lesson Nineteen, we'll describe **G72** and **G73**. You will see many similarities among these multiple repetitive cycles. All roughing multiple repetitive cycles are followed by the multiple repetitive cycle for finishing. **G70** specifies the finishing cycle.

With roughing multiple repetitive cycles, you only need to describe the finish pass of the workpiece surface being roughed. Based on this *finish pass definition* and one very simple command, the control will completely rough the entire workpiece.

The two phases of G71

G71 and **G72** commands will be completed in two *phases*. In the phase one, the machine will make a series of roughing passes based on a specified depth-of-cut. But as soon as the tool comes close to an end point for a roughing pass, the machine will immediately retract the tool for another pass. Figure 6.7 shows the motions generated during the first phase of the **G71** command.

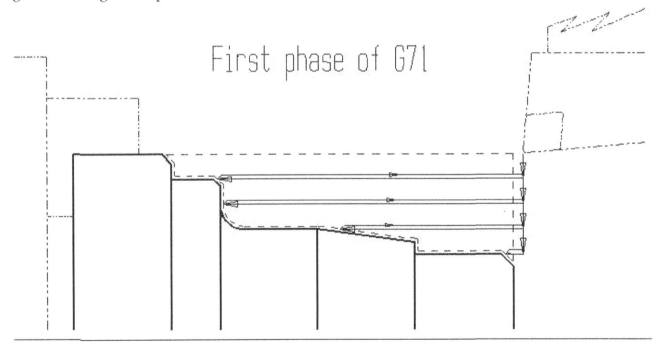

Figure 6.7 – Motions during phase one of a G71 command

After the control finishes the first phase of **G71**, the workpiece will have a series of steps (like a staircase) and it will not be close to its proper size (it will not have the appropriate amount of finishing stock on all surfaces). None of the steps left by the first phase of **G71** will be larger than the depth-of-cut specified in the **G71** command. Figure 6.8 shows what the workpiece will look like *after* the first phase of the **G71** command.

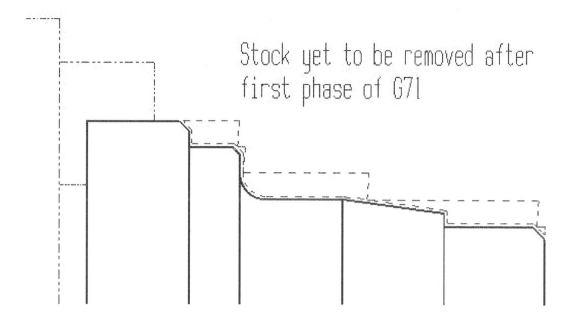

Figure 6.8 – What workpiece will look like after phase one of G71 command

In the second phase of **G71**, the machine will return to the starting point of the finish pass definition and make one sweeping semi-finish pass over the entire workpiece. It will stay away from the finished surface by the finishing stock values specified in the **G71** command. Figure 6.9 shows the motions generated during the second phase of the **G71** command.

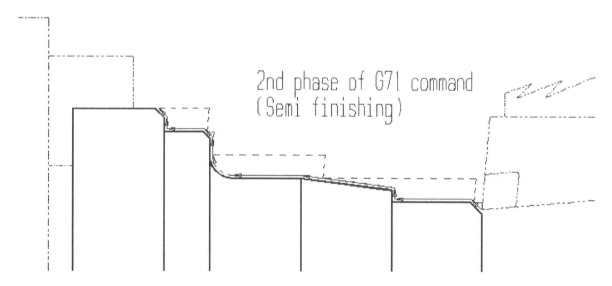

Figure 6.9 – Motions during phase two of G71 command

This is all accomplished by one command (the **G71** command) together with the finish pass definition. **G71** makes it very easy to program even very lengthy and complex rough turning and boring operations. In essence, the machine will figure out how to make all of the roughing passes for you, in much the same way a computer aided manufacturing (CAM) system does.

There are three steps to programming a roughing operation with **G71**.

First, position the tool to the convenient starting position. For rough turning (external work), this position must be flush with the stock diameter in X and clear of the workpiece in Z. For rough boring (internal work), the convenient starting position must be flush with the current hole size in X and clear of the workpiece in Z.

Second, specify the G71 command. This command tells the machine *how* to perform the rough turning or boring operation. A series of special words acts as *variables* to specify the required information.

Third, specify the finish pass definition. The finish pass definition begins with a rapid positioning movement in X to the first diameter to be machined. It ends with the tool's movement to the diameter of the convenient starting position. Two of the words in the G71 command (one for diameters and one for faces) specify how much stock is to be left for finishing. All movements during the finish pass definition are specified with finished workpiece dimensions.

At the completion of the G71 command, the workpiece will be entirely rough turned or bored. There will still be stock left on all surfaces for finishing. After a turret index to the finishing tool, another command is used to finish turn or bore after a G71 has been used for roughing (G70).

Understanding G71 command words

Again, a G71 command tells the control how to perform the rough turning or boring operation. Here are the words included within the G71 command.

P word

This word points to the sequence number of the command that starts the finish pass definition. The finish pass definition will always begin with the command immediately following the G71 command. For example, if the G71 command is specified in line N020 (and you skip five numbers for each sequence number), the finish pass definition will begin in line N025. In this case, the word P025 will be included in the G71 command.

Q word

This word points to the sequence number of the command that ends the finish pass definition. The machine will use the commands specified from the P word through the Q word in order to understand what the finished workpiece looks like. As you actually write the G71 command, you won't know how many commands will be in the finish pass definition, meaning you won't know the value of the Q word (yet). So leave the Q word blank, circling it to help you remember to come back and fill it in when the finish pass definition is completed.

Most controls require the P and Q word values to perfectly match the sequence numbers being specified. If N025 is being specified by a P word, for example, the correct word is P025, not P25 or P0025. Also note that the sequence numbers specified by P and Q must only occur once in the program.

U word

The U word tells the control how much stock is to be left for finishing on all diameters. The value U0.040, for example, tells the control to leave 0.040 in on all diameters (0.020 on the side). The U word also points in the direction of stock to be left. For external diameter machining (turning), the U word must be positive. For internal diameter machining (boring), the U word must be negative.

W word

The W word tells the control how much stock is to be left for finishing on all faces. The word W0.005 will leave 0.005 in on all faces. As long as turning or boring is done toward the chuck (in the minus Z direction), as most turning and boring is, the W word value must be positive. If the cutting tool is back turning (machining in the Z plus direction), the W word value must be negative.

D word

The D word tells the control how much stock to remove (on the side) per pass. This is the *depth-of-cut* for the roughing operation. All but the most recent controls do not allow a decimal point to be used with this word. Instead, a four place fixed format must be used in the inch mode (a three place fixed format must be used in the metric mode). For example, the word D2500 specifies a 0.250 inch depth-of-cut. D1250 specifies a 0.125 inch depth-of-cut. D0500 specifies a 0.050 inch depth-of cut. And so on. (In the metric mode, D3000 specifies a three millimeter depth-of-cut.) If you want to know whether a decimal point is allowed with the D word for your particular control, reference the control manufacturer's programming manual in the description of G71.

F word

This is the feedrate for the entire roughing operation. Any feedrate word specified during the finish pass definition will be ignored by during the roughing operation (but will be used during the finishing operation).

What about finishing?

Before we can show a complete example, we need to introduce the G70 finishing command. As stated, at the completion of the G71, the machine will have completely roughed the workpiece based on the G71 command and what's in the finished pass definition (between the P and Q blocks). But as you know, the workpiece will not yet be finished.

To keep from having to repeat the commands given during the finish pass definition, the machine will allow you to specify the finishing operation with a G70 command. Like G71, the G70 uses a P word to point to the first command of the finish pass definition and a Q word to point to the last command of the finish pass definition. The P and Q of the G70 command will always match the P and Q of the related G71 command. When G70 is executed, the control will simply go back to the sequence number specified by the P word and do through the sequence number specified by the Q word. This will cause the cutting tool to finish the workpiece. At the completion of the finishing cycle, the tool will be left at the convenient starting position.

Example showing G71 for rough turning and G70 for finish turning

Admittedly, the G71 command probably sounds a little complicated at this point. When you see an example, things should clear up a bit. And regardless of how difficult you find this to be, it is well worth your time to study until you thoroughly understand the G71 and G70 commands. You will save countless programming hours when you master them.

Figure 6.10 shows the workpiece to be used for our first example program. Tool number one is the rough turning tool and tool two is the finish turning tool. The end of this workpiece has been previously faced to size.

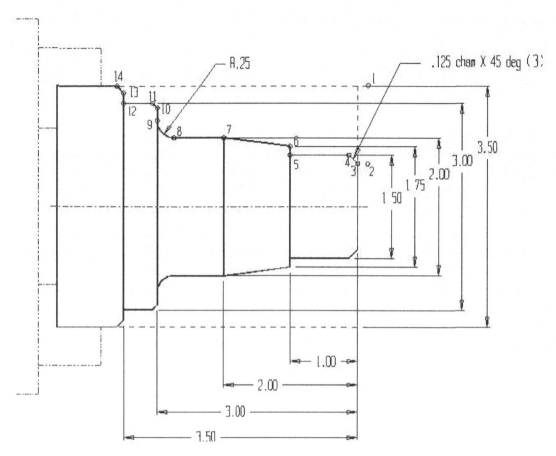

Figure 6.10 – Drawing for first G71 example program

Program:

O0010 (Program number)

(ROUGH TURNING TOOL)

N003 G99 G20 G23 (Ensure that initialized states are still in effect)

N004 G50 S4000 (No need for limiting, limit to machine's maximum speed)

N005 T0101 M41 (Select roughing tool and low spindle range)

N010 G99 G96 S400 M03 (Turn spindle on fwd at 400 sfm)

N015 G00 X3.5 Z0.1 M08 (Rapid to convenient starting position, point 1, start coolant)

N020 **G71 P025 Q085 U0.040 W0.005 D1250 F0.015** (Rough turn entire workpiece based on what is between lines N025 and N085)

N025 G00 X1.25 (First block of finish pass definition, rapid to point 2)

N030 G01 Z0 **F0.008** (Feed to point 3)

N035 X1.5 Z-0.125 (Feed to point 4)

N040 Z-1.0 (Feed to point 5)

N045 X1.75 (Feed to point 6)

N050 X2.0 Z-2.0 **F0.005** (Feed to point 7)

N055 Z-2.75 **F.008** (Feed to point 8)

N060 G02 X2.5 Z-3.0 R.25 (Circular move to point 9)

N065 G01 X2.75 (Feed to point 10)

N070 X3.0 Z-3.125 (Feed to point 11)

N075 Z-3.5 (Feed to point 12)

N080 X3.25 (Feed to point 13)

N085 X3.5 Z-3.625 (Feed to point 14)

N090 G00 X6.0 Z6.0 (Rapid to safe index position)

N095 M01 (Optional stop)

(FINISH TURNING TOOL)

N100 T0202 M42 (Select finish turning tool and high spindle speed range)

N105 G96 S600 M03 (Turn spindle on fwd at 600 sfm)

N110 G00 **G42** X3.5 Z0.1 M08 (Rapid to the same convenient starting position as used for the rough turning operation, start coolant)

N115 **G70 P025 Q085 F0.008** (Go back to line N025 and do through line N085, finish turning workpiece)

N120 G00 **G40** X6.0 Z6.0 (Rapid to safe index point)

N125 M30 (End of program)

Notes about the example program:

1) At the completion of the G71 command (line N020), the workpiece will be entirely rough turned. The control will continue with the program's execution at line N090, after the last command of the finish pass definition. For this reason, it may help to think of all of the commands from N020 through N085 as one command.

2) As you write line N020, you will not know which command ends the finish pass definition. For this reason, the Q word must be left blank until the finish pass definition is completed (in line N085). The finish pass definition will end when the tool reaches the stock diameter (the same X position as in the convenient starting position). In our example, notice that the tool comes up to a 3.5 position in X in line N085, which is the same X position as is in line N015.

3) Notice that the feedrates specified during the finish pass definition (in lines N030, N050, and N055) will be ignored during the rough turning operation. The entire rough turning operation will be performed at 0.015 ipr (as is commanded in line N020). However, these feedrates within the finish pass definition *will* be used during the finishing operation (commanded by line N115). This allows you to specify different feedrates (meaning different surface finishes) for different surfaces of the workpiece. In our case, the tapered surface will be machined at a slower feedrate than the balance of the workpiece. If no feedrate commands are given during the finish pass definition, the control will use the currently instated feedrate at the time the G70 command is executed. If you wish the entire finishing pass to be made at the same feedrate, you can simply include this feedrate in the G70 command.

4) Notice that the first command of the finish pass definition (line N025) is a motion at rapid to a position flush with the first X surface to be machined (the lower right corner of the chamfer). **There is a special rule about this command.** It goes like this: *The block specified by the P word of a G71 command must include either a G00 or G01 and **only** an X word to the starting diameter of the finish pass definition.* If you leave out the G00 (or G01) or if you include a Z word in this command, the control will generate a format alarm.

5) Notice (in line N110) that the finishing tool is sent to precisely the same convenient starting location as the rough turning tool. Since all canned cycles leave the tool back at the convenient starting position, this is where the finishing tool will go when the G70

command is finished. If you keep the same convenient starting position for the finishing tool as you used for the roughing tool, you can rest assured that there will be no interference when the finish turning tool returns to its convenient starting position. On the other hand, if you rapid the finish turning tool closer to the starting diameter prior to the G70 command (if the convenient starting position is closer to the first diameter to machine), it is likely that the finish turning tool will crash through the workpiece on its return to the convenient starting position at the completion of the finishing operation.

6) Think of how easy it will be to change the cutting conditions at the machine when G71 is used. If, for example, the depth of cut is too deep and the spindle stalls, only one word (the D word) has to be changed in order to generate a new set of rough turning movements.

7) Notice that it is still possible to use tool nose radius compensation in conjunction with the G70 command. Since tool nose radius compensation is desired for the finishing tool, the proper word (G42 in this case) is included in the tool's first approach to the workpiece (to the convenient starting position) in line N110. It is also being canceled (with G40) on the tool's return to the safe index point in line N120.

Lesson 20

G76 – Threading Command

This command is extremely helpful for machining threads. It machines external threads and internal threads with equal ease. Only one command is needed to machine the thread – regardless of how many passes are required.

Threading involves *chasing* a thread on the workpiece with a single point threading tool. By chasing, we mean the threading tool will make a series of synchronized passes to machine the thread to its required depth. Thread depth will increase with each pass. The threading tool has the form of the thread ground into its insert.

G76 is an extremely helpful multiple repetitive cycle that will machine an entire thread in one command. Like the other multiple repetitive cycles, the G76 command allows you to specify the criteria for the thread to be machined with a series of *variables* that are included right in the G76 command.

Like the G71 command, G76 can be used for outside diameters as well as inside diameters. Also like the G71, the programmer will rapid the threading tool to a *convenient starting position*. This position for threading will be clear of the surface to be threaded in the X axis (by at least 0.03 inch or so), and well off the face of the diameter to be threaded in the Z axis (we recommend a distance of four times the pitch or 0.2 inch, *whichever is smaller*). This large approach distance in Z is required to allow the threading tool to accelerate to the required motion rate (feedrate) before contacting the workpiece. If it does not, the thread's pitch will not be correct at the beginning of the thread.

Figure 6.17 shows the words related to the G76 threading cycle. Note that an external thread is being shown.

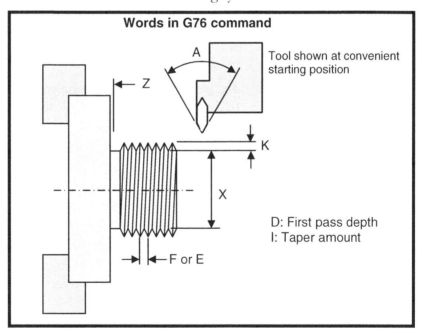

Figure 6.17 – Words related to G76 command

X word

The X word in a G76 command specifies the *final threading diameter*. For external threads, the X word is the minor diameter of the thread. For internal threads, the X word is the major diameter of the thread.

Z word

The Z word in a **G76** specifies the end point for every threading pass in the Z axis. When threading into a recess (as Figure 6.17 shows), the Z word value will be in the middle of the recess the recess. When threading a solid surface, the Z word value will be the ending point of the thread. As discussed in Lesson Ten, we recommend programming the leading edge of the threading insert in the Z axis (as opposed to the point of the insert). This makes it easy for the setup person to determine the program zero assignment values for the tool – and when threading up to a shoulder (as is done in Figure 6.17), it allows you to specify a Z word that is very close to the shoulder without fear of crashing the threading tool.

What is thread chamfering?

There is a feature called *thread chamfering* that affects how close to the Z word value the threading tool will actually come before chamfering begins. If thread chamfering is turned on, the tool will come to within about one pitch (1 divided by the number of threads per inch) of the Z value. At this point, the threading tool will begin feeding away in the X axis while it continues moving in the Z axis (forming a forty-five degree angle of pull-out). This causes the threading tool to gradually retract from the thread.

Thread chamfering is most helpful when there is no thread relief groove at the end of the thread. With this kind of thread – and especially with older machines - the thread will *look better* if threading is done with thread chamfering turned on.

If thread chamfering is off (which is the initialized state for most turning centers), the tool will feed precisely to the Z word value of the **G76** command and pull directly away from the thread. This is required when threading into a thread relief.

How thread chamfering is turned on and off is determined by the machine tool builder. Some builders provide two M codes, one to turn thread chamfering on and another to turn it off. Others require that a control parameter be changed in order to activate thread chamfering. Again, the initialized state for thread chamfering is usually off.

Since newer machines provide very accurate threading motions, the need for thread chamfering has been reduced over the years. Most current machines can provide accurate and nice looking threads with thread chamfering turned off – even when there is no relief groove at the end of the thread.

K word

The K word specifies the thread depth (on the side). Unfortunately, most design engineers do not provide the thread depth as part of the thread specification. To determine this value, you will need to reference a machining handbook (like Machinery's Handbook). Look for the major and minor diameters of the thread. The thread depth (K word value) can be calculated by subtracting the minor diameter from the major diameter – and then dividing the result by two.

Some programmers approximate thread depth by using 75% of the thread's pitch. This works quite well for National Standard (60 degree included angle) threads. If the thread depth is approximated, the threading tool's wear offset can be used to fine-tune the depth of the thread when the first workpiece is run.

D word

The D word specifies the *depth of the first pass* (on the side). The control will automatically reduce each subsequent threading pass depth from this value – so the depth of successive passes will get progressively shallower.

The D word value controls how many passes will be made. Generally speaking, the larger the D word, the fewer the number of threading passes. And actually determining the D word value will take some practice. If you're in doubt, we recommend starting with a small value (forcing more passes than necessary) and working your way up. Generally speaking, coarse threads will allow a larger D word than fine threads. As a starting point, we recommend a D word of about 15% of the pitch for external threads and about 10% of the pitch for internal threads. For example, if machining a 1"-8 external thread (0.125 pitch), start with a D word of about 0.019 (which is fifteen percent of 0.125).

As with **G71** and **G72**, the D word in **G76** can help with optimizing. If the threading cycle is making too many passes, increase the D word. If too few passes are being made, reduce the D word.

Also like **G71** and **G72**, most controls do not allow a decimal point to be used with the D word used with **G76**. In the inch mode, a four place fixed format must be used. **D0150** specifies a 0.015 inch depth of cut. **D0100** is 0.010 inch, and so on. In the metric mode, a three place format must be used. **D500** specifies a 0.5 mm depth-of-cut.

A word

The A word specifies the *tool angle*. The A word does not allow a decimal point. Many threading tools are designed to machine with only the leading edge of the tool. If A is specified in the **G76** command (and not set to 0), the threading tool will machine with only its leading edge.

For National Standard threads, the included angle for the thread is sixty degrees. With an A word specified as **A60** in the **G76** command, the threading tool will in-feed with each pass at an angle of just under thirty degrees.

Older Fanuc controls require that you specify the A word from these choices: **A80**, **A60**, **A55**, **A30**, **A29**, and **A0**. Newer controls allow any tool angle to be specified. An A word of **A29**, for example, will be used to machine an **ACME** thread.

Some threading tools are designed to machine on both sides of the threading insert. These threading tool inserts are commonly referred to as *cresting inserts*. If you use a cresting insert, simply leave the A word out of the **G76** command – or set it to a value of zero (**A0**). This will cause the threading tool to approach in only the X axis for each threading pass. That is, three will be no angular in-feed.

F word and E word

As you know, the F word specifies feedrate in feed per revolution (always thread in feed per revolution mode – **G99**). And of course, the pitch of any thread is the amount of Z axis motion needed during each spindle revolution – meaning the F word will be equal to the pitch of the thread.

With threads specified in the inch mode, pitch is equal to one divided by the number of threads per inch. With metric threads, the pitch is provided right in the threads specification.

For example, an external metric thread specified as 50-1.5 has a fifty millimeter major diameter and a 1.5 millimeter pitch. An external inch thread specified as 2.0-12 has a two inch major diameter and twelve threads per inch. The pitch for this thread is 0.08333333333333... (1/12).

In the inch mode, you know the F word has a four-place decimal format. This means you can only program the F word to four places. A sixteen-threads-per-inch thread has a pitch of precisely 0.0625, meaning you can perfectly specify this pitch with the F word (with **F0.0625**). But with a twelve threads per inch thread, the pitch is 0.083333333333... The best you can do with an F word is **F0.0833**.

This is usually acceptable (as long as mating parts are programmed with the same pitch value). However, you can gain two more digits of pitch accuracy by using the E word instead of the F word in the **G76** command. That is, the E word allows a six-place decimal format. With the E word, you can specify the pitch of a twelve threads per inch thread as **E0.083333**. Again, in most applications, you won't notice the difference. But consider workpieces that require a perfect pitch value for threading. If you are machining a lead screw, for example, it is quite important to make the pitch of the lead screw as perfect as possible.

I word

The I word is only used for *taper threading* to specify the amount of taper in the X axis. We'll show how taper threading is programmed a little later in this lesson.

Q word

Some newer Fanuc controls use the Q word to specify the angular position of the thread start when machining multiple start threads. We'll show how multiple start threads are programmed a little later in this lesson.

Example program for threading

Figure 6.18 shows the workpiece to be threaded. All machining (other than threading) has been previously done. Only the threading tools are shown (one external threading tool and one internal threading tool).

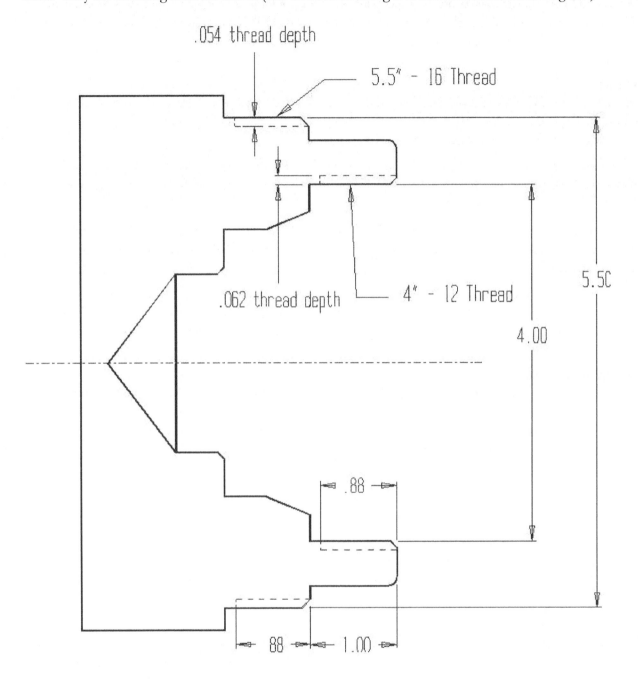

Figure 6.18 – Drawing for threading example program

Program:

```
O0016 (Program number)
N003 G99 G20 G23 (Ensure that initialized states are still in effect)
N004 G50 S4000 (No need for limiting, limit to machine's maximum speed)
N005 T0505 M41 (Select external threading tool and low spindle range)
N010 G97 S500 M03 (Start spindle fwd at 500 rpm)
N015 G00 X5.7 Z-0.8 M08 (Rapid to convenient starting position, start coolant)
N020 G76 X5.392 Z-1.88 K0.054 D0100 A60 F0.0625 (Chase 5.5-16 thread)
```

N025 G00 X8.0 Z6.0 (Rapid to safe index position)

N030 M01 (Optional stop)

N035 T0707 M41 (Select internal threading tool and low spindle range)

N040 G97 S700 M03 (Start spindle fwd at 700 rpm)

N045 G00 X3.8 Z0.2 M08 (Rapid to convenient starting position, start coolant)

N050 **G76 X4.124 Z-0.88 K0.062 D0080 A60 F0.0833** (Chase 4-12 thread)

N055 G00 X8.0 Z6.0 (Rapid to safe index point)

N060 M30 (End of program)

Fundamentals of CNC

INDEX

Made in the USA
Monee, IL
28 October 2020